# Gradient Estimation via Perturbation Analysis

**THE KLUWER INTERNATIONAL SERIES IN ENGINEERING AND COMPUTER SCIENCE**

DISCRETE EVENT DYNAMIC SYSTEMS

*Consulting Editor*

**Yu - Chi Ho**
Harvard University

# GRADIENT ESTIMATION VIA PERTURBATION ANALYSIS

*by*

**Paul Glasserman**
*AT&T Bell Laboratories*

*forword by*

**Yu-Chi Ho**
*Harvard University*

KLUWER ACADEMIC PUBLISHERS
BOSTON/DORDRECHT/LONDON

**Distributors for North America:**
Kluwer Academic Publishers
101 Philip Drive
Assinippi Park
Norwell, Massachusetts 02061 USA

**Distributors for all other countries:**
Kluwer Academic Publishers Group
Distribution Centre
Post Office Box 322
3300 AH Dordrecht, THE NETHERLANDS

**Library of Congress Cataloging-in-Publication Data**

Glasserman, Paul.
Gradient estimation via perturbation analysis / by Paul Glasserman ; forward by Yu-Chi Ho.
p. cm. -- (The Kluwer international series in engineering and computer science. Discrete event dynamic systems)
Includes bibliographical references and index.
ISBN 0-7923-9095-4 (alk. paper)
1. System analysis. 2. Control theory. 3. Perturbation (Mathematics) I. Title II. Series.
QA402.G57 1991
003--dc20 90-48166
CIP



*Printed on acid-free paper.*

Printed in the United States of America

# Contents

# Series Foreword

Consider the daily operation of a large international airport. On the air side, there are a multitude of aircraft, jumbo or small, international or domestic, private or commercial, including helicopters, using airspace, sectors, runways, and gates. On the land side, there are all kinds of vehicles, buses or cars, rental, private or service, taxis or limos, crowding the access ways and parking spaces. Inside the airport building, passengers and luggage require services from ticket counters, security gates, and transport mechanisms. All of these *jobs* requiring different kinds of *services* from a multitude of *resources* interact with each other continuously and dynamically at every instant. Anyone who has experience with such a system on a hot Sunday afternoon in July at the Kennedy airport in NYC or a stormy weekend in January at O'Hare Airport of Chicago can easily visualize the complexity and challenge in the management of such a system. Slightly less familiar but equally real is the example of a large automated manufacturing plant whether it is involved in the making of microchips or jet engines. Here again, the system evolves dynamically in time depending on a host of events such as the arrival of a batch of parts to be machined, the breakdown of a piece of equipment, the outcome of a test, etc. Finally, a more abstract but no less important example is the functioning of many worldwide telecommunication/computer networks. The fact that we can place a phone call to almost anyone in the world within range of a telephone while flying 30,000 feet above in a plane simultaneously with millions of other calls is a technological feat worth appreciating. Other examples of such Discrete Event Dynamic Systems (DEDS) are almost endless: military C3I / logistic systems, emergency ward of a metropolitan hospital, back offices of large insurance and brokerage firms, service and spare part operations of multinational firms ... the point is the pervasive nature of such systems in the daily life of human beings. Yet DEDS are a relatively new phenomenon in dynamic system studies. From the days of Galileo to Newton to quantum mechanics and cosmology of the present, dynamic systems in nature are primarily differential equation based and time driven. A large literature and endless success stories have been built up on such Continuous Variable Dynamic Systems (CVDS). It is, however, equally clear that DEDS are fundamentally different from CVDS. They are event driven, asynchronous, mostly man-made and only became significant during the past generation. Increasingly however, it can be argued that in the modern world our lives are being impacted by and dependent upon the efficient operations of such DEDS. Yet compared to the successful paradigm of differential equations for CVDS, the mathematical modelling of DEDS is in its infancy. Nor are there as many successful and established techniques for their analysis and synthesis.

The purpose of this series is to promote the study and understanding of the modelling, analysis, control, and management of DEDS. The idea of the series came from editing a special issue of the *Proceedings of the IEEE* on DEDS during 1988. The kind reception by the scientific public of that special issue suggests that a more substantial publication effort devoted to DEDS may be in order. I am delighted that Kluwer has decided to undertake such an effort. As readers of the foregoing paragraph can infer, our intention is to publish both engineering and mathematically oriented books. We firmly believe in the synergistic interplay between the serendipitous application of the pure and the discovery of new knowledge by the applied.

Perturbation Analysis (PA) is a technique for the efficient performance analysis of DEDS trajectories. PA had its beginning in a real world application which is more engineering oriented and less formal. Although it was recognized at the beginning as a general approach by its proponent, it wasn't until much later in the work of Cao, Zazanis, Suri, and Glasserman that the foundation of infinitesimal perturbation analysis (IPA) was established. This particular book by Paul Glasserman is an outgrowth of the author's Ph.D. thesis work and its immediate extensions. It collects in one place in a consistent set of notation the rigorous basis of the technique of IPA as well as the structural conditions on DEDS for IPA applicability. Some more recent work on IPA consistency and unbiasedness by Gong, Strickland and Hu in collaboration with Glasserman is also included. In future years this book undoubtedly will become the standard reference on IPA for theoreticians. For application oriented readers, a companion and more informal book on PA, both infinitesimal and finite, by Ho and Cao will also make its appearance in this series at about the same time.

The succeeding volumes of this series will deal with other topics of DEDS.

*Y.C. Ho*
*Cambridge, Massachusetts*

# Preface

In analyzing a stochastic system, such as a network of queues, one is often interested in how system performance depends on system parameters. Gradients provide useful information on this dependence. If the system in question is simulated (or perhaps just observed) one may therefore be interested in estimating gradients from sample paths.

This monograph brings together a circle of ideas on the validation and implementation of a class of gradient estimates, those based on *infinitesimal perturbation analysis* (IPA). IPA is easy to use, but it does not always yield correct results. Our purpose here is to bring together fairly simple, *structural* conditions under which IPA works, and to stretch its scope to make it work on as many problems as possible.

Since this book only considers IPA and some of its extensions, it takes a rather narrow view of gradient estimation. The treatment is, in this sense, more method-oriented than problem-oriented. Techniques requiring an altogether different analysis—in particular estimation via likelihood ratios or finite differences—are not covered, though they can (and should!) be used to solve similar problems.

The core of the theory developed here is in Chapters 1-3: Chapter 1 motivates the topic and reviews some basic tools; Chapter 2 shows how to compute derivatives from sample paths; Chapter 3 provides conditions under which these derivative estimates are unbiased.

Subsequent chapters branch off into more specialized topics. Chapter 4 applies the results of previous chapters to networks of queues; some readers may want to read that chapter together with Chapters 2 and 3 to see what the general theory leads to. Chapters 5 and 6 are something of a digression into another set of structural conditions for IPA and can be skipped by a reader interested only in the basic theory. Chapter 7 develops extensions of IPA based on *smoothing*; these are applicable to a class of performance measures different from those covered in Chapters 2 and 3. Chapter 8 returns to the setting of Chapter 3 and extends results for finite time horizons to the problem of steady-state derivative estimation. Since in many settings steady state is of primary interest, some readers may want to skip directly to Chapter 8 from Chapter 3.

There are few specific prerequisites for reading this book. Basic knowledge of probability and stochastic processes is assumed, as is some familiarity with interchanging limits and integrals. An interest in queueing and in simulation is probably essential.

I am happy to thank here some of the people who have helped in this effort. Foremost among them is Professor Y. C. Ho; this book would not exist without his efforts and ideas before, during and after my years as his student.

Chapter 7 of this book is based on joint work with Wei-Bo Gong, Chapter 8 on work with Jian-Qiang Hu and Steve Strickland. The corresponding results are theirs as well as mine.

Several people read parts of a draft of this book, or drafts of some of the papers on which it is based, and provided corrections and suggestions at all levels. These include, in addition to those noted above, Aseem Chandawarkar, Youyi Feng, Bennett Fox, Michael Fu, Peter Glynn, Youghua Pan, David Yao, Bing Zhao, and Shaohui Zheng.

Finally, I thank AT&T Bell Laboratories, especially Hanan Luss and C. J. McCallum, Jr., for supporting this project.

*Paul Glasserman*
*Holmdel, New Jersey*

# Selected Notation

Numbers indicate sections in which the notation is defined or used in a special way. The symbol 2.3.1 refers to Section 3.1 of Chapter 2, and so on.

| | |
|---|---|
| $\mathbf{1}\{\cdot\}$ | The indicator function of the set $\{\cdot\}$. |
| $a_n$ | The $n$th event of a GSMP; 2.2. |
| $\underline{a}_n$ | The vector $(a_1, \ldots, a_n)$ of the first $n$ events; 7.3.1, 7.4.2. |
| $\mathbf{A}$ | The set of events of a GSMP; 2.2. |
| $\mathcal{A}_n$ | The set of vectors $(a'_1, \ldots, a'_n)$ obtained through selected changes in $\underline{a}_n$; 7.3.1, 7.4.2. |
| $A(s)$ | The set of immediate successors of state $s$; 6.3. |
| $A(x)$ | The set of "arcs" emanating from a state $x$ of a Markov chain; 5.2. |
| $\alpha, \beta$ | Typical events of a GSMP; 2.2. |
| $c_n$ | The vector of clock readings following the $n$th transition of a GSMP; 2.1, 2.3.1. Similar but slightly different definitions are used in 5.2, 6.3. |
| $C_n$ | The set of assigned clocks following the $n$th transition of a Markov chain; 5.2. |
| $D_{n_0}, D_T$ | Discontinuous additive performance measures; 7.1.3, 7.3.5, 7.4.2, 7.5. |
| $\delta_n(\alpha)$ | Perturbation in scheduled occurrence of $\alpha$; 2.3.1, 8.3.1. |
| $\Delta(\cdot\|\cdot)$ | The expected size of a discontinuity; 7.1.2, 7.3.1, 7.3.5, 7.4.2. |
| $e_i, e_0$ | $e_i$ is the $i$th unit vector and $e_0$ is the vector of all zeros; 2.1, 4.2.1. |
| $\varepsilon_s(s')$ | The event $\alpha$ for which $p(s'; s, \alpha) > 0$; 6.3,6.4. |
| $\mathcal{E}(s)$ | The event list for a state $s$ of a GSMP; 2.1. |
| $F_\alpha$ | The distribution of clock samples for event $\alpha$; 2.1. |
| $\eta$ | Triggering indicator; 2.2, 2.3.1. |
| $\mathcal{H}_n$ | History of a GSMP, up to its $n$th transition; 7.3.1, 7.4.2. |
| $J_n$ | Index of clock that triggers $n$th transition of a Markov chain; 5.2. |
| $K(x, y)$ | Common successor of two states in a Markov chain (5.1.2) or GSMP ($K(s, s')$, 6.3). |
| $K_P$ | Common successor matrix for routing matrix $P$; 5.4, 6.5.2; |
| $\lambda_\alpha(s)$ | The clock speed for event $\alpha$ in state $s$; 3.2.1, 4.2.5. |

| | |
|---|---|
| $L_T, L_n$ | Cumulative performance measures; 2.1, 2.3.1, 4.1.2, 8.1, 8.3. |
| $\mu_\alpha$ | The hazard rate for event $\alpha$; 6.1, 7.2. |
| $\mu_\alpha^*$ | The *parametric* hazard rate for event $\alpha$; 6.2.1. |
| $M(c,t,\theta)$ | A shifted cumulative hazard function; 6.2. |
| $\mathbf{n}$ | A vector of queue lengths; 4.1.1, 4.2.1. |
| $N(t)$ | The number of state transitions in $[0,t]$; 2.1, 2.2. |
| $N(\alpha,n)$ | The number of occurrences of $\alpha$ among $a_1,\ldots,a_n$; 2.2. |
| $p(s';\alpha,s)$ | The GSMP transition probability from $s$ to $s'$ on the occurrence of $\alpha$; 2.1, 2.2. |
| $P_{ij}$ | Routing matrix for a queueing network; 2.1, 4.1.1, 4.2.1, 6.5.2. |
| $\phi$ | The mapping that determines the next state, given the current state, the current event and a routing indicator; 2.2, 4.1.1. |
| $\Phi_{ij}$ | The state transition mapping that effects the movement of a job from node $i$ to node $j$; 4.1.1, 4.2. |
| $\psi_\alpha$ | A function such that $dX_\theta(\alpha,k)/d\theta = \psi_\alpha(X_\theta(\alpha,k),\theta)$; 7.2, 8.3.1. |
| $\Psi_\alpha$ | The solution to $dx/d\theta = \psi_\alpha(x,\theta)$; 7.2. |
| $Q_\theta, \mathbf{Q}_\theta$ | Generator matrix of a continuous time Markov chain; 5.1, 5.2, 7.5. |
| $r_n$ | The $n$th event-order pair, i.e., $(a_n, N(a_n,n))$; 2.2. |
| $R_n$ | Scheduled occurrence of $\alpha$ ($R_n(\alpha)$, 3.2.2), of clock $j$ ($R_n(j)$, 5.3.1), of transition to $s'$ ($R_n(s')$, 6.3). |
| $R_T, R_{n_0}$ | Terminal reward performance measures; 7.1.3, 7.3, 7.5. |
| $\mathbf{S}$ | State space of a GSMP, Markov chain, etc.; 2.1, 5.2. |
| $S_n(x)$ | The set of states, $y$, for which a clock assigned to $(Y_{n-1},y)$ is reassigned to $(Y_n,x)$ following the $n$th transition; 5.3.2, 6.4.1. |
| $S(\alpha,k)$ | Epoch of the $k$th setting of a clock for $\alpha$; 7.3.2. |
| $\tau_n$ | Epoch of the $n$th state transition of a GSMP, Markov chain, etc.; 2.2. |
| $t_n(\alpha)$ | Age of $\alpha$-clock at $n$th transition; 6.3, 7.3.2, 7.4.2. |
| $T(\alpha,k)$ | Epoch of the $k$th occurrence of event $\alpha$; 2.2. |
| $U(\alpha,k)$ | The $k$th routing indicator for $\alpha$; 2.2. |
| $X(\alpha,k)$ | The $k$th clock sample for $\alpha$; 2.2. |
| $Y_n$ | The $n$th state of a GSMP or Markov chain; 2.1, 2.2, 5.2. |
| $Z_t$ | The state, at time $t$, of a GSMP, Markov chain, etc.; 2.1, 2.2, 5.2. |

# Gradient Estimation via Perturbation Analysis

# Chapter 1

# Introduction

## 1.1 The Gradient Estimation Problem

The title of this book demands immediate explanation of what this book is *not* about. This book is not about approximating derivatives by differences. While finite difference approximation will always be the most general way of estimating derivatives, the (numerical) issues inherent to that method are quite different from those we consider. Nor does this book have anything to do with *time* derivatives of sample paths of stochastic systems; indeed, virtually all the processes we consider have piecewise constant sample paths, so their derivatives in time are zero wherever they exist. Rather, this book is about using *stochastic gradients* to estimate gradients of expectations. All derivatives we consider are with respect to *parameters* of a system, in a sense to be made clear. The unifying methodological theme throughout this book is *the continuous construction of parametric families of stochastic processes.*

An abstract statement of the simplest form of the gradient estimation problem is the following: Given a function $\ell$, differentiable in the vector argument $\theta$, find a vector-valued random function $\xi(\theta)$ with expectation $\mathbf{E}[\xi(\theta)]$ equal to $\nabla_\theta \ell(\theta)$, the gradient of $\ell$. In the problems we consider, $\ell(\theta)$ arises as the expectation, $\mathbf{E}[L(\theta)]$, of an almost surely locally differentiable random function $L(\theta)$; we study properties of $\nabla_\theta L(\theta)$ as a candidate $\xi(\theta)$. Conditions for a key property, the *unbiasedness* of $\nabla_\theta L(\theta)$, are closely tied to conditions for the continuity of $L$ in $\theta$. We explain why after developing some background.

### Motivation

By a stochastic *discrete event system* we mean, broadly, a stochastic process that changes state only at discrete, random time instants. The state of the process represents the physical configuration of a system, and the epochs of state transitions typically correspond to the occurrence of physically meaningful *events*. We think primarily of queueing networks; the "state" is the vector of queue lengths at the various nodes, possibly supplemented with information about different job classes, blocking and routing mechanisms, etc. The events are the different kinds of arrivals and service completions supported by the network. The timing of events is determined by random *lifetimes* or *clock samples*; these are, for example, interarrival times and service times. Models of this type are widely used in the study of computer, communications and manufacturing systems.

One studies discrete event systems to understand their performance; but no one interested in performance is interested in the performance of just one system. Competing alternatives, even if never fully articulated, always influence the analysis of a system—often, by determining what aspects to analyze. To put it another way, performance *analysis* is a step towards performance *improvement*, and improvement takes place within a limited (though perhaps loosely defined) class of systems.

In the setting we consider, the (neatly defined) set of alternatives is a parametric family of stochastic processes. In most cases, the parameters will be parameters of the clock *distributions*; e.g., the distributions of service times, interarrival times, etc. (A simple example, the family of M/M/1 queues, is parameterized by an arrival and a service rate.) If the performance of this family of systems changes "smoothly" with the parameter, then information about the *gradient* of the performance with respect to the parameter is valuable in comparing the alternatives and in finding the best one.

Let us develop this framework in more detail. Denote a parametric family of processes by $\{Z_t(\theta), t \geq 0, \theta \in \Theta\}$, where $t$ is the time index and $\Theta$ is the parameter space; thus $Z_t(\theta)$ is the state of the system at time $t$ under parameter value $\theta$. A performance measure $L$, say, is a mapping assigning to each sample path $\{Z_t(\theta), t \geq 0\}$ a real number. Since $Z$ depends on $\theta$, $L$ does, too; we write $L(\theta)$ for the performance at parameter $\theta$. Each $L(\theta)$ is a random variable because each $\{Z_t(\theta), t \geq 0\}$ is stochastic. We are primarily interested in comparing (and improving) the *expected* performance $\mathbf{E}[L(\theta)]$, viewed as a function of $\theta$. To do so, we would like to know something about $\nabla_\theta \mathbf{E}[L(\theta)]$, the gradient of expected performance with respect to the parameter $\theta$.

With this set-up, the problem of gradient estimation becomes one of finding, for each $\theta$, a sequence $\{D_n(\theta), n = 1, 2, \ldots\}$ of random vectors such that $\lim_{n\to\infty} D_n(\theta) = \nabla_\theta \mathbf{E}[L(\theta)]$, the limit taken in the almost sure sense. A particularly convenient case is one in which there is a random vector $\xi(\theta)$ for which $\mathbf{E}[\xi(\theta)] = \nabla_\theta \mathbf{E}[L(\theta)]$; for in this case we may take a sequence $\{\xi_i(\theta), i = 1, 2, \ldots\}$ of independent random vectors, each distributed like $\xi(\theta)$, and set $D_n(\theta) = \sum_{i=1}^n \xi_i(\theta)/n$. This $\{D_n(\theta)\}$ converges to $\nabla_\theta \mathbf{E}[L(\theta)]$ by the law of large numbers.

For this idea to be of any practical value, two fundamental conditions must be met: there should be a fairly general mechanism for finding a candidate estimator $\xi(\theta)$, and the estimator should be computable with reasonable effort from available data. One method that meets these conditions for a broad class of discrete event system uses a *stochastic gradient* to estimate the gradient of an expectation. This is the method of *infinitesimal perturbation analysis* (IPA) pioneered by Y.C. Ho and co-workers. Simply put, IPA chooses $\xi(\theta)$ to be (the random vector) $\nabla_\theta L(\theta)$. Just what this means requires further explanation, which we provide shortly. But we stress immediately that $\nabla_\theta L(\theta)$ is, in great generality, easily calculated from knowledge of $\{Z_t(\theta), t \geq 0\}$ at the single value of $\theta$ at which the gradient is evaluated. In particular, $\nabla_\theta L(\theta)$ is calculated *exactly* without any change in $\theta$ ever being introduced. Given that this estimator can be computed, the harder question becomes the validity

$$\mathbf{E}[\nabla_\theta L(\theta)] = \nabla_\theta \mathbf{E}[L(\theta)]; \tag{1.1}$$

i.e., the *unbiasedness* of the stochastic gradient. The left side is what we obtain by averaging i.i.d. copies of the stochastic gradient; the right side is what we want. Conditions for the validity of (1.1) are the main subject of this book.

On the surface, (1.1) is a routine problem in real analysis, the interchange of an integral and a derivative. It is easy to write down purely analytic, necessary and sufficient conditions for (1.1) to hold. Such conditions turn out to be of little use in practice, and this point of view completely misses the central role played by *structural* properties of a system in ensuring (1.1). An appreciation of this point requires a closer look at the meaning of the symbol $\nabla_\theta L(\theta)$.

### The Simulation View

A standard, intuitive explanation of $\nabla_\theta L(\theta)$ helps motivate a more detailed discussion and is relevant in practice. Suppose we simulate the processes $\{Z_t(\theta), t \geq 0, \theta \in \Theta\}$. Our computer has an idealized random number generator which, given an ideal *seed* (a number uniformly distributed on the unit interval), generates an infinite sequence of independent, likewise uniformly distributed random variables. Given a parameter value $\theta$, our simulation algorithm transforms the sequence into a sample path of $\{Z_t(\theta), t \geq 0\}$.

Let $\theta$ be an $n$-vector with $i$th component $\theta_i$; let $e_i$ be the $i$th unit $n$-vector. One way to estimate the change in $L$ under a change in $\theta$ runs a "nominal" simulation at parameter setting $\theta$ using seed $U_0$, then runs $n$ "perturbed" simulations using independent seeds $U_1, \ldots, U_n$, with the $i$th simulation run at a parameter value $\theta + h \cdot e_i = (\theta_1, \ldots, \theta_i + h, \ldots, \theta_n)$. The runs produce outputs $L(\theta, U_0), L(\theta + he_1, U_1), \ldots, L(\theta + he_n, U_n)$. The $n$ differences

$$L(\theta + he_i, U_i) - L(\theta, U_0), \; i = 1, \ldots, n, \tag{1.2}$$

give an indication of the effect of $\theta$ on $L$. By averaging over $N$ such differences using seeds $U_i^{(j)}$, $j = 1, \ldots, N$, $i = 0, \ldots, n$ and letting $N \to \infty$ we indeed obtain the differences $\mathbf{E}[L(\theta + he_i)] - \mathbf{E}[L(\theta)]$.

A better way to estimate these differences uses the same seeds for nominal and perturbed runs to get

$$L(\theta + he_i, U_0) - L(\theta, U_0),\ i = 1, \ldots, n. \tag{1.3}$$

This is the method of *common random numbers.* Intuitively, (1.3) should be preferred to (1.2) because it puts the nominal and perturbed systems on an equal footing, comparing their responses to exactly the same inputs. A more precise statement is that (1.3) typically has lower variance than (1.2); see, e.g., Bratley, Fox, and Schrage [3] for a discussion of common random numbers.

IPA can be viewed as a limiting case of (1.3) in which one computes

$$\nabla_\theta L(\theta, U_0) = (\partial_{\theta_1} L(\theta, U_0), \ldots, \partial_{\theta_n} L(\theta, U_0)), \tag{1.4}$$

where

$$\partial_{\theta_i} L(\theta, U_0) = \lim_{h \to 0} \frac{L(\theta + he_i, U_0) - L(\theta, U_0)}{h}. \tag{1.5}$$

(The limit in (1.5) is a definition, not an algorithm; IPA evaluates the derivative exactly without approximating it through a convergent sequence.) When (1.1) holds, (1.4) is certainly a better gradient estimate than (1.2) and (1.3), which are approximations at best. Moreover, (1.4) evaluates all $n$ components of the gradient from a *single* simulation (at $\theta$) while the others require $n + 1$ runs to get all $n$ differences.

### The Mathematical View

A short step takes us from simulation to the mathematical setting. Suppose the family of processes $\{Z_t(\theta), t \geq 0, \theta \in \Theta\}$ is defined on a probability space $(\Omega, \mathcal{F}, P)$. Thus, $Z$ is really $\{Z_t(\theta, \omega)\}$, a function of three arguments, in which $\omega \in \Omega$ represents "randomness". (Think of the outcome of the simulation seed.) We view the composition of $L$ with $Z$ as a function of two arguments, $L(\theta, \omega)$; now (1.5) becomes

$$\partial_{\theta_i} L(\theta, \omega) = \lim_{h \to 0} \frac{L(\theta + he_i, \omega) - L(\theta, \omega)}{h}, \tag{1.6}$$

with the requirement that the limit exist for almost all $\omega$; that is, with probability one. This is a stochastic partial derivative. (It is a random variable.) The vector of such partial derivatives is the stochastic gradient.

But this formulation, as it stands, overlooks an important point at the heart of our subject, which has to do with what it means to specify a stochastic process. We assumed that $Z_t(\theta, \cdot)$ is a function on a space $(\Omega, \mathcal{F}, P)$; one might well ask *which* function on *which* space. Ordinarily it does not matter. So long as the functions $\{Z_t(\theta), t \geq 0, \theta \in \Theta\}$ accurately represent, say, the family of M/M/1 queues we do not ask how they do so, we do not ask about the explicit dependence of $Z_t(\theta, \cdot)$ on $\omega$. The value of any $\mathbf{E}[L(\theta)]$ must be the same for all legitimate representations of the family of processes. Indeed, probability theory is properly concerned only with those aspects of a process that do not depend on the choice of probability space. Nevertheless, choosing a particular space is a powerful tool—for our purposes, an essential one. The partial derivatives in (1.6) are not well-defined until we have specified explicitly how $Z$ depends on $\theta$ and $\omega$.

What uniquely characterizes a process is its *law*, the measure it induces on a space of sample paths. Suppose that for every $\theta \in \Theta$, every sample path of $\{Z_t(\theta), t \geq 0\}$ is an element of a space $\mathcal{Z}$. Let $\mathcal{Z}$ be endowed with a $\sigma$-algebra $\mathcal{A}$ of sets declared measurable. Each (measurable) process $\{Z_t(\theta), t \geq 0\}$, induces a probability measure $\mathcal{P}_\theta$ on $(\mathcal{Z}, \mathcal{A})$ via

$$\mathcal{P}_\theta(A) = P(\{Z_t(\theta), t \geq 0\} \in A), \; A \in \mathcal{A}.$$

Any function on any probability space inducing the same measure $\mathcal{P}_\theta$ on $(\mathcal{Z}, \mathcal{A})$ is a legitimate representation of the process represented by $\{Z_t(\theta), t \geq 0\}$. Rather than start from a parametric family of processes, we should start from the family $\{\mathcal{P}_\theta, \theta \in \Theta\}$ of probability measures, which minimally characterize the processes under consideration. Now $Z$ could be any family of processes (with sample paths in $\mathcal{Z}$) on any probability space for which $\{Z_t(\theta), t \geq 0\}$ induces $\mathcal{P}_\theta$ on $(\mathcal{Z}, \mathcal{A})$, for every $\theta \in \Theta$. Notice that this is a *marginal* constraint on the choice of $Z$, in the sense that it constrains the measure of each $Z(\theta) = \{Z_t(\theta), t \geq 0\}$ separately, but says nothing about the *joint* dependence of $Z(\theta)$'s at different $\theta$'s. In contrast, the existence of (1.6) is a property of the dependence of $Z$ across $\theta$'s for fixed $\omega$. We are free to choose the dependence across $\theta$'s as we wish, so long as the marginal constraint for each $\theta$ is met. In particular, we can (try to) choose it so that (1.6) exists almost surely.

Within this more elaborate setting, we can pose the problem of unbiased gradient estimation using stochastic gradients in the following way: Given a specification of a parametric family of processes (i.e., a family of probability measures $\{\mathcal{P}_\theta, \theta \in \Theta\}$) construct, on a single probability space $(\Omega, \mathcal{F}, P)$ a family $\{Z_t(\theta), t \geq 0, \theta \in \Theta\}$ of processes such that *(i)* each $\{Z_t(\theta), t \geq 0\}$ has law $\mathcal{P}_\theta$; *(ii)* the limit in (1.6) almost surely exists for the performance measure $L$ of interest; and *(iii)* equation (1.1) holds. Step *(i)* is the mathematical equivalent of finding a simulation algorithm for $\{\mathcal{P}_\theta\}$. Step *(ii)* is rarely an obstacle, and when it fails it is easy to see why. By far the most involved issue is step *(iii)*. It requires a deeper look at the structure of a system.

### The Role of System Structure and Continuity

The conditions for interchanging derivatives and integrals that one finds in textbooks aim for generality. But in studying the performance of discrete event systems one does not encounter "arbitrary" functions. For reasonable constructions of reasonable systems with reasonable parameters and performance measures, some basic properties of $L(\theta)$ can be said to hold quite broadly. These arise from the structure and *dynamics* of discrete event systems. An important property is this: For each $\theta$, there is, almost surely, a neighborhood of $\theta$ in which $L$ is "well-behaved"—i.e., continuously differentiable. Typically, $L$ has this property when (for almost all $\omega$) a sufficiently small change in $\theta$ changes the *timing* but not the *order* of events on any finite time interval $[0, t]$. This behavior is natural in queueing systems: a sufficiently small change in service and interarrival times will not change the order of service completions and arrivals, only their timing.

Throughout a neighborhood of $\theta$ in which events do not change order, the dependence of $L$ on $\theta$ is often easy to describe. It is this property which makes it possible to evaluate a stochastic gradient at $\theta$ from observation of $\{Z_t(\theta), t \geq 0\}$ at $\theta$ only. For example, Figure 1.1 illustrates the effect of small perturbations in service times on a sample path of a single-server queue. A small increase in a service time delays the departure of the job in service. It also delays the service initiation of any waiting jobs. Thus, in the figure, the departure of the first job in a busy period is delayed by one perturbation, that of the second by two, and so on. Service perturbations are *propagated* to subsequent jobs in the same busy period. A job arriving to find the system empty, enters service immediately and is unaffected by earlier delays. By analyzing the system in this way, then passing to the limit in which the perturbations become "infinitesimally small", we develop an algorithm for calculating the *derivatives* of the departure epochs in terms of the *derivatives* of the service times, with respect to a parameter.

While there is typically a neighborhood in which $L$ is smooth, it is not much of an overstatement to say that cases in which $L$ is almost surely continuously differentiable *throughout* $\Theta$ are trivial. (If $\Theta$ is an open set in Euclidean space, if $L$ is a.s. $C^1$, and if $\mathbf{E}[|\nabla_\theta L(\theta)|] < \infty$ for all $\theta \in \Theta$, then (1.1) holds throughout $\Theta$.) If (with positive probability) a sufficiently large change in $\theta$ *does* change the order of events, then $L$ may fail to be differentiable on the (random) set of $\theta$'s at which events change order. (In Figure 1.1, sufficiently large service time perturbations would cause the two busy periods to merge by changing the order of the third departure and the fourth arrival.) If no change in $\theta$ changes the order of events, then $\theta$ affects only the *time scale* of the system, and not its dynamics; this case is of limited interest. For discrete event systems, the study of conditions for (1.1) is the study of the behavior of $L$ at points where there is a change in the order of events.

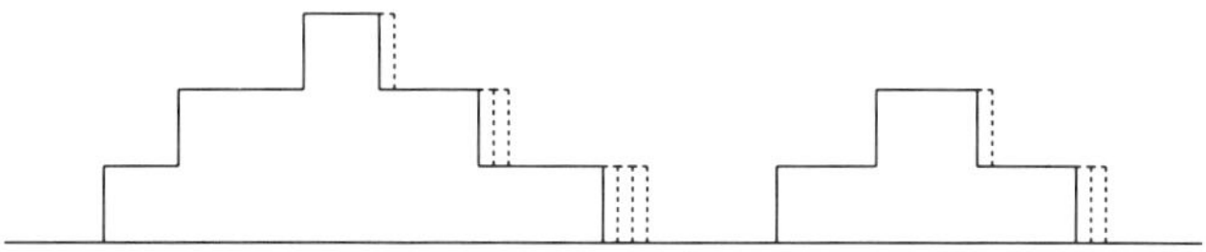

Figure 1.1: Sample path of a queue with perturbed service times

At this point, the "typical" behavior of discrete event systems splits—two cases are commonly encountered. In one, $L$ has jump discontinuities at some $\theta$'s at which order changes occur. In the other, $L$ has at worst a "kink" (a discontinuity in its gradient) at an order change, but is almost surely continuous throughout $\Theta$. This dichotomy is fundamental: if $L$ is discontinuous with non-negligible probability, we cannot expect the interchange in (1.1) to hold. Intuitively, $\nabla_\theta L(\theta)$ reflects only the local behavior of $L$, and if $L$ is not continuous then its local behavior is not indicative of that of $\mathbf{E}[L(\theta)]$. Continuity of $L$ is not sufficient (or strictly necessary) for (1.1); but given continuity we can impose reasonable additional hypotheses under which (1.1) holds. In practice, possible discontinuities of $L$ are the limiting factor in using $\nabla_\theta L(\theta)$ as an estimate of $\nabla_\theta \mathbf{E}[L(\theta)]$.

For reasonable choices of $L$, continuity across order changes depends on the *structure* of the system modeled by $\{Z_t(\theta), t \geq 0, \theta \in \Theta\}$. Consider a queueing network. A change in the order of events might correspond to a change in the order in which two different jobs arrive at a queue. If the jobs are of different classes with different service requirements and different routes through the network, then a change in their order of arrival may significantly alter the future evolution of the process. In this case, $L$ might well have a discontinuity at any $\theta$ at which such an order change occurs. On the other hand, if the jobs are indistinguishable then just after both arrive the state of the system is independent of their order of arrival. The future evolution of the process is unaffected, and we can expect $L$ to be continuous even at an order change. Thus, some order changes are significant and others are not. Whether or not a system is vulnerable to significant order changes is a property of the logical structure of its dynamics, rather than its detailed stochastic characteristics. Much of this book is devoted to making these ideas precise and finding easily verified conditions under which they can be applied.

## Summing Up

It may be helpful to tie together the bits of the theory sketched above, and to relate these to the chapters ahead. Our starting point is always a specification of a parametric family of processes. Just what specification we start from depends on the setting; a Markov chain, for example, is (notationally) easier to characterize than a network of queues. We identify structural conditions under which it is possible to construct the family of processes on a common probability space to satisfy almost sure differentiability (as in (1.6)) and a continuity condition across $\theta$'s. The particular structure needed varies to some extent with the kind of system considered, but the intuitive argument of the preceding paragraph is reflected in every case. Once we have established structural conditions for continuity, we add regularity conditions for unbiasedness—that is, for (1.1).

Our analysis treats only the case of scalar $\theta$ explicitly. A stochastic gradient is unbiased if and only if every one of its component partial derivatives is unbiased, so in this sense the scalar case suffices. One may think of a scalar $\theta$ as a component of a vector parameter.

Throughout much of this book, we model discrete event systems as *generalized semi-Markov processes*, or GSMPs. These processes allow a unified treatment of a wide diversity of systems. More importantly, they separate the detailed stochastic properties from the structure of an event-driven system. For example, as GSMPs, all infinite buffer, single-server queues look essentially alike—as they should—regardless of their service and interarrival time distributions. For this reason, the GSMP framework allows a succinct statement of a key condition for continuity.

Chapter 2 introduces GSMPs, carries out a detailed construction of a parametric family of GSMPs on a single probability space, then develops the resulting stochastic derivatives, culminating in a generic algorithm to compute these derivatives. Chapter 3 identifies structural conditions for continuity, and, under these conditions, verifies unbiasedness for a class of finite-horizon performance measures. It then develops an analogous theory for more general GSMPs, those with speeds.

Figure 1.2 illustrates the interdependence of the chapters. As the figure indicates, Chapters 2 and 3 lay groundwork on which the remaining chapters depend. Chapters 4-8 develop new directions, but all have roots, to varying degrees, in the conditions of Chapter 3, Section 1.1. Briefly, these conditions assert that changing the order of events cannot change the state reached—that events "commute." Chapter 4 applies and interprets these conditions in the setting of queueing systems. These applications are of practical value, and also help make vivid the results of Chapters 2 and 3.

Chapter 5 exploits special properties of Markov chains to find unbiased stochastic derivatives. These derivatives result from an unusual construction of a parametric family of specially structured chains. Though Markov chains form a restricted class of GSMPs, their special properties and wide use makes them worth considering separately.

Chapter 6 shows that, in some ways, Markov chains are not so special after all. It takes the condition and construction of Chapter 5 and applies them to

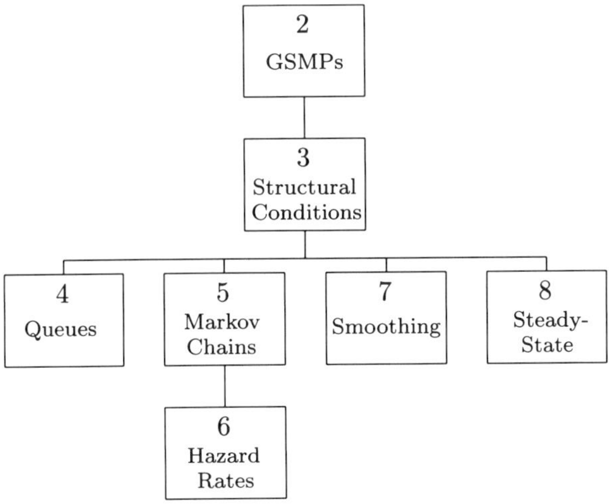


Figure 1.2: Dependence among the chapters

(non-Markovian) GSMPs. This leads to conditions and derivative estimates different from those of Chapter 2 and 3. The principle tool of this chapter is the *hazard rate* which permits transformation of exponential random variables to more general random variables, thus transforming a Markovian setting to a non-Markovian one.

Chapter 7 is the only one that focuses on derivative estimation in the absence of continuity. Smoothing techniques are developed to get around a special class of discontinuities. Chapter 7 takes as its starting point the framework of Chapters 2 and 3.

Finally, Chapter 8 considers the problem of *steady-state* derivative estimation; previous chapters consider only finite-horizon performance. As one might expect, the steady-state estimation problem is more difficult. The theory for that case is not as fully developed. Chapter 8 studies a class of GSMPs for which a neat theory exists. Convergence of steady-state derivative estimates is established by supplementing the conditions of Chapter 3 with an explicit regenerative structure.

Almost all the techniques and analyses presented here apply whether the processes $\{Z_t(\theta), t \geq 0, \theta \in \Theta\}$ are simulated or just *observed*, as in the case of data from a real system. (In the second case, the construction on a common probability space is a "thought experiment" carried out merely to derive the form of the derivative estimator.) In a few places, we point out methods that apply only to simulation—uniformization, in Chapter 5, is one example. Usually, the distinction is obvious. It is likely that some bias towards the simulation setting has crept in; it is the one with which we have practical experience.

The topics in this book are linked as much by a common methodology as by the problems they address. Several other methods of gradient estimation—methods that solve similar problems using an altogether different approach—are not covered. The most glaring omission is the absence of any treatment of the *likelihood ratio method.* In any practical setting where gradient estimation seems a relevant tool this method should be considered together with those of this book. Indeed, the likelihood ratio method is broadly applicable without the kinds of structural restrictions we must impose here. Its principle drawback is a typically large variance. For background on this method see Glynn [38,39], Reiman and Weiss [66], and Rubinstein [69].

## 1.2 Applications

A gradient is such a useful piece of information that it hardly seems necessary to justify estimating one. Nevertheless, a brief discussion of some stochastic applications helps motivate the methods of this book.

### Optimization

Gradient estimates can be used to optimize analytically intractable systems through *stochastic approximation.* The situation we have in mind is the following: One simulates or observes a process $\{Z_t(\theta), t \geq 0\}$. The parameter $\theta$ can be controlled. The performance criterion is $\mathbf{E}[L(\theta)]$. From $\{Z_t(\theta), t \geq 0\}$, one simultaneously estimates both the performance $\mathbf{E}[L(\theta)]$ and its gradient. Based on the gradient estimate, $\theta$ is adjusted in the direction of improved performance.

In more detail, we start from an initial parameter setting $\theta_1$. After observing $Z_t(\theta)$ over some time interval, we update $\theta$ according to

$$\theta_{n+1} = \theta_n + a_n g_n \tag{1.7}$$

where $\{a_n, n = 1, 2, \ldots\}$ is a sequence of positive step sizes and $\{g_n\}$ estimates $\nabla_\theta \mathbf{E}[L(\theta)]$. For example, the $g_n$'s could take the form $\nabla_\theta L(\theta_n)$. (The recursion (1.7) searches for a maximum of $\mathbf{E}[L(\theta)]$; to minimize, replace addition with subtraction.) Intuitively, (1.7) drives the sequence $\{\theta_n\}$ to a zero of the gradient $\nabla_\theta \mathbf{E}[L(\theta)]$ (a possible optimum) if $\{g_n\}$ consistently estimates $\nabla_\theta \mathbf{E}[L(\theta)]$.

There is an extensive literature on conditions for convergence of approximation schemes like (1.7); we will not attempt to review it. However, a few remarks on conditions for convergence to an optimal $\theta$ are relevant. In general, three sorts of conditions are imposed. First, one needs a condition on the objective $\mathbf{E}[L(\theta)]$ as a function of $\theta$. For example, one might require that the equation $\nabla_\theta \mathbf{E}[L(\theta)] = 0$ have just one root, which is, in fact, a global maximum. (Weaker assumptions are possible.) Next, one needs conditions on the step-size sequence $\{a_n\}$. A typical condition is

$$\sum_{n=1}^{\infty} a_n = \infty \quad \text{and} \quad \sum_{n=1}^{\infty} a_n^2 < \infty.$$

(An example is $a_n = 1/n$.) The second of these is needed for the eventual convergence of the algorithm: if $\{\theta_n\}$ is to converge, the step sizes must eventually become small. On the other hand, the first condition accelerates convergence by ensuring that the steps do not become small too fast. The divergence of the step sizes partly compensates for the error in the gradient estimate.

Finally, one must impose conditions stating that the error in the gradient estimate is not too bad. A natural condition is that the estimator bias $\mathbf{E}[g_n(\theta_n)] - \nabla_\theta \mathbf{E}[L(\theta_n)]$ vanish as $n \to \infty$, but this by itself is not enough. One also needs a bound on the "noise" in the estimator as measured by the variance of the error. A strong condition of this type imposes a uniform bound on the covariance matrices of $\{g_n(\theta)\}$.

Stochastic approximation was introduced by Robbins and Monro [67] as a way of finding the zero of a function defined as an expectation with respect to a distribution depending on a parameter. Kiefer and Wolfowitz [60] modified the Robbins-Monro procedure to find the extremum of a function using finite difference approximations to the derivative of a function. Not surprisingly, algorithms of the Robbins-Monro type converge faster and under weaker conditions than those of the Kiefer-Wolfowitz type. When the gradient estimator is unbiased, (1.7) is a Robbins-Monro algorithm. It directly approximates the zero of a function—a function which happens to be a gradient. Thus, in using an unbiased gradient estimate, we can typically expect substantial improvement over optimization driven by finite differences.

### Sensitivity Analysis

The simplest and most immediate application of gradient estimation is to sensitivity analysis—to getting a sense of how the performance measure $\mathbf{E}[L(\theta)]$ varies as $\theta$ varies. In the absence of information about the form of $\mathbf{E}[L(\theta)]$ (e.g., whether it is roughly linear, unimodal, etc.) this analysis may be largely heuristic. Given an estimate of $\mathbf{E}[L(\theta)]$ and a gradient estimate $g(\theta)$, one might expect that for perturbations $\delta$ within, say, 10% of $\theta$, the change in performance from $\theta$ to $\theta + \delta$ is well approximated by $\delta g(\theta)$. In some cases, just getting a good estimate of the *sign* of a derivative can be useful, particularly when the performance of the system is not well understood.

A more formal approach to sensitivity analysis attempts to model explicitly the response of outputs to changes in inputs. For example, one might measure performance at a set of parameters and then fit a polynomial to the data. If the performance estimates are good, and if $\mathbf{E}[L(\theta)]$ is well behaved, the polynomial fit allows interpolation and extrapolation of performance estimates to other parameter values. Here, too, gradient estimates are valuable. If one samples at just two points, performance estimates alone can only produce a straight line model. But when the two function values are supplemented with two gradient estimates they allow a cubic fit, or an improved linear or quadratic fit. Of course, the gradient estimate at a point may be correlated with the performance estimate at that point, and this correlation must be accounted for (and ordinarily estimated) in making a least-squares fit.

Holtzman [55] describes a different setting for sensitivity analysis. Suppose that $\theta$ is a (scalar) parameter of a real system—e.g., the rate of arrivals to a facility. We do not know the true value of $\theta$, but from statistics of the real system we can form an estimate $\tilde{\theta}$ of $\theta$; $\tilde{\theta}$ is a random variable. We assume it has expectation $\theta$ and finite variance $\sigma^2$. If we now simulate the real system, using the estimate $\tilde{\theta}$ for $\theta$, we would like to have a way of mapping our uncertainty in the input $\theta$ to the appropriate uncertainty in the simulation output.

Holtzman discusses the following approach: Let $\ell(\theta) = \mathbf{E}[L(\theta)]$. Assume that $\ell$ is reasonably smooth; in particular, suppose it is well approximated by its second-order Taylor polynomial. The random variable $\ell(\tilde{\theta})$ is then approximated by

$$\ell(\tilde{\theta}) \approx \ell(\theta) + (\tilde{\theta} - \theta)\ell'(\theta) + \frac{1}{2}(\tilde{\theta} - \theta)^2\ell''(\theta).$$

Taking expectations and using the assumption $\mathbf{E}[\tilde{\theta}] = \theta$, we get

$$\mathbf{E}[\ell(\tilde{\theta})] \approx \ell(\theta) + \frac{1}{2}\ell''(\theta)\sigma^2.$$

By squaring the second-order approximation to $\ell(\tilde{\theta})$, we obtain, similarly,

$$\mathbf{Var}[\ell(\tilde{\theta})] \approx [\ell'(\theta)]^2\sigma^2.$$

These approximations give an indication of the uncertainty in $\ell(\tilde{\theta})$. The second, in particular, is useful in bounding the likely values of $\ell(\theta)$, given an estimate of the derivative $\ell'(\theta)$ and a ballpark figure for the variance $\sigma^2$ of $\tilde{\theta}$.

In contrasting this approach based on a Taylor expansion with one based on a (Stirling) difference expansion, Holtzman argues effectively for advantages of differences over derivatives in sensitivity analysis. Even so, when derivative estimates can be obtained at substantially lower computational cost than difference estimates, an analysis based on derivatives may be preferred.

### Simulation Output Analysis

Glynn [37] discusses the role of gradient estimates in simulation output analysis, particularly in forming confidence intervals for functions of estimates. Suppose, as above, that $\ell(\theta) = \mathbf{E}[L(\theta)]$, and that $\theta$ is unknown. Let $\{\tilde{\theta}_n, n = 1, 2, \ldots\}$ be a consistent estimator of $\theta$; i.e., $\tilde{\theta}_n$ converges to $\theta$ as $n \to \infty$, in probability or perhaps almost surely. A natural estimator for $\ell(\theta)$ is, then, $\ell(\tilde{\theta}_n)$.

In a broad range of circumstances, $\{\tilde{\theta}_n\}$ satisfies a central limit theorem, to the effect that as $n \to \infty$, $n^{1/2}(\tilde{\theta}_n - \theta)$ has a normal distribution with mean zero and some variance $\sigma^2$. This is the basis of most techniques for estimating a confidence interval for a point estimate. We would also like a confidence interval for our estimate of $\ell(\theta)$; hence, we need an estimate of the asymptotic variance of $\{\ell(\tilde{\theta}_n), n = 1, 2, \ldots\}$. If $\sigma^2 < \infty$ and $\ell$ is differentiable, then the distribution of $n^{1/2}[\ell(\tilde{\theta}_n) - \ell(\theta)]$ is asymptotically normal with mean zero and standard deviation $\ell'(\theta)\sigma$. Thus, a derivative estimate is needed to form a confidence interval for $\ell(\theta)$. In practice, one might well use an estimate of $\ell'(\tilde{\theta}_n)$ (such as $L'(\tilde{\theta}_n)$) to get at $\ell'(\theta)$.

Glynn and Iglehart [41] provide several central limit theorems that support this analysis. In one example, (their Theorem 1) we can take $\tilde{\theta}_n$ to be a sample mean of $n$ i.i.d. copies of $Y$, an unbiased, finite mean square estimator of $\theta$. The parameter $\theta$ could be a vector. If $\ell$ is differentiable, $\ell(\tilde{\theta}_n)$ satisfies a central limit theorem with asymptotic variance $\nabla\ell(\theta)C\nabla\ell(\theta)^t$, where the superscript $t$ denotes transpose, and $C$ is the covariance matrix of $Y$, $C = \mathbf{E}[Y^tY] - \mathbf{E}[Y^t]\mathbf{E}[Y]$.

## 1.3 The Finite Dimensional Setting

In this section, we consider a special case of the gradient estimation problem, a simplified version of the topic of later chapters. We take the "underlying system" to be a family of random vectors $\{\underline{X}(\theta) = (X_1(\theta), \ldots, X_n(\theta)), \theta \in \Theta\}$, rather than a family of stochastic processes. There are several reasons for considering this finite dimensional setting. Reason enough is that it contains in microcosm many of the essential aspects of the more general theory. It is also a necessary prelude to evaluating stochastic derivatives for functionals of stochastic processes: just as, say, queue length statistics are built up from service times and interarrival times, *derivatives* of queue length statistics are built up from *derivatives* of service times and interarrival times. In addition, random vectors are, for some applications, studied through simulation, so the topic of this section has practical value as well.

### 1.3.1 Unbiased Stochastic Derivatives

Let us assume that the vectors $\{\underline{X}(\theta), \theta \in \Theta\}$ are *a priori* defined on a common probability space. Below, we take up the question of how to effect such a construction. For simplicity, let $\theta$ be scalar and let $\Theta$ be a bounded open interval of the real line. A function $f : \mathbf{R}^n \to \mathbf{R}$ is a performance measure for $\underline{X}$. We want to estimate the derivative of $\ell(\theta) = \mathbf{E}[f(\underline{X}(\theta))]$.

We need to impose some conditions. Let us suppose that

$$\text{for all } \theta \in \Theta \text{ and } i = 1, \ldots, n,\ X_i \text{ is a.s. differentiable at } \theta. \tag{1.8}$$

Define $D_f$ to be the subset of $\mathbf{R}^n$ on which $f$ is continuously differentiable. Suppose that

$$\text{for all } \theta \in \Theta,\ P(\underline{X}(\theta) \in D_f) = 1. \tag{1.9}$$

Under these conditions, the stochastic derivative of $f(\underline{X}(\cdot))$ exists almost surely at every $\theta \in \Theta$ and is given by

$$f'(\underline{X}(\theta)) = \frac{d}{d\theta} f(\underline{X}(\theta)) = \nabla f \cdot \underline{X}'(\theta) = \sum_{i=1}^{n} \partial_i f(\underline{X}(\theta)) X_i'(\theta).$$

Under (1.8)-(1.9), derivative estimation via stochastic derivatives is well-posed.

When is $f'(\underline{X}(\theta))$ an unbiased estimate of $\ell'(\theta)$? Let us review some material on interchanging limits and expectations (see, e.g., Chung [12], especially p.97, for more detail). Recall the definition of

**Uniform Integrability.** A family of random variables $\{Y_h, h \in H\}$, indexed by an arbitrary set $H$, is *uniformly integrable* if

$$\sup_h \mathbf{E}[|Y_h| \cdot \mathbf{1}\{|Y_h| > x\}] \to 0 \quad \text{as } x \to \infty.$$

(The notation $\mathbf{1}\{\cdot\}$ denotes the indicator of $\{\cdot\}$.)

Uniform integrability is a necessary and sufficient condition for the interchange of limit and expectation, in the following sense: If $\mathbf{E}[|Y_n|] < \infty$, $n = 1, 2, \ldots$, and $Y_n \to Y$ in probability, then $\lim_{n\to\infty} \mathbf{E}[|Y_n|] = \mathbf{E}[|Y|]$ if and only if $\{Y_n, n = 1, 2 \ldots\}$ is uniformly integrable. The following is an immediate consequence:

**Theorem 1.1.** Suppose (1.8) and (1.9) hold and that $\mathbf{E}[|f(X(\theta))|] < \infty$ throughout $\Theta$. Then $\mathbf{E}[f'(\underline{X}(\theta))] = \ell'(\theta)$ if and only if for some $h_1 > 0$, the family $\{h^{-1}[f(\underline{X}(\theta + h)) - f(\underline{X}(\theta))], 0 < |h| < h_1\}$ is uniformly integrable.

While uniform integrability provides the most general means of establishing the interchange of derivative and expectation, it is not necessarily the most convenient. Our proofs rely instead on the

**Dominated Convergence Theorem.** If $\lim_{n\to\infty} Y_n = Y$, a.s., and if, for all $n$, $|Y_n| \le X$, a.s., with $\mathbf{E}[X] < \infty$, then $\lim_{n\to\infty} \mathbf{E}[Y_n] = \mathbf{E}[Y]$.

Thus, for the "if" half of Theorem 1.1 it is enough to show that the difference quotients $h^{-1}[f(\underline{X}(\theta+h)) - f(\underline{X}(\theta))]$ are almost surely bounded by an integrable random variable. We use dominated convergence in conjunction with continuity and the following

**Generalized Mean Value Theorem.** Let $G : \mathbf{R} \to \mathbf{R}$ be a continuous function which is differentiable on a finite interval $[a, b]$, except possibly on a set $\tilde{D}$ of countably many points. Then for all $x$ and $x + h$ in $(a, b)$

$$\left| \frac{G(x+h) - G(x)}{h} \right| \leq \sup_{a \leq y \leq b, y \notin \tilde{D}} |G'(y)|.$$

(See Dieudonné [18], p.160.)

Impose the following continuity condition:

$$f(\underline{X}(\cdot)) \text{ is a.s. continuous and piecewise differentiable throughout } \Theta. \quad (1.10)$$

This does not follow from (1.8) and (1.9). Those conditions imply that, for each $\theta$, the set of $\omega$'s at which differentiability fails has probability zero. The set of $\omega$'s for which continuity fails at *some* $\theta$ could still have positive probability. Condition (1.10) rules out this possibility. We now have

**Theorem 1.2.** Suppose (1.8)-(1.10) hold on a compact interval $\Theta$. Let $\tilde{D} = \tilde{D}(\omega)$ be the (random) subset of $\Theta$ at which $f(\underline{X}(\cdot))$ fails to be differentiable. If $\mathbf{E}[\sup_{\theta \notin \tilde{D}} |f'(\underline{X}(\theta))|] < \infty$, then $\mathbf{E}[f'(\underline{X}(\theta))] = \ell'(\theta)$ on $\Theta$.

**Proof.** From the generalized mean value theorem we obtain

$$\left| \frac{f(\underline{X}(\theta + h) - f(\underline{X}(\theta))}{h} \right| \leq \sup_{\theta \notin \tilde{D}} |f'(\underline{X}(\theta))|.$$

The right side is integrable, by hypothesis; thus, the dominated convergence theorem applies, and we have

$$\begin{aligned} \mathbf{E}[f'(\underline{X}(\theta))] &= \mathbf{E}\left[\lim_{h \to 0} \frac{f(\underline{X}(\theta + h)) - f(\underline{X}(\theta))}{h}\right] \\ &= \lim_{h \to 0} \mathbf{E}\left[\frac{f(\underline{X}(\theta + h)) - f(\underline{X}(\theta))}{h}\right] \\ &= \ell'(\theta). \end{aligned}$$

That is, the derivative on the right exists and is equal to the expectation on the left. □

## 1.3.2 Constructions of Parametric Families

We now address the construction of parametric families of random variables on a common probability space. Such a construction is presupposed by the results of Section 1.3.1. We consider only families of positive real-valued random variables depending on a scalar parameter $\theta$. Our starting point is a family of distributions $\{F(x,\theta), x \geq 0, \theta \in \Theta\}$ on the positive real line. We are to construct random variables $\{X(\theta), \theta \in \Theta\}$ under the constraint that each $X(\theta)$ have distribution $F(\cdot,\theta)$. In addition, we want $X(\cdot)$ to be a.s. differentiable at every $\theta \in \Theta$, and a.s. continuous throughout $\Theta$. Our discussion draws on Suri and Zazanis [79], Section 4, and Glynn [37].

With no essential loss of generality we may take our probability space $(\Omega, \mathcal{F}, P)$ to be the unit interval $[0,1]$ with the $\sigma$-algebra of Borel sets and Lebesgue measure. To put it another way, we may construct all $X(\theta)$'s from a single random variable, $U$, uniformly distributed on $[0,1]$. An algorithm for sampling from $F(\cdot,\theta)$ given $U$ is a function $g(\cdot,\theta)$ satisfying

$$P(g(U,\theta) \leq x) = F(x,\theta), \quad \text{for all } x \geq 0. \tag{1.11}$$

By defining $X(\theta,u) = g(u,\theta)$, for $u \in \Omega = [0,1]$ and $\theta \in \Theta$, we obtain a legitimate construction of $\{X(\theta), \theta \in \Theta\}$. We omit the argument $u$ except to emphasize a particular construction.

### The Inversion Representation

Books on simulation catalog a variety of algorithms for transforming uniformly distributed variates to more general variates. The most general such algorithm is the method of *inversion*. It sets

$$g(u,\theta) = F^{-1}(u,\theta) \equiv \sup\{x \geq 0 : F(x,\theta) \leq u\}.$$

This choice of $g$ always satisfies (1.11), as is easily checked.

Suri [76] recognized inversion as a mechanism for modeling the change in $X$ under a change in $\theta$. He noted that if $X(\theta,u) = F^{-1}(u,\theta)$, then the value of $X(\theta+h)$ corresponding to the same outcome, $u$, is given by $F^{-1}(F(X(\theta,u),\theta),\theta+h)$. If $F^{-1}$ is differentiable in $\theta$, we directly obtain $X'(\theta,u) = \partial_\theta F^{-1}(u,\theta)$. For example, if $F(\cdot,\theta)$ is the exponential distribution with mean $\theta$, $F(x,\theta) = 1-\exp(-x/\theta)$, then $F^{-1}(x,\theta) = -\theta\ln(1-u)$. Setting $X(\theta,u) = -\theta\ln(1-u)$ we get $dX(\theta)/d\theta = -\ln(1-u) = X(\theta)/\theta$.

Under stronger but reasonable conditions on $F(x,\theta)$, the inversion representation leads to an important simplification, embodied in

**Theorem 1.3.** (Suri and Zazanis [79], Glynn [37].) Let $\Theta$ be an open interval. Suppose that *(i)* for each $\theta \in \Theta$, $F(\cdot,\theta)$ has a density $\partial_x F(\cdot,\theta)$ which is strictly positive on an open interval $I_\theta$ and zero elsewhere; and (*ii*) $F$ is continuously differentiable on $\{(x,\theta) : x \in I_\theta, \theta \in \Theta\}$. Let $X(\theta,u) = F^{-1}(u,\theta)$. Then for $0 < u < 1$ and $\theta \in \Theta$,

$$\frac{\partial X(\theta,u)}{\partial\theta} = -\frac{\partial_\theta F(X(\theta),\theta)}{\partial_x F(X(\theta),\theta)}. \tag{1.12}$$

The numerator on the the right side is the partial derivative of $F$ with respect to its second argument only.

**Proof** ([37]). Condition *(ii)* implies that for $0 < u < 1$ and $\theta \in \Theta$, $X(\theta, u)$ is uniquely determined by $F(X(\theta, u), \theta) = u$; hence, for $\theta + h \in \Theta$,

$$F(X(\theta + h, u), \theta + h) = u = F(X(\theta, u), \theta). \tag{1.13}$$

By the (ordinary) mean value theorem,

$$F(X(\theta + h, u), \theta + h) = \\ F(X(\theta, u), \theta) + \partial_\theta F(\hat{x}, \hat{\theta})h + \partial_x F(\hat{x}, \hat{\theta})[X(\theta + h, u) - X(\theta, u)],$$

where $(\hat{x}, \hat{\theta})$ lies on the line segment joining $(X(\theta, u), \theta)$ and $(X(\theta+h, u), \theta+h)$. Cancelling according to (1.13) and rearranging terms, we get

$$X(\theta + h, u) - X(\theta, u) = -\partial_\theta F(\hat{x}, \hat{\theta})h/\partial_x F(\hat{x}, \hat{\theta}).$$

Dividing both sides by $h$ and letting $h \to 0$ yields (1.12) via continuity of the partial derivatives of $F$. □

It is easy to check that for the exponential distribution (1.12) indeed simplifies to $dX(\theta)/d\theta = X(\theta)/\theta$. The significance of (1.12) is that it expresses $dX(\theta)/d\theta$ as a function of $X(\theta)$. Thus, in simulating, one may generate $X(\theta)$ using any algorithm, then generate its derivative using the expression (1.12), which is merely *derived* from inversion. There is also an important implication for observed systems. If $X(\theta)$ is, say, a service time at a queue, one would typically be able to observe $X(\theta)$ itself, but not the (hypothetical) random outcome of $U$ that determines $X(\theta)$. Thus, when $dX(\theta)/d\theta$ depends directly on the outcome of $U$, it cannot be observed from a sample path of the queue. But if it depends only on $X(\theta)$ it can be "observed" indirectly. This leads us, in several places, to impose the more general assumption (more general than (1.12)) that there is *some* function $\psi$ for which $dX(\theta)/d\theta = \psi(X(\theta), \theta)$.

### Scale and Location Parameters

Two special classes of distributions for which a particularly simple such $\psi$ is available are the *scale* and *location* families: $\theta$ is a scale parameter if $F(x, \theta) = \hat{F}(x/\theta)$, for all $\theta$, (for some distribution $\hat{F}$) and a location parameter if $F(x, \theta) = \hat{F}(x - \theta)$ (for some $\hat{F}$). When $F(x, \theta)$ satisfies the conditions for Theorem 1.3, (1.12) becomes $dX(\theta)/d\theta = X(\theta)/\theta$ for a scale parameter, and $dX(\theta)/d\theta = 1$ for a location parameter. (The mean of the exponential distribution is a scale parameter: if $X$ is exponential, then so is $\theta X$.) In fact, these relations hold more generally than the conditions of Theorem 1.3. If $\theta$ is a scale parameter, then there is a random variable $Y$ (not depending on $\theta$) such that $\theta Y$ has distribution $F(\cdot,\theta)$, for all $\theta$. For a location parameter, $Y$ may be chosen so that $Y+\theta$ has distribution $F(\cdot,\theta)$. Thus, in the first case we may take $X(\theta) = \theta Y$ and in the second $X(\theta) = Y + \theta$. From these representations of $X$ the expressions for $dX(\theta)/d\theta$ follow.

Scale and location parameters are convenient for implementation and broadly applicable in practice. Many natural parameters of standard distributions are of one of these forms. One might also take a hybrid of these cases as a starting point, postulating that the effect of a parameter on a random variable has, for example, the form $X(\theta) = (aY + b)\theta$, for some $Y$, and some constants $a$ and $b$.

### Discrete Distributions

Consider the distribution of a random variable that takes on only finitely many values $x_1, \dots, x_n$ with respective probabilities $p_1, \dots, p_n$, $p_i > 0$, $i = 1, \dots, n$, $\sum_i p_i = 1$. This distribution is given by

$$F(x) = \sum_{i=1}^{n} p_i \mathbf{1}\{x_i \le x\}.$$

Its inverse, defined on $[0,1]$, is

$$F^{-1}(u) = x_{N(u)}, \quad \text{where} \quad N(u) = \min\{n : \sum_{i=1}^{n} p_i \ge u\}.$$

A parametric family of such distributions can be defined by allowing the atoms $x_i$, $i = 1, \dots, n$, or the masses $p_i$, $i = 1, \dots, n$, or both, to vary with a parameter $\theta$. Let us assume, in every case, that the $x_i$'s and $p_i$'s are continuously differentiable in $\theta$ if they depend on $\theta$. Suppose, first, that only the atoms vary. Write $F(x,\theta)$ to reflect the dependence on $\theta$, and define $X(\theta, u) = F^{-1}(u,\theta)$. Then, for all $\theta \in \Theta$ and $u \in [0,1]$, $\partial_\theta X(\theta,u) = x'_{N(u)}(\theta)$. To put it another way, $X'(\theta) = x'_i(\theta)$ whenever $X(\theta) = x_i(\theta)$; and $X$ is continuously differentiable in $\theta$ for every $u$.

If, on the other hand, only the masses vary, the situation is quite different. When the $p_i$'s depend on $\theta$, $N \equiv N(\theta,u)$ does, too. However, it is easy to see that, for each $u \in [0,1]$, $N(\cdot,u)$ is piecewise constant in $\theta$; hence, $X(\theta,u) \equiv x_{N(\theta,u)}$ is also piecewise constant. It jumps from one $x_i$ to another at those $(\theta,u)$ values at which, for some $1 \le k \le n$, $\sum_{i=1}^{k} p_i(\theta) = u$; elsewhere, it is constant. It follows that $\partial_\theta X(\theta,u)$ is zero, wherever it exists. Since, in general, $\partial_\theta \sum_{i=1}^{n} x_i p_i(\theta) \ne 0$, we have $\mathbf{E}[X'(\theta)] \ne \mathbf{E}[X(\theta)]'$. In fact, we can make a stronger claim: if only the $p_i$'s vary with $\theta$ (and they are not constant in $\theta$) then *no* construction of $\{X(\theta), \theta \in \Theta\}$ is almost surely continuous in $\theta$. For this reason, parametric families of discrete distributions in which the masses vary with the parameter are generally not suited to derivative estimation via stochastic derivatives.

### From Perturbed Variables to Perturbed Processes

We end this section with a brief description of how derivatives of parametric random variables translate into derivatives of performance measures for parametric stochastic processes. Think of a (parametric) simulation algorithm as generating a sequence $\{X_i(\theta), i = 1, 2, \ldots\}$ of random variables (representing, e.g., service times, interarrival times, etc.) and transforming these into a sample path of $\{Z_t(\theta), t \geq 0\}$ (representing, e.g., a queue-length process). A perturbation analysis derivative estimation algorithm evaluates the effect of an "infinitesimal" change in $\theta$ on the $X_i(\theta)$'s, and the effect of changes in the $X_i(\theta)$'s on the sample path of $\{Z_t(\theta), t \geq 0\}$. Thus, it consists of two parts; the first is commonly called *perturbation generation*, the second, *perturbation propagation.*

An instance of propagation is illustrated in Figure 1.1. As noted earlier, a small perturbation in a service time is propagated to subsequent jobs in the same busy period. But Figure 1.1 says nothing of how perturbations in service times are introduced in the first place. If the service times depend on the parameter $\theta$, and the perturbations in service times arise from perturbations in $\theta$, then we need to know how a small $\Delta\theta$ maps to each $\Delta X_i(\theta)$. In other words, we need to know $dX_i(\theta)/d\theta$. The examples above show how to construct $X(\theta)$ and evaluate its derivative in a variety of cases.

Consider the system of Figure 1.1 with exponential, mean $\theta$ service times. Let $X(\theta)$ be a generic service time. Then

$$\frac{dX}{d\theta} = \frac{X}{\theta}; \tag{1.14}$$

$X/\theta$ is the perturbation *generated* during the service time $X$. (If $\theta$ is the service *rate*, the reciprocal of the mean, then $dX/d\theta = -X/\theta$.) Each service time generates such a perturbation, which is propagated as in the figure. By tracking perturbations along a sample path, we evaluate the derivatives of the transition epochs—i.e., the epochs of arrivals and departures. From derivatives of transition epochs, we evaluate derivatives of performance measures.

*　　　*　　　*

## Notes and Comments

The first "perturbation analysis"-like algorithm was formulated in Ho, Eyler, and Chien [50]. Its pupose was to determine the effect of perturbing a discrete parameter, a buffer size; it estimated differences rather than derivatives. Hence, this early ancestor is quite different from its descendants. The first algorithm to evaluate derivatives directly is the one in Ho and Cao [51], which estimates derivatives of throughput with respect to service rates in Jackson networks. In conjuction with this algorithm, Suri [76] introduced the inversion model (described in Section 1.3) for the effect of parameter changes on sampled random variables. (Glasserman [23] considers perturbations in non-inversion methods.) Suri and Zazanis [79], in developing a derivative estimation algorithm for the single-server queue, also laid some of the early groundwork. A 1987 overview of the subject can be found in Ho [49]

The interchange of derivative and expectation as a fundamental issue was formulated in Cao [5], where the role of continuity is also discussed. A general discussion of continuity and its significance for derivative estimation is given in Glasserman [26], from which Theorem 1.2 is taken. The role of the construction of a family of processes in ensuring continuity (implicit in earlier work) is pointed out explicitly in Glasserman [25] and Glynn [37].

The use of gradient estimates in stochastic optimization is an active area of research. A standard reference for stochastic approximation in general is Kushner and Clark [61]. Theoretical and empirical work linking general results with gradient estimation includes Glynn [38], Suri and Leung [78], Fu [22], and L'Ecuyer, Giroux, and Glynn [62]. It is likely that renewed interest in stochastic approximation stimulated through results in gradient estimation will lead to improved optimization algorithms.

# Chapter 2

# Generalized Semi-Markov Processes

Generalized semi-Markov processes (GSMPs) provide a framework for studying stochastic discrete event systems. In this chapter, we define and give examples of GSMPs, then give a detailed construction of their sample paths. Using this construction, we consider families of GSMPs depending on a parameter and introduce general expressions for stochastic derivatives with respect to the parameter. This leads to derivative estimates for a broad class of performance indices associated with GSMPs, and a simple algorithm that implements these estimates. The crucial question of when the estimates are unbiased is postponed to Chapter 3.

## 2.1 Examples and Definition

We begin with a verbal description of generalized semi-Markov processes. The states of a GSMP represent possible "physical" configurations of a system, which need not be states in the Markovian sense. In a queueing context, the state may be simply a vector of queue lengths, perhaps supplemented by information about the classes of jobs in queue, which servers are blocked, etc. The process jumps from state to state upon the occurrence of *events*; typical events are departures from and external arrivals to queues. The state to which the process moves when an event occurs is governed by a set of transition probabilities. In a queueing network, these determine the routing of jobs. Just when events occur is determined by random *clocks* associated with the possible events in a state. Each clock represents the time remaining until the associated event occurs, so the event with the shortest remaining clock time is the next to occur.

When, for example, the events are arrivals to and departures from a queue, the initial settings of the respective clocks are simply interarrival times and service times. After being set, clocks run down at unit rate. When a clock runs out, the corresponding event occurs, the process changes state, and new clocks may be set for new events possible in the new state.

A GSMP is therefore characterized by the following elements:

**S** A state space (finite or countable) representing the set of physical states of a system.

**A** A finite set events. Typical events will be denoted by $\alpha, \beta$.

$\mathcal{E}(s)$ The set of possible events, the *event list*, in state $s$. For example, service completion at a queue is only a possible event in those states in which the server is busy. We never allow $\mathcal{E}(s)$ to be empty.

$p(s'; s, \alpha)$ The probability of jumping to state $s'$ when event $\alpha$ occurs in state $s$.

$\{F_\alpha, \alpha \in \mathbf{A}\}$ The set of distributions (one for each $\alpha \in \mathbf{A}$) of new clock samples for events. If, for example, $\alpha$ is an external arrival to a queue, then $F_\alpha$ is the interarrival time distribution, while if $\alpha$ is the departure from a server then $F_\alpha$ is the service time distribution.

The precise definition of a GSMP as a stochastic process is somewhat involved, though the intuition behind it is quite natural. For this reason, we first give three examples of how GSMPs are used to model familiar systems, postponing the definition of a GSMP until after the examples. These examples will also be useful later.

**Example 1.** *GI/G/1 queue.* Take **S** to be the non-negative integers (the set of possible queue lengths). Let $\alpha$ denote arrival and $\beta$ denote service completion; then $F_\alpha$ and $F_\beta$ are the interarrival and service time distributions. An arrival is a possible event in every state, but a service completion is possible only when the server is busy; hence, $\mathcal{E}(s) = \{\alpha, \beta\}$ if $s > 0$, and $\mathcal{E}(0) = \{\alpha\}$. Upon the occurrence of an arrival, the state increases by one, while at a departure it decreases by one. This is captured by setting $p(s+1; s, \alpha) = 1$ and $p(s-1; s, \beta) = 1$ (for $s > 0$), and all other transition probabilities equal to zero.

**Example 2.** *Jackson-like networks.* By a *Jackson-like* network we mean a network of single server, infinite capacity, first come, first served queues, with a single class of jobs whose transitions are governed by a Markov routing matrix

$P$. Thus, upon completion of service at queue $i$, a job moves immediately to queue $j$ with probability $P_{ij}$, independent of everything else. (An ordinary Jackson network is the special case in which all service and interarrival times are exponential.) To consider the open and closed cases together easily, we consider only one (renewal) stream of arrivals. We include a fictitious node 0, which is never idle, from which all arrivals to the network originate and to which all departures from the network go. In a closed network, $P_{00} = 1$ and all other $P_{i0}$ and $P_{0i}$ are zero.

To obtain a GSMP description, let $\mathbf{S}$ be the set of possible queue-length vectors $s = (n_1, \ldots, n_M)$ where $n_i \geq 0$ is the number of jobs at queue $i$, and $M$ is the number of nodes. In the closed case, $\mathbf{S}$ is restricted to $\{s : \sum_i n_i = K\}$, where $K$ is the network population. Let $\beta_i$ denote service completion at node $i$, and let $\beta_0$ denote an external arrival (in the open case). Then $F_{\beta_i}$ is the service time distribution at server $i$ (and $F_{\beta_0}$ is the interarrival time). The event lists, $\mathcal{E}(s)$, are characterized by $\beta_i \in \mathcal{E}(s)$, $i = 1, \ldots, M$, if and only if $n_i > 0$ in $s$ (and $\beta_0 \in \mathcal{E}(s)$ for every $s$ in the open case). The transition probabilities are given by $p(s - e_i + e_j; s, \beta_i) = P_{ij}$, where $e_i$ is the $i$th unit vector and $e_0$ is the vector of all zeros.

**Example 3.** *Cyclic server queues.* In cyclic server models, a single resource provides service to $M$ queues, one at a time. There are many variants; we consider only one. Service is *exhaustive*, meaning that the server continues to serve a queue until it empties. At that point, it immediately begins serving the next queue. To get a GSMP description, we represent a typical state by $(n_1, \ldots, n_M, J)$ where $n_i$ is the number in queue $i$, and $J$ is the index of the queue being served. (Set $J = 0$ if all queues are empty.) Let $\alpha_i$ be arrival at queue $i$ and $\beta_i$ service completion at queue $i$, $i = 1, \ldots, M$; $\beta_i \in \mathcal{E}(s)$ if and only if $J = i$ in $s$. If $J = i$ and $n_i > 1$, then

$$p((n_1, \ldots, n_i - 1, \ldots, n_M, i); (n_1, \ldots, n_i, \ldots, n_M, i), \beta_i) = 1;$$

while if $n_i = 1$, upon occurrence of $\beta_i$, the server moves to the next non-empty queue. Thus, if $n_{i+1} > 0$,

$$p((n_1, \ldots, 0, n_{i+1}, \ldots, n_M, i + 1); (n_1, \ldots, 1, n_{i+1}, \ldots, n_M, i), \beta_i) = 1.$$

If $i = M$, then the "next" queue is understood to be queue 1. The occurrence of $\alpha_i$ increments $n_i$ by one, but does not change $J$ unless $J = 0$, in which case

$$p((e_i, i); (0, \ldots, 0, 0), \alpha_i) = 1.$$

We now turn to the precise definition of a GSMP, adopting the approach in Whitt [84]. We define a (continuous time) GSMP via a discrete time, general state space Markov chain. This discrete time process records the state and the clock readings at each transition.

For $s \in \mathbf{S}$, define the set $C_s$ of possible clock readings in state $s$ by

$$C_s = \{(c(\alpha), \alpha \in \mathbf{A}) : c(\alpha) \geq 0, \text{ and } c(\alpha) > 0 \text{ only if } \alpha \in \mathcal{E}(s)\}. \tag{2.1}$$

Since $\mathbf{A}$ is a finite set, we may (and do) identify a sequence $(c(\alpha), \alpha \in \mathbf{A})$ with an $|\mathbf{A}|$-dimensional vector, and view $C_s$ as a subset of $|\mathbf{A}|$-dimensional Euclidean space. For $s \in \mathbf{S}$ and $c \in C_s$, let

$$t^*(s, c) = \min\{c(\alpha) : \alpha \in \mathcal{E}(s)\}; \tag{2.2}$$

then $t^*(s, c)$ represents the time until the next event occurs, when the state is $s$ and the vector of clock readings is $c$. Let the elements of $\mathbf{A}$ be ordered in some way and let

$$\alpha^*(s, c) = \min\{\alpha : c(\alpha) = t^*(s, c)\}; \tag{2.3}$$

then $\alpha^*$ is the event that triggers the transition out of $s$, because it has the shortest remaining clock time. We use the "min" as an arbitrary way to break ties: this definition picks out a unique triggering event even when more than one clock runs out at the same time.

For $s, s' \in \mathbf{S}$ and $\alpha \in \mathbf{A}$ define the sets $\mathcal{N}(s'; s, \alpha)$ and $\mathcal{O}(s'; s, \alpha)$ of *new* and *old* events by

$$\mathcal{N}(s'; s, \alpha) = \mathcal{E}(s') \backslash (\mathcal{E}(s) - \{\alpha\}) \tag{2.4}$$

and

$$\mathcal{O}(s'; s, \alpha) = \mathcal{E}(s') \cap (\mathcal{E}(s) - \{\alpha\}). \tag{2.5}$$

At a transition from $s$ to $s'$ triggered by $\alpha$, new clocks are set for events in $\mathcal{N}(s'; s, \alpha)$; clocks for events in $\mathcal{O}(s'; s, \alpha)$ continue to run down. Clocks for events in $\mathcal{E}(s) \backslash \mathcal{O}(s'; s, \alpha)$ are *interrupted* and set to zero; the associated event ceases to be a "possible" next event.

We can now define a discrete time Markov process $\{(Y_n, c_n), n \geq 0\}$ with the interpretation that $Y_n$ is the $n$th (GSMP) state and $c_n$ is the vector of clock readings just after the $n$th transition. This process has state space

$$\Sigma = \bigcup_{s \in \mathbf{S}} (\{s\} \times C_s).$$

It is characterized by the following transition probability: For $A \subseteq \Sigma$ of the form

$$A = \{s'\} \times \{c' \in C_{s'} : c'(\alpha) \leq x(\alpha), \alpha \in \mathcal{E}(s')\}$$

for some fixed real numbers $(x(\alpha), \alpha \in \mathbf{A})$, the transition probability from $(s, c)$ to $A$ is

$$P((s, c), A) = p(s'; s, \alpha^*) \prod_{\alpha \in \mathcal{N}} F_\alpha(x(\alpha)) \prod_{\alpha \in \mathcal{O}} \mathbf{1}\{c(\alpha) - t^* \leq x(\alpha)\},$$

where $t^* = t^*(s, c)$, $\alpha^* = \alpha^*(s, c)$, $\mathcal{N} = \mathcal{N}(s'; s, \alpha^*)$ and $\mathcal{O} = \mathcal{O}(s'; s, \alpha^*)$.

A GSMP is a continuous time process with right-continuous, piecewise constant sample paths, which stays in state $Y_n$ for an interval of length $t^*(Y_n, c_n)$. Let $\tau_0 = 0$ and

$$\tau_n = \sum_{i=0}^{n-1} t^*(Y_n, c_n),$$

so that $\tau_n$ is the epoch of the $n$th transition. Also let

$$N(t) = \sup\{n \geq 0 : \tau_n \leq t\};$$

and suppose that $N(t)$ is a.s. (almost surely) finite for finite $t$ (we will impose conditions that ensure this). Then the GSMP $Z_t$ associated with $\{(Y_n, c_n)\}$ is defined by

$$Z_t = Y_{N(t)};$$

i.e., $Z_t$ is the state of the GSMP at time $t$.

If in every state there is just one active event, the GSMP is an ordinary semi-Markov process. If all clock distributions are exponential, the process is Markov.

We introduce GSMPs to study their performance, so some mention of the kind of performance measures we consider is in order. We focus almost exclusively on performance measures of the general form

$$L = \int_0^T f(Z_t)dt,$$

where $f$ is a bounded, real-valued function on $\mathbf{S}$, and $T$ is in a restricted class of *stopping times.* (Among other things, $T$ could be either a fixed time or the epoch of the $k$th transition.) This represents a fairly broad class of finite-horizon performance measures, from which even more general measures can be obtained by taking differences and ratios. For example, if $Z$ is a queueing system and $f(s)$ is the queue length at some queue, then $L/T$ is the mean queue length in the interval $[0, T]$. If $T$ is the epoch of the $k$th departure from some queue, and $f \equiv 1$, then $k/L$ is the throughput of the queue over the horizon $[0, T]$. More generally, if $f(s)$ represents the rate at which "cost" is incurred in state $s$, then $L$ is the total cost incurred up to time $T$. While most of our development focuses on finite-horizon performance, this analysis is relevant to the infinite-horizon setting. Steady-state quantities are often estimated by the limit of $L/T$ as $T \to \infty$, and also by the *regenerative method* which uses ratios of quantities like $L$, but with the integral taken over a regenerative *cycle.* Steady-state performance and derivative estimation is the subject of Chapter 8.

## 2.2 Construction of GSMPs

In order to develop derivative estimates for GSMPs depending on a parameter, we need an explicit construction of their sample paths. This section carries out a detailed construction for a single GSMP, which is easily extended to the parametric case in Section 2.3.

The construction we undertake is the "obvious" one corresponding to the definition of a GSMP given in Section 2.1. It can be thought of as a general simulation algorithm for a GSMP. Our probability triple $(\Omega, \mathcal{F}, P)$ is a generic space for a simulation driven by independent, uniformly distributed random numbers. A typical element $\omega \in \Omega$ is a sequence

$$\omega = \{(u(\alpha, k), v(\alpha, k)), \alpha \in \mathbf{A}, k = 1, 2, \ldots\},$$

where every $u(\cdot, \cdot)$ and $v(\cdot, \cdot)$ is in $[0, 1]$. For every $\alpha$ and $k$, define *coordinate random variables* $U(\alpha, k)$ and $V(\alpha, k)$, mapping $\Omega$ to $[0, 1]$, by $U(\alpha, k) = U(\alpha, k)(\omega) = u(\alpha, k)$, $V(\alpha, k) = V(\alpha, k)(\omega) = v(\alpha, k)$. The probability, $P$, is the product measure that makes all $U$'s and $V$'s independent and gives each a uniform distribution on $[0, 1]$: For all $n \geq 1$, for all $\alpha_i, k_i$ and $\alpha'_i, k'_i$, $i = 1, \ldots, n$, and all $0 \leq x_i, x'_i \leq 1$, $i = 1, \ldots, n$,

$$P(U(\alpha_i, k_i) \leq x_i, V(\alpha'_i, k'_i) \leq x'_i, i = 1, \ldots, n) = \prod_{i=1}^{n} x_i \prod_{i=1}^{n} x'_i.$$

The $\sigma$-algebra $\mathcal{F}$ plays no explicit role in our discussion; for concreteness, let it be the one generated by products of Borel sets on $[0, 1]$; cf. Chung [12], p.57.

In our construction, $U(\alpha, k)$ determines the state transition at the $k$th occurrence of event $\alpha$, and $V(\alpha, k)$ determines the $k$th sample from the clock-setting distribution $F_\alpha$. The sample itself is denoted by $X(\alpha, k)$. There is an extensive literature on transforming uniform random variables to sample from other distributions. We take as given, for each $\alpha \in \mathbf{A}$, a function $g_\alpha : [0, 1] \to \mathbf{R}$ with the property that if $V$ is uniformly distributed on $[0, 1]$, then, for all $x$,

$$P(g_\alpha(V) \leq x) = F_\alpha(x). \tag{2.6}$$

We then define $X(\alpha, k) = g_\alpha(V(\alpha, k))$ for every $\alpha$ and $k$. One way to choose $g_\alpha$, called *inversion* (and discussed in Chapter 1) takes $g_\alpha = F_\alpha^{-1}$. This choice always satisfies (2.6)

Transforming $U(\alpha, k)$ to generate transitions means sampling from a distribution $p(\cdot; s, \alpha)$. We assume we are given a mapping $\phi : \mathbf{S} \times \mathbf{A} \times [0, 1] \to \mathbf{S}$ satisfying, for all $s, s'$ and $\alpha \in \mathcal{E}(s)$

$$P(\phi(s, \alpha, U) = s') = p(s'; s, \alpha),$$

when $U$ is uniform on $[0,1]$. Then if the $k$th occurrence of $\alpha$ occurs in state $s$, the next state is given by $\phi(s,\alpha,U(\alpha,k))$. We call the sequence $\{X(\alpha,k),\alpha\in\mathbf{A},k=1,2,\ldots\}$ the *clock samples*, and the sequence $\{U(\alpha,k),\alpha\in\mathbf{A},k=1,2,\ldots\}$ the *routing indicators*. It is significant that we assign separate streams of random numbers to state transitions due to different event types (but the significance may not be clear until Section 3.1.2).

Examples 1-3 of Section 2.1 share a property which is important in the construction of sample paths: they are *non-interruptive* GSMPs, meaning that the occurrence of one event never interrupts clocks for other events. At every transition, all clocks from the previous state (except the one that ran out) continue to run down. A clock, once set, eventually runs out and causes the associated event to occur. We will need this condition to obtain unbiased derivative estimates; our construction is greatly simplified if we impose it from the outset:

**(C1). Non-interruption Condition.** For every $s,s'\in\mathbf{S}$ and $\alpha\in\mathbf{A}$, if $\alpha\in\mathcal{E}(s)$ and $p(s';s,\alpha)>0$, then $\mathcal{E}(s)-\{\alpha\}\subseteq\mathcal{E}(s')$.

We now take up the construction. As in Section 2.1, we denote the GSMP itself by $\{Z_t,t\geq 0\}$. We denote additional sample path characteristics as follows:

$$\begin{aligned}
\tau_n &= \text{the epoch of the } n\text{th state transition;}\\
a_n &= \text{the } n\text{th event;}\\
Y_n &= \text{the } n\text{th state visited by the process;}\\
c_n &= \text{the vector of clock readings just after the } n\text{th transition;}\\
c_n(\alpha) &= \text{at } \tau_n\text{, the time remaining until } \alpha \text{ occurs, if } \alpha\in\mathcal{E}(Y_n);\\
N(\alpha,n) &= \text{number of instances of } \alpha \text{ among } a_1,\ldots,a_n.
\end{aligned}$$

### Recursive Construction

Let an initial state $Y_0$ be given. Initialize by setting $\tau_0=0$ and every $N(\alpha,0)=0$, and by setting clocks for the possible events: if $\alpha\in\mathcal{E}(Y_0)$ then set $c_0(\alpha)=X(\alpha,1)\ (=g_\alpha(V(\alpha,1)))$. Now repeat the following recursions (using $t^*$ and $\alpha^*$ as defined in (2.2) and (2.3)):

$$\tau_{n+1}=\tau_n+t^*(Y_n,c_n); \tag{2.7}$$

$$a_{n+1}=\alpha^*(Y_n,c_n); \tag{2.8}$$

$$N(\alpha,n+1)=\begin{cases} N(\alpha,n)+1, & \alpha=a_{n+1},\\ N(\alpha,n), & \text{otherwise;}\end{cases} \tag{2.9}$$

$$Y_{n+1}=\phi(Y_n,a_{n+1},U(a_{n+1},N(a_{n+1},n+1))). \tag{2.10}$$

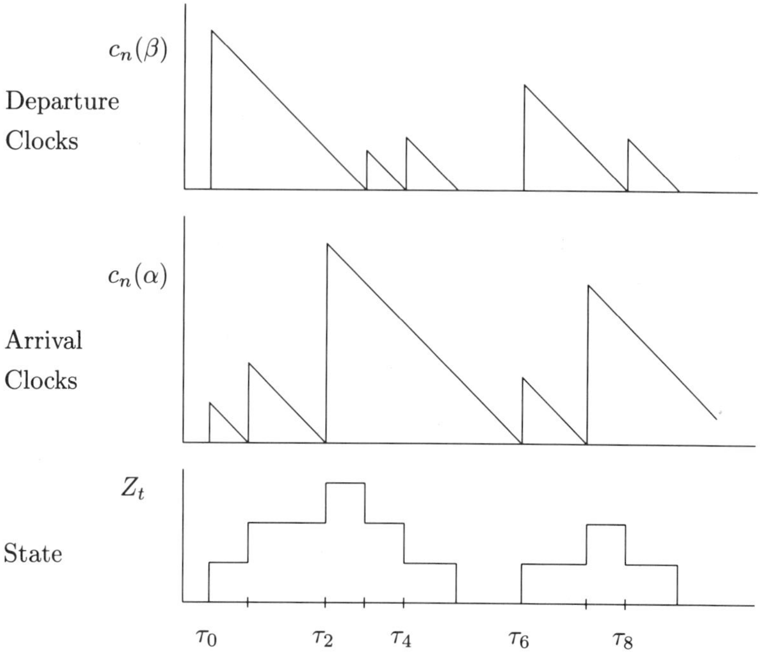


Figure 2.1: GSMP view of the GI/G/1 queue

At each state transition, adjust the clock readings by setting clocks for any new events and reducing the time left on any old clocks by the time since the last transition. Thus, if $\alpha \in \mathcal{E}(Y_n)$ and $\alpha \neq a_{n+1}$, then, under (C1), $\alpha \in \mathcal{E}(Y_{n+1})$ and

$$c_{n+1}(\alpha) = c_n(\alpha) - t^*(Y_n, c_n); \tag{2.11}$$

if $\alpha \in \mathcal{E}(Y_{n+1})$ and either $\alpha \notin \mathcal{E}(Y_n)$ or $\alpha = a_{n+1}$, then

$$c_{n+1}(\alpha) = X(\alpha, N(\alpha, n+1) + 1). \tag{2.12}$$

From these recursions, we obtain $Z$ from $Y$ as before. In particular, if every $t^*(Y_n, c_n) > 0$, then $Z_t = Y_n$ on $[\tau_n, \tau_{n+1})$.

Figure 2.1 illustrates how clock samples for arrivals and departures drive the state of a single server queue under this construction. The vertical jumps of the clock processes correspond to the setting of new clocks; thus, the $k$th jump of $\{c_n(\alpha)\}$ would have height $X(\alpha, k)$. When a departure clock runs out, the queue length, $Z_t$, jumps down one unit (corresponding to a departure); and when an arrival clock runs out, it jumps up one unit.

To further specify characteristics of the sample paths of $\{Z_t, t \geq 0\}$, we define $T(\alpha, k)$ to be the epoch of the $k$th occurrence of event $\alpha$. That is,

$T(\alpha, k)$ is equal to $\tau_{n^*}$ if

$$n^* = \min\{n \geq 0 : N(\alpha, n) = k\}.$$

If the event $\alpha$ does not occur $k$ times, then $T(\alpha, k) = \infty$. Every $\tau_n$ is equal to some $T(\alpha, k)$—in particular, with $\alpha = a_n$ and $k = N(a_n, n)$. Thus, if we define $r_n = (a_n, N(a_n, n))$, we may write $\tau_n = T(r_n)$. We call the $r_n$'s defined in this way the sequence of *event-order pairs.*

**Remark.** Knowing that $T(\alpha, k) < \infty$ a.s. is useful but verifying it can be difficult. An important property in ensuring that an event $\alpha$ occurs $k$ times is *irreducibility.* A GSMP is irreducible if for every pair of states $s$ and $s'$ there is a sequence of events $\alpha_0, \ldots, \alpha_m$ and a sequence of states $s_1, \ldots, s_m$ such that $\alpha_0 \in \mathcal{E}(s)$, $\alpha_i \in \mathcal{E}(s_i)$, $i = 1, \ldots, m$, and

$$p(s_1; s, \alpha_0)p(s_2; s_1, \alpha_1) \cdots p(s_m; s_{m-1}, \alpha_{m-1})p(s'; s_m, \alpha_m) > 0.$$

In a finite-state, irreducible GSMP, if every clock-setting distribution $F_\alpha$ has a finite mean and a density which is continuous and *strictly* positive on $(0, \infty)$, then, over an infinite horizon, every event $\alpha$ occurs infinitely many times with probability one (see the proof of a closely related result in Shedler [75] pp.50-51). Thus, at least for finite-state, irreducible GSMPs, the assumption that $T(\alpha, k) < \infty$ is fairly general. Virtually all GSMPs we consider are irreducible.

### Triggering Indicators

It is convenient to work with a more explicit representation of the jump epochs $\tau_n$ and $T(\alpha, k)$ than that given by (2.7). In obtaining such a representation, the key observation is that every $\tau_n$ and every $T(\alpha, k)$ is the sum of a subset of $\{X(\beta, j), \beta \in \mathbf{A}, j = 1, 2, \ldots\}$. In (2.7), the clock that runs out and determines the jump at $\tau_{n+1}$ was set at some earlier transition $\tau_m$, $m \leq n$, to some clock sample $X(\beta, j)$ (in particular, with $\beta = a_{n+1}$ and $j = N(a_{n+1}, n + 1)$) so that $\tau_{n+1} = \tau_m + X(\beta, j)$. In general, working backwards in this way from any $\tau_n$ we can find a sequence of events and indices $(\beta_1, j_1), \ldots, (\beta_{m_n}, j_{m_n})$ such that

$$\tau_n = X(\beta_1, j_1) + \cdots + X(\beta_{m_n}, j_{m_n}),$$

and such that the $j_i$th occurrence of $\beta_i$ triggers the setting of the $j_{i+1}$th clock of event $\beta_{i+1}$. If $\tau_n$ is the epoch of the transition caused by the $k$th occurrence of $\alpha$ (i.e., $a_n = \alpha$ and $N(\alpha, n) = k$), call this the *triggering sequence* for $(\alpha, k)$. To pick out which event-order pairs are in the triggering sequence for $(\alpha, k)$ we use the following:

**Definition 2.1.** Suppose (C1) holds. The *triggering indicators* $\eta(\cdot, \cdot; \cdot, \cdot)$ are equal to zero except as follows:

**(i)** For every $\alpha$ and $k$, $\eta(\alpha, k; \alpha, k) = 1$;
**(ii)** if the $k$th clock for $\alpha$ is set at the $j$th occurrence of $\beta$, $\eta(\alpha, k; \beta, j) = 1$;
**(iii)** if $\eta(\alpha, k; \beta, j) = 1$ and $\eta(\beta, j; \beta', j') = 1$, then $\eta(\alpha, k; \beta', j') = 1$.

The following is an immediate consequence of this definition:

**Lemma 2.1.** Suppose (C1) holds. For every $\alpha \in \mathbf{A}$ and every $k > 0$, if $T(\alpha, k) < \infty$, then

$$T(\alpha, k) = \sum_{\beta,j} X(\beta, j)\eta(\alpha, k; \beta, j); \tag{2.13}$$

for every $n \geq 0$,

$$\tau_n = \sum_{i=1}^{n} X(r_i)\eta(r_n; r_i). \tag{2.14}$$

(If, as we always require, every $\mathcal{E}(s)$ is non-empty and no clock sample takes the value infinity, then every $\tau_n$ is defined and finite. Thus we do not need to assume $\tau_n < \infty$ in the lemma.)

In Figure 2.1 we see that every arrival ($\alpha$) triggers the setting of the next arrival clock; hence, the only $\eta(\alpha, k; \cdot, \cdot)$'s equal to unity are of the form $\eta(\alpha, k; \alpha, j)$ with $j < k$. But departures ($\beta$) may have both arrivals and departures in their triggering sequences. For example, the departure clock that runs out at $\tau_9$ was set at $\tau_8$; the clock that runs out at $\tau_8$ was set at $\tau_6$; and the clock that runs out at $\tau_6$ was set at $\tau_2$. Thus, we see that the triggering sequence for the departure at $\tau_9$ consists of the departures at $\tau_9$ and $\tau_8$, the arrival at $\tau_6$ and all previous arrivals, so

$$\tau_9 = X(\alpha, 1) + X(\alpha, 2) + X(\alpha, 3) + X(\beta, 4) + X(\beta, 5).$$

That this is correct is easily verified in the figure by noting that the "height" of a new clock sample is equal to the length of the interval from the time it is set to the time it runs out: $X(\alpha, 1) = \tau_1 - \tau_0$, $X(\alpha, 2) = \tau_2 - \tau_1$, $X(\alpha, 3) = \tau_6 - \tau_2$, and so on.

An important observation is that the triggering sequences and indicators are determined by the *order* in which events occur, but do not depend on the particular epochs of their occurrence.

## 2.3 Derivative Estimation

This section develops derivative estimates for GSMPs depending on a parameter. We consider a situation in which the clock-setting distributions $\{F_\alpha, \alpha \in \mathbf{A}\}$ depend on a parameter $\theta \in \Theta$. *We do not allow the transition probabilities*

$p(\cdot;\cdot,\cdot)$ *or the initial state* $Y_0$ *to depend on* $\theta$. For each value of $\theta$, $\{Z_t(\theta), t \geq 0\}$ is a GSMP with performance given by the random variable $L(\theta)$. Our goal is to obtain an unbiased estimate of $d\mathbf{E}[L(\theta)]/d\theta$, the derivative of the expected performance.

To obtain derivative estimates, we construct the parametric family of processes $\{Z_t(\theta), t \geq 0, \theta \in \Theta\}$ on a common probability space in such a way that *(i)* $L$ is almost surely differentiable in $\theta$, and *(ii)* $dL/d\theta$ evaluated at, say, $\theta_0$ can be calculated from $\{Z_t(\theta_0), 0 \leq t \leq T\}$—that is, without ever changing $\theta$. The stochastic derivative $dL/d\theta$ provides an unbiased estimate if

$$\mathbf{E}[\frac{dL}{d\theta}] = \frac{d\mathbf{E}[L]}{d\theta}.$$

Conditions for unbiasedness will be taken up in Section 3.1.1. Here, we adapt the construction of Section 2.2 to the parametric case to satisfy *(i)*, then show how to accomplish *(ii)*.

### 2.3.1 Stochastic Derivatives for GSMPs

Using the construction of the previous section, we obtain, for each $\theta \in \Theta$, a GSMP $\{Z_t(\theta), t \geq 0\}$ corresponding to $(\mathbf{S}, \mathbf{A}, \mathcal{E}, p, \{F_\alpha(\cdot, \theta), \alpha \in \mathbf{A}\})$. All of these processes are driven by the same input sequences $\{U(\alpha, k), V(\alpha, k), \alpha \in \mathbf{A}, k = 1, 2, \ldots\}$ which *do not depend on* $\theta$. In other words, the processes $\{Z_t(\theta), t \geq 0, \theta \in \Theta\}$ are defined on a common probability space.

To make this explicit, we need only modify each $g_\alpha$ to correctly transform uniform random variables to random variables with distribution $F_\alpha(\cdot, \theta)$, for each $\theta \in \Theta$. Thus, let, now, every $g_\alpha$ map $[0,1] \times \Theta$ to $\mathbf{R}$ in such a way that, for all $x$,

$$P(g_\alpha(V, \theta) \leq x) = F_\alpha(x, \theta),$$

when $V$ is uniform on $[0,1]$. Define $X_\theta(\alpha, k) = g_\alpha(V(\alpha, k), \theta)$.

To consider derivatives, we need some conditions on the clock samples and their distributions:

**(A1).** For each $\theta \in \Theta$ and each $\alpha \in \mathbf{A}$, $F_\alpha(x, \theta)$ is continuous in $x$ and zero at $x = 0$.

**(A2).** For every $\alpha$ and $k$, $X_\theta(\alpha, k)$ is, with probability one, a continuously differentiable function of $\theta$ on $\Theta$.

Condition (A1) allows us to work with only strictly positive $X_\theta(\alpha, k)$. It is also sufficient to ensure that, for *each* $\theta$, two or more events never occur at the same time. We still need the generality of (2.3) because as $\theta$ is varied with $\omega$ held fixed, it becomes possible for clocks to run out simultaneously. In other words, if we let $\{\tau_n(\theta)\}$ be the sequence of event epochs obtained at $\theta$, then (A1) ensures that, for *each* $\theta$,

$$\{\omega : \tau_{n+1}(\theta) = \tau_n(\theta) \text{ for some } n\}$$

has probability zero; but

$$\{\omega : \tau_{n+1}(\theta) = \tau_n(\theta) \text{ for some } n \text{ and some } \theta\}$$

may have positive probability.

Condition (A2) requires that the function $g_\alpha(v, \theta)$ be continuously differentiable in $\theta$ for almost every $v \in [0, 1]$—"almost every" with respect to the uniform distribution.

Once the clock samples depend on $\theta$, so do all the sample path characteristics defined in (2.7)-(2.12), and we write, for example, $Y_n(\theta)$ to denote the $n$th state under the parameter value $\theta$. In this regard, we make the following notational convention: Whenever a sample path characteristic appears without a parameter argument, it is understood to be evaluated at a fixed, nominal value of $\theta$; thus, $a_i = a_i(\theta)$ and $Y_i = Y_i(\theta)$. When we need to emphasize a small change in $\theta$ we write, for example, $\tau_n(\theta + h)$ and $T_{\theta+h}(\alpha, k)$.

## Derivatives of Event Epochs

We now develop expressions for derivatives, with respect to $\theta$, of performance measures of the general type described at the end of Section 2.1. Once we have all $\{Z_t(\theta), t \geq 0, \theta \in \Theta\}$ defined on a common probability space, the performance measures, $\{L(\theta), \theta \in \Theta\}$ become random variables on the same space, and we may consider the derivative of $L$, with respect to $\theta$, for fixed $\omega$. The first step in showing that $L$ is differentiable, and explicitly evaluating its derivative, is calculating $d\tau_n/d\theta$ for each $n > 0$, and $dT(\alpha, k)/d\theta$ for each $\alpha \in \mathbf{A}$ and each $k > 0$.

Under (A1), for each $\theta$, we may assume events occur singly so that the inequalities (implicit in $t^*$ and $\alpha^*$) that determine $\tau_1, \ldots, \tau_n$ and $a_1, \ldots, a_n$ are strict. Since the clock samples $\{X_\theta(\alpha, k)\}$ change smoothly in $\theta$, this implies that throughout a sufficiently small neighborhood of $\theta$, these inequalities retain their sense and remain strict. Throughout such a neighborhood, the $\tau_i$'s change continuously—and, under (A2), differentiably—in $\theta$; the $a_i$'s, and hence the $Y_i$'s, remain constant. A potential discontinuity in some $\tau_i$, $a_i$ or $Y_i$ can only occur where the change in $\theta$ is large enough to change the argument of minimization in (2.7) or (2.8)—i.e., large enough to change the event that triggers the transition out of some state.

Observe, next, that so long as the $a_i$'s and $Y_i$'s, $i \leq n$ remain unchanged, so does the triggering sequence for each $\tau_i$, $i \leq n$. Writing $\eta_\theta(r_i; r_j)$ to emphasize the dependence of the triggering indicators on $\theta$, we conclude that, for sufficiently small $h$, $\eta_{\theta+h}(r_i; r_j) = \eta_\theta(r_i; r_j)$ for all $i, j \le n$. Using (2.14), we find that for sufficiently small $h$,

$$\tau_n(\theta + h) - \tau_n(\theta) = \sum_{i=1}^{n} [X_{\theta+h}(r_i) - X_\theta(r_i)]\eta_\theta(r_n; r_i).$$

Furthermore, if $T_\theta(\alpha, k) < \infty$, then so is $T_{\theta+h}(\alpha, k)$ for all sufficiently small $h$, and

$$T_{\theta+h}(\alpha, k) - T_\theta(\alpha, k) = \sum_{\beta,j} [X_{\theta+h}(\beta, j) - X_\theta(\beta, j)]\eta_\theta(\alpha, k; \beta, j).$$

Combining these observations, we have

**Lemma 2.2.** Suppose (C1), (A1) and (A2) hold. For each $\theta$ and every $n$, with probability one, $a_n$ and $Y_n$ are constant in a neighborhood of $\theta$; $\tau_n$ is differentiable at $\theta$ with

$$\frac{d\tau_n}{d\theta} = \sum_{i=1}^{n} \frac{dX_\theta(r_i)}{d\theta}\eta_\theta(r_n; r_i); \tag{2.15}$$

and if $T_\theta(\alpha, k) < \infty$, then $T_\theta(\alpha, k)$ is differentiable with

$$\frac{dT_\theta(\alpha, k)}{d\theta} = \sum_{\beta,j} \frac{dX_\theta(\beta, j)}{d\theta}\eta_\theta(\alpha, k; \beta, j). \tag{2.16}$$

Lemma 2.2 states, roughly, that the "perturbations" $dX(r_i)/d\theta$ propagate along the triggering sequences. This is illustrated in Figure 2.2 in the case of a GI/G/1 queue. The figure shows the effect of small increases in the departure clocks. Each increase in a departure clock delays the occurrence of the associated departure, and also delays the setting of the next departure clock—i.e., delays the next service initiation—unless the next departure clock is set by an arrival. The arrival epochs are unaffected because a departure never triggers the setting of an arrival clock. We saw earlier that the epoch, $\tau_9$, of the last departure shown in Figures 2.1 and 2.2 has the triggering sequence representation

$$\tau_9 = X(\alpha, 1) + X(\alpha, 2) + X(\alpha, 3) + X(\beta, 4) + X(\beta, 5). \tag{2.17}$$

Figure 2.2 illustrates that for sufficiently small perturbations,

$$\Delta\tau_9 = \Delta X(\beta, 4) + \Delta X(\beta, 5).$$

Since the interarrival times do not change, for every $k$, $\Delta X(\alpha, k) = 0$, so we also have

$$\Delta\tau_9 = \Delta X(\alpha, 1) + \Delta X(\alpha, 2) + \Delta X(\alpha, 3) + \Delta X(\beta, 4) + \Delta X(\beta, 5),$$

which is what we get by applying (2.15) to (2.17).

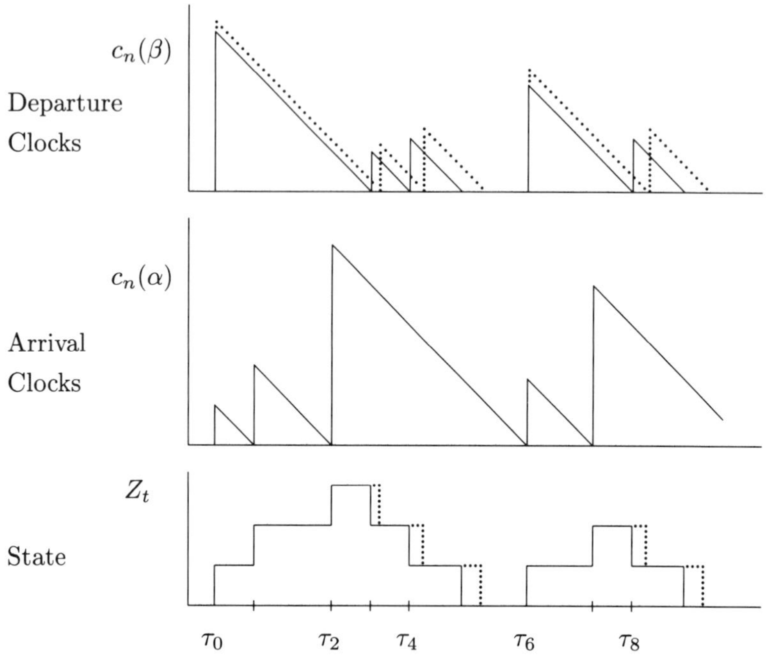


Figure 2.2: GI/G/1 queue with perturbed service times

### Tracking Perturbations

Lemma 2.2 may create the mistaken impression that it is necessary to evaluate the triggering indicators in order to evaluate the derivatives of the event epochs. In fact, a simple scheme allows us to track perturbations, using (2.15) only implicitly. To each $\alpha \in \mathbf{A}$, associate a sequence $\{\delta_n(\alpha), n \geq 0\}$ of "perturbation accumulators". Initialize the sequence by setting $\delta_0(\alpha) = dX_\theta(\alpha, 1)/d\theta$ if $\alpha \in \mathcal{E}(Y_0)$, and $\delta_0(\alpha) = 0$ otherwise. If, at the $n$th transition, a clock is set for $\alpha$ for the $k$th time, then update by setting $\delta_n(\alpha) = \delta_{n-1}(a_n) + dX_\theta(\alpha, k)/d\theta$. With this scheme, it is not hard to see (from Lemma 2.2 and Definition 2.1) that $d\tau_n/d\theta = \delta_{n-1}(a_n)$, for all $n \geq 1$. This is the basis of Algorithm 2.1, below.

### Derivatives of Performance Measures

From the derivatives of the state transition epochs we can build up expressions for derivatives of a general class of finite-horizon performance measures. We define this class of performance measures more precisely. Let $f$ be a *bounded*, real-valued function on the state space $\mathbf{S}$ of the GSMPs $\{Z_t(\theta)\}$. For any (deterministic) real $T > 0$ and any integer $n_0 > 0$ define

$$L_T(\theta) = \int_0^T f(Z_t(\theta))dt, \tag{2.18}$$

and

$$L_{n_0}(\theta) = \int_0^{\tau_{n_0}} f(Z_t(\theta))dt. \tag{2.19}$$

If $T_\theta(\alpha, k) < \infty$, also define

$$L_{\alpha,k}(\theta) = \int_0^{T_\theta(\alpha,k)} f(Z_t(\theta))dt. \tag{2.20}$$

Examples of different applications of these performance measures were given at the end of Section 2.1. From the persective of simulation, $L_T$, $L_{n_0}$ and $L_{\alpha,k}$ differ in how they terminate a run. With $L_T$, the number of events is random and the time is fixed; with $L_{n_0}$, the number of events is fixed but the time is random; and with $L_{\alpha,k}$, both the number of events and the time are random. *Henceforth, whenever we refer to $L_{\alpha,k}$ we assume that $T_\theta(\alpha, k) < \infty$ with probability one, for all $\theta \in \Theta$.*

For the following, recall that $N(t)$ is the number of events in $[0, t]$; i.e.,

$$N(t) = \sup\{k : \tau_k \leq t\}.$$

**Lemma 2.3.** Under (C1) and (A1), $N(t)$ is, with probability one, finite for finite $t$.

**Proof.** Define, for each $\alpha \in \mathbf{A}$,

$$N_\alpha(t) = \sup\{k \geq 0 : \sum_{i=1}^{k} X(\alpha, i) \leq t\}.$$

Under (C1), a clock, once set, is never interrupted, which implies that $N(t) \leq \sum_\alpha N_\alpha(t)$. Each $N_\alpha$ is a renewal process, and under (A1) every $X(\alpha, i) > 0$, a.s. A result from renewal theory (Theorem 5.2.1 of Prabhu [65]) therefore implies that every $N_\alpha(t)$ is finite a.s. Since $\mathbf{A}$ is finite, we conclude that $N(t)$ is too. □

**Lemma 2.4.** Under (C1), (A1) and (A2), for each $\theta \in \Theta$, $L_T$, $L_{n_0}$ and $L_{\alpha,k}$ are, with probability one, differentiable at $\theta$, with

$$\frac{dL_T}{d\theta} = \sum_{i=1}^{N(T)} \frac{d\tau_i}{d\theta}[f(Y_{i-1}) - f(Y_i)], \tag{2.21}$$

$$\frac{dL_{n_0}}{d\theta} = \sum_{i=0}^{n_0-1} f(Y_i)[\frac{d\tau_{i+1}}{d\theta} - \frac{d\tau_i}{d\theta}], \tag{2.22}$$

and

$$\frac{dL_{\alpha,k}}{d\theta} = \sum_{i=0}^{N(T(\alpha,k))-1} f(Y_i)[\frac{d\tau_{i+1}}{d\theta} - \frac{d\tau_i}{d\theta}]. \tag{2.23}$$

**Proof.** Note, first, that because $Z_t$ is constant (and equal to $Y_i$) on $[\tau_i, \tau_{i+1})$, we may write

$$L_T = \sum_{i=1}^{N(T)} f(Y_{i-1})[\tau_i - \tau_{i-1}] + (T - \tau_{N(T)})f(Y_N(T)), \tag{2.24}$$

$$L_{n_0} = \sum_{i=0}^{n_0-1} f(Y_i)[\tau_{i+1} - \tau_i], \tag{2.25}$$

and, since $T(\alpha, k) = \tau_{N(T(\alpha,k))}$,

$$L_{\alpha,k} = \sum_{i=0}^{N(T(\alpha,k))-1} f(Y_i)[\tau_{i+1} - \tau_i]. \tag{2.26}$$

Starting with $L_{n_0}$, recall from Lemma 2.2 that, for each $\theta$, with probability one there exists a neighborhood of $\theta$ throughout which $Y_1, \ldots, Y_{n_0}$ are constant (and $\tau_1, \ldots, \tau_{n_0}$ are differentiable). Thus, with probability one, for sufficiently small $h$,

$$\begin{aligned} L_{n_0}(\theta + h) - L_{n_0}(\theta) = \\ \sum_{i=0}^{n_0-1} f(Y_i(\theta)) \{[\tau_{i+1}(\theta + h) - \tau_i(\theta + h)] - [\tau_{i+1}(\theta) - \tau_i(\theta)]\} \\ = \sum_{i=0}^{n_0-1} f(Y_i(\theta)) \{[\tau_{i+1}(\theta + h) - \tau_{i+1}(\theta)] - [\tau_i(\theta + h) - \tau_i(\theta)]\}. \end{aligned}$$

Dividing by $h$ and letting $h \to 0$ we obtain (2.22). The same argument yields (2.23) because $N(T(\alpha, k))$ is, with probability one constant throughout a neighborhood of $\theta$.

Differentiating (2.24), we get

$$\frac{dL_T}{d\theta} = \sum_{i=1}^{N(T)} f(Y_{i-1})[\frac{d\tau_i}{d\theta} - \frac{d\tau_{i-1}}{d\theta}] + (-\frac{d\tau_{N(T)}}{d\theta})f(Y_N(T)),$$

because $T$ is constant, and $N(T)$ (hence, $Y_{N(T)}$) is a.s. constant throughout a neighborhood of $\theta$. Rearranging the terms in the sum yields (2.21). □

The derivatives (2.21), (2.22) and (2.23) admit a simple interpretation: sufficiently small perturbations in the clock samples that drive $Z$ introduce small changes in the state transition epochs without changing the sequence of states visited (at least over a finite horizon). As $\theta$ is varied over a small interval, the sample paths of $Z$ are deformed by stretching or contracting the state holding times $\tau_{i+1} - \tau_i$ without changing the basic shape as given by the $Y_i$. This is depicted in Figure 2.2.

## 2.3.2 A Generic Derivative Estimation Algorithm

The expression (2.15) for $d\tau_n/d\theta$, while convenient for analysis, is cumbersome to implement because the triggering indicators, $\eta$, are defined by working *backwards* from $\tau_n$. But as already noted, it is not necessary to evaluate the $\eta$'s to calculate $d\tau_n/d\theta$ or the performance derivatives (2.21), (2.22) and (2.23). Algorithm 2.1 below meshes the steps needed to generate a GSMP (essentially following the recursions (2.7)-(2.12)), with the propagation of perturbations using the sequences $\{\delta_n(\alpha)\}$. Most of the steps of the algorithm would be needed even without the derivative calculations.

Algorithm 2.1 takes as inputs sequences of uniformly distributed random variables, used to generate clock samples $X_\theta(\alpha)$ (using $g_\alpha$) and to generate state transitions (using $\phi$). The algorithm generates as output the sequences $\{(Y_n, a_n, \tau_n)\}$ of states, events and transition epochs of a GSMP, and also the sequence $\{d\tau_n/d\theta\}$ of derivatives of transition epochs (in the algorithm, the sequence of values taken on by the variable $\delta^*$). The derivatives of performance measures are not explicitly generated by the algorithm. These are easily evaluated using (2.21)-(2.23).

In the algorithm, $Y$ denotes a generic state, $\tau$ an event epoch, $c(\alpha)$ a clock reading for $\alpha$, and $X_\theta(\alpha)$ a sample from $F_\alpha(\cdot, \theta)$; $\delta^*$ is the derivative of the current transition epoch. The accumulator $\delta(\alpha)$ represents, roughly, the accumulated infinitesimal delay in the epoch of the next occurrence of $\alpha$.

**Algorithm 2.1: Perturbation Propagation in a GSMP**

0. Initialization.
    a. $\tau := 0$;
    b. Set $Y$ to initial state;
    c. For every $\alpha \in \mathcal{E}(Y)$
        Generate $X_\theta(\alpha)$ from $F_\alpha(\cdot, \theta)$;
        Generate $dX_\theta(\alpha)/d\theta$;
        $c(\alpha) := X_\theta(\alpha)$;

$\delta(\alpha) := dX_\theta(\alpha)/d\theta$;

d. For every $\alpha \notin \mathcal{E}(Y)$,

$c(\alpha) := 0$;

$\delta(\alpha) := 0$;

1. Next Transition.

a. Find $\alpha^* \in \mathcal{E}(Y)$ with smallest $c(\alpha)$;

b. $t^* := c(\alpha^*)$;

c. $\tau := \tau + t^*$;

d. $\delta^* := \delta(\alpha^*)$;

e. Generate $Y'$ from $p(\cdot; Y, \alpha^*)$;

f. For every $\alpha \in \mathcal{E}(Y') \backslash (\mathcal{E}(Y) - \{\alpha^*\})$

Generate $X_\theta(\alpha)$ from $F_\alpha(\cdot, \theta)$;

Generate $dX_\theta(\alpha)/d\theta$;

$c(\alpha) := X_\theta(\alpha)$;

$\delta(\alpha) := \delta^* + dX_\theta(\alpha)/d\theta$;

g. For every $\alpha \in \mathcal{E}(Y') \cap (\mathcal{E}(Y) - \{\alpha^*\})$

$c(\alpha) := c(\alpha) - t^*$;

h. $Y := Y'$;

2. Check Stopping Condition.

If stopping condition is met, STOP; else go to 1.

The stopping condition in Step 2 is deliberately left vague; there are many reasonable choices that do not alter the algorithm in any substantive way. Each of $L_T$, $L_{n_0}$ and $L_{\alpha,k}$ implicitly prescribes a stopping rule.

Algorithm 2.1 shows, among other things, that it is possible to compute stochastic derivatives from *observation* (as opposed to *simulation*) of $Z$ and its events under one additional condition: namely, that for each $\alpha \in \mathbf{A}$ there exist a function $G_\alpha$ such that, for all $k$,

$$\frac{dX_\theta(\alpha, k)}{d\theta} = G_\alpha(X_\theta(\alpha, k), \theta).$$

Since every $X_\theta(r_i)$, $i = 1, \ldots, n$, can be "observed" from the sequence of states, events, and event epochs $\{(Y_i(\theta), a_i(\theta), \tau_i(\theta)), i = 1, \ldots, n\}$, the existence of $G_\alpha$ makes it possible to "observe" $dX_\theta(r_i)/d\theta$ as well. The condition that such a $G$ exist is broadly, though not universally, applicable; see Section 1.3.2 of Chapter 1.

In Algorithm 2.1, as well as in Sections 2.2 and 2.3, we have taken the initial state, $Y_0$, to be fixed. All our results go through if, instead, $Y_0$ is drawn from a distribution on $\mathbf{S}$, provided the distribution does not depend on $\theta$. In this case, Step 0.b is changed to generate a state from the initial distribution.

* * *

## Notes and Comments

Generalized semi-Markov processes were introduced by Matthes [64] to study the phenemenon of *insensitivity* of steady-state distributions. They are well-suited to that purpose for the same reason they are suited to ours: they separate the structure of a discrete event system from the distributions that drive it. Applications of GSMPs to insensitivity are further developed in Schassberger [72,73,74]. The event list formulation used here is from Whitt [84]. GSMPs as models for simulation are discussed in Glynn and Iglehart [41] and Shedler [75]. The construction in Section 2.2 is from Glasserman [24,27].

The first infinitesimal perturbation analysis derivative estimation algorithm was formulated in Ho and Cao [51]. That algorithm (for the sensitivity of throughput in Jackson networks) and the one in Suri and Zazanis [79] (for the sensitivity of mean waiting time in a queue) contain the essential ingredients of more general algorithms that followed. The general algorithm in Suri [77] is similar to Algorithm 2.1, though Suri does not work with GSMPs *per se.*

An entirely different approach to derivative estimation for GSMPs, based on *likelihood ratios*, is studied in Glynn [38,39]. See also Reiman and Weiss [66] and Rubinstein [69]. Generally speaking, likelihood ratio estimators depend in a less intricate way on the particular GSMP and performance measure under consideration. In addition, they are applicable without the special structure developed here in subsequent chapters: they impose (mild) regularity conditions on the distributions that drive a system, rather than on the performance measure or the system itself. This increased generality comes, however, at the expense of large (often prohibitively large) variance.

# Chapter 3

# Structural Conditions for GSMPs

This chapter continues the line of development of the previous one, turning now to conditions on the structure of a GSMP that ensure that the derivative estimates we have derived are unbiased. We then extend the derivative estimation technique of Chapter 2 to GSMPs with *speeds*, and provide conditions for these estimates to be unbiased as well.

## 3.1 Continuity and Unbiasedness

In Chapter 2 we established the existence of, and derived expressions for, almost sure derivatives of a class performance measures. We now turn to the key question of whether or not these stochastic derivatives are unbiased estimators of the derivative of expected performance—that is, whether or not

$$\mathbf{E}[\frac{dL}{d\theta}] = \frac{d\mathbf{E}[L]}{d\theta}, \tag{3.1}$$

where $L$ is any of the performance measures introduced in Chapter 2.

We first review briefly why, in practice, (3.1) may fail to hold; see also the discussion in Chapter 1. Lemmas 2.2 and 2.4 rely on the fact that for each $\theta$ and almost every $\omega$, there is a neighborhood of $\theta$ throughout which the $\tau_n$'s and $L$'s are continuous, and, in fact, differentiable. However, the size of that neighborhood may depend on $\omega$; and for each *fixed* $h > 0$, the set of $\omega$ for which a particular $\tau_n$ or $L$ has a discontinuity somewhere in $(\theta, \theta + h)$ may have positive probability. These potential discontinuities arise when perturbations in $\theta$ introduce changes in the clock samples large enough to change the event that triggers the transition out of a state (in (2.8)).

Consider the case of $L_{n_0}$. Let $\mathcal{D}_h$ be the set of $\omega$ for which at least one of the first $n_0$ events $a_1, \ldots, a_{n_0}$ changes somewhere in $(\theta, \theta + h)$. We may write

$$\begin{aligned} \frac{d\mathbf{E}[L_{n_0}]}{d\theta} &= \lim_{h \to 0} \mathbf{E}\left[\frac{L_{n_0}(\theta+h) - L_{n_0}(\theta)}{h} \mathbf{1}\{\Omega \backslash \mathcal{D}_h\}\right] \\ &+ \lim_{h \to 0} \mathbf{E}\left[\frac{L_{n_0}(\theta+h) - L_{n_0}(\theta)}{h} \mathbf{1}\{\mathcal{D}_h\}\right]. \end{aligned} \quad (3.2)$$

It is typically the case that

$$\mathbf{E}[\frac{dL_{n_0}}{d\theta}] = \lim_{h \to 0} \mathbf{E}\left[\frac{L_{n_0}(\theta+h) - L_{n_0}(\theta)}{h} \mathbf{1}\{\Omega \backslash \mathcal{D}_h\}\right],$$

since $dL_{n_0}/d\theta$ is evaluated by ignoring the dependence of $a_1, \ldots, a_{n_0}$ (and of $Y_1, \ldots, Y_{n_0}$) on $\theta$. Hence, for (3.1) to hold, it is typically necessary that (3.2) vanish as $h \to 0$. A prerequisite for this is that

$$\lim_{h \to 0} \mathbf{E}\left[(L_{n_0}(\theta+h) - L_{n_0}(\theta))\mathbf{1}\{\mathcal{D}_h\}\right] = 0.$$

This holds under mild additional conditions if $L_{n_0}$ is continuous in $\theta$ even for $\omega \in \mathcal{D}_h$—i.e., if $L_{n_0}$ is continuous *even when the change in $\theta$ is large enough to change the order of events.*

### 3.1.1 Commuting Conditions

With these points in mind, we introduce conditions on a GSMP that guarantee the continuity of the $L$'s even at points where events change order. As functions of $\theta$, the $L$'s may have "kinks" where order changes occur, but they will still be continuous.

We first state the main condition in provisional form, in terms of the state transition mapping $\phi$. We then give a more appropriate, though less intuitive, statement of the condition in terms of the transition probabilities $p(\cdot; \cdot, \cdot)$. Recall from Section 2.2 that $\phi(s, \alpha, u)$ is the state reached from $s$ under routing indicator $u$ when event $\alpha$ occurs.

**(C2'). Provisional Commuting Condition.** The mapping $\phi$ can be chosen so that for every $s \in \mathbf{S}$, every $\alpha, \beta \in \mathbf{A}$ and every $u_1, u_2 \in [0,1]$, if $\{\alpha, \beta\} \subseteq \mathcal{E}(s)$ then

$$\phi(\phi(s, \alpha, u_1), \beta, u_2) = \phi(\phi(s, \beta, u_2), \alpha, u_1).$$

This condition says that when two events are possible, the state reached through the occurrence of both is independent of their order—provided the same routing indicators are used. Under our construction of a GSMP, routing indicators are assigned to event types, so that if events of different types change order they keep their routing indicators; cf. (2.10). Part of the requirement in (C2') is that both sides of the equation be defined: $\{\alpha, \beta\} \subseteq \mathcal{E}(s)$ must imply that $\beta \in \mathcal{E}(\phi(s, \alpha, u_1))$ and $\alpha \in \mathcal{E}(\phi(s, \beta, u_2))$. This is just condition (C1).

As a simple illustration of (C2'), we mention

**Example 1.** *(continued) GI/G/1 queue.* Recall that $s$ records the number in the system, $\alpha$ denotes arrival and $\beta$ denotes departure. Since the state changes deterministically, the routing indicators actually play no role: an arrival changes $s$ to $s+1$, a departure to $s-1$. Thus, for any $s > 0$, and any $u_1, u_2$,

$$\begin{aligned} \phi(\phi(s, \alpha, u_1), \beta, u_2) &= \phi(s+1, \beta, u_2) \\ &= s \\ &= \phi(s-1, \alpha, u_1) \\ &= \phi(\phi(s, \beta, u_2), \alpha, u_1). \end{aligned}$$

Condition (C2') makes no requirement of $s = 0$ because $\mathcal{E}(0)$ contains only one event ($\alpha$) whereas (C2') is a condition on states with at least two possible events.

As stated, (C2') has the shortcoming that it depends on $\phi$, an object we introduced for our construction, and not solely on the basic GSMP data $\mathbf{S}, \mathbf{A}, \mathcal{E}, p$ and $\{F_\alpha, \alpha \in \mathbf{A}\}$. Given the specification of a GSMP, it should be possible to check the condition without reference to the way the process is constructed or simulated. Furthermore, there are generally many equally valid choices of $\phi$ for the same GSMP, and some may satisfy (C2') while others do not. We improve on (C2') by giving a condition in terms of the transition probabilities $p$. (We introduced (C2') because it is sometimes easier to work with.) The proof of the proposition that follows shows how to define $\phi$ to satisfy (C2') when the condition on $p$ (plus another minor condition) is satisfied.

**(C2). Commuting Condition.** For any $s_1$, if $\{\alpha, \beta\} \subseteq \mathcal{E}(s_1)$, and $s_2$ and $s_3$ satisfy $p(s_2; s_1, \alpha)p(s_3; s_2, \beta) > 0$, then there is an $s_4$ such that

$$p(s_4; s_1, \beta) = p(s_3; s_2, \beta)$$

and

$$p(s_3; s_4, \alpha) = p(s_2; s_1, \alpha).$$

This condition says that if it is possible to go from $s_1$ to $s_3$ through the occurrence of $\alpha$, then $\beta$, it must also be possible through the occurrence of $\beta$,

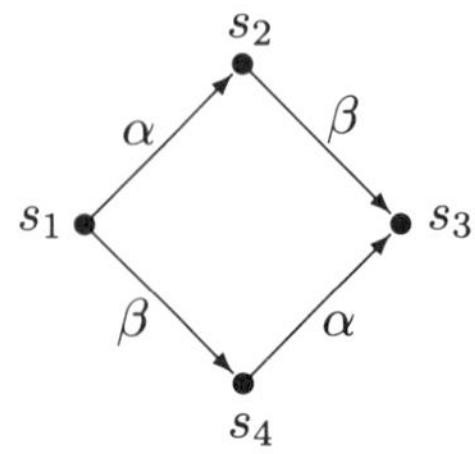


Figure 3.1: Illustrating the commuting condition (C2)

then $\alpha$, so that each of the transitions triggered by the same event has the same probability. This situation is depicted in Figure 3.1; transitions represented by opposite sides of the parallelogram are required to have the same probability.

To get $\phi$ from (C2) for arbitrary GSMPs, we will impose one additional condition on $p$; namely, that for any $s, s', s''$ and $\alpha$, if $p(s'; s, \alpha) > 0$ then

$$p(s'; s, \alpha) = p(s''; s, \alpha) \Rightarrow s'' = s'. \tag{3.3}$$

In words, no two possible transitions from the same state due to the occurrence of the same event can have exactly the same probability. In practice, this is not much of a restriction since the difference could be arbitrarily small. In the queueing examples of Chapter 4, we will be able to define $\phi$ to satisfy (C2') without recourse to (3.3).

**Proposition 3.1.** If $p$ satisfies (C2) and (3.3), $\phi$ can be chosen to satisfy (C2').

**Proof.** We choose a $\phi$ that generates transitions by *inversion.* Starting with any state $s$ and any event $\alpha$, let $s_1, s_2, \ldots$ be the states for which $p(s_i; s, \alpha) > 0$, ordered so that, for all $i$, $p(s_{i+1}; s, \alpha) < p(s_i; s, \alpha)$. For $u \in [0, 1]$ let

$$m^* \equiv m^*(u; s, \alpha) = \min\{m > 0 : u \leq \sum_{k=1}^{m} p(s_k; s, \alpha)\}, \tag{3.4}$$

and define $\phi(s, \alpha, u) = s_{m^*}$. Clearly, the set of $u \in [0, 1]$ for which $\phi(s, \alpha, u) = s_k$ is an interval of length $p(s_k; s, \alpha)$; so if $U$ is uniformly distributed on the unit interval,

$$P(\phi(s, \alpha, U) = s') = p(s'; s, \alpha).$$

Therefore, (3.4) defines a legitimate choice for $\phi$. We now show that it satisfies (C2').

Suppose $\{\alpha, \beta\} \subseteq \mathcal{E}(s)$ and $p(s'; s, \alpha) > 0$; then (C2) implies that for every $s_i'$ with $p(s_i'; s', \beta) > 0$ there is some $s_i$ with $p(s_i; s, \beta) = p(s_i'; s', \beta)$. Condition

(3.3) implies this $s_i$ is unique. Thus, the states reachable from $s$ and $s'$ via $\beta$ are in one-to-one correspondence, with corresponding states determined by having the same transition probabilities under $\beta$. It follows, then, from (3.4) that

$$\phi(s', \beta, u_2) = s'_i \text{ if and only if } \phi(s, \beta, u_2) = s_i.$$

Reversing the roles of $\alpha$ and $\beta$, we get the analogous condition for states reachable via $\alpha$ from $s$ and $s''$ whenever $p(s''; s, \beta) > 0$.

Now suppose $s_2 = \phi(s, \alpha, u_1)$ and $s_3 = \phi(s_2, \beta, u_2)$. Then $s_4 = \phi(s, \beta, u_2)$ and $s_5 = \phi(s_4, \alpha, u_1)$ satisfy $p(s_4; s, \beta) = p(s_3; s_2, \beta)$ and $p(s_5; s_4, \alpha) = p(s_2; s, \alpha)$. Comparing with (C2) and invoking the uniqueness in (3.3), we find that $s_5$ must, in fact, be $s_3$; that is,

$$\phi(\phi(s, \alpha, u_1), \beta, u_2) = \phi(\phi(s, \beta, u_2), \alpha, u_1),$$

which is (C2'). □

### 3.1.2 Unbiasedness

We now show that the commuting conditions imply the continuity of the $L$'s, via Lemma 3.1 below. To motivate the lemma and the continuity that follows, we consider the simple example illustrated in Figure 3.2. This figure illustrates the effect of a single change in the order of events on a sample path of a GSMP satisfying (C2'). Initially, the sequence of events is $a_1(\theta), a_2(\theta), \ldots$ with, say, $a_2(\theta) = \alpha$ and $a_3(\theta) = \beta$. A change in $\theta$ causes $\beta$ to occur before $\alpha$. Under our conditions and construction, just at the point where $a_2$ becomes $\beta$, the residual clock time for $\alpha$ is zero, so $a_3$ becomes $\alpha$. Moreover, after both $\alpha$ and $\beta$ have occurred, the same state is reached regardless of the order, because of (C2'). Right at the point where the events change order no time elapses between them, so all residual clock times are unchanged. Hence, the future evolution of the sample path is unaffected by the order change. In particular, the event epochs are continuous, even across the order change. Furthermore, since the holding time $\tau_3 - \tau_2$ in $Y_2$ is zero right at the point where $\alpha$ and $\beta$ change order, the area under the sample path $\{Z_t(\theta), t \geq 0\}$ is continuous (in $\theta$) at that point. The same argument applied to sample paths of $f(Z_t(\theta))$ shows that the $L$'s are continuous. We now formalize these ideas.

#### Continuity

**Lemma 3.1.** Suppose (C2') holds (e.g., (C2) and (3.3) hold) and that (A1) and (A2) hold throughout $\Theta$. Then

**(i)** Every $\tau_i$ and every finite $T(\alpha, k)$ is a.s. continuous in $\theta$ throughout $\Theta$.

**(ii)** At a discontinuity of $Y_i$, $\tau_{i+1} = \tau_i$.

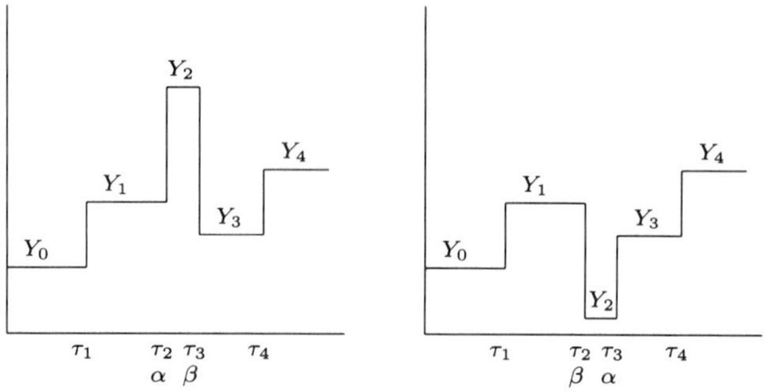


Figure 3.2: Before and after $\alpha$ and $\beta$ change order

**(iii)** At a discontinuity of $N(T)$, $\tau_{N(T)} = T$.
**(iv)** At a discontinuity of $N(T(\alpha, k))$, $T(\alpha, k) = \tau_{N(T(\alpha,k))-1}$.

**Proof, part (i).** From (2.7)-(2.10) it is clear that, for each $i$, $\tau_1, \ldots, \tau_i$ are continuous wherever $a_1, \ldots, a_i$ are. Suppose, then, that at some $\theta$ some $a_j$ is discontinuous, and let $j$ be, in fact, the smallest index of a discontinuous event. For $a_j$ to be discontinuous, we see from (2.8) that clocks for two or more events in $\mathcal{E}(Y_{j-1})$ must run out simultaneously; and any of these clocks potentially determines $a_j$. But if, say, clocks for $\alpha$ and $\beta$ run out simultaneously, then $c_{j-1}(\alpha) = c_{j-1}(\beta)$. That is, $c_{j-1}(a_j)$ is continuous at $\theta$ even if $a_j$ is not. Thus, we see that $\tau_j \equiv \tau_{j-1} + c_{j-1}(a_j)$ is also continuous at $\theta$.

Now suppose, for simplicity, that there are exactly two events $\alpha$ and $\beta$ in $\mathcal{E}(Y_{j-1})$ whose clocks run out at the same time. (The case where many clocks run out together works the same way; considering only two makes the argument clearer.) If $\alpha$ occurs first then (from (2.11)) $c_j(\beta) = c_{j-1}(\beta) - c_{j-1}(\alpha) = 0$ and, similarly, if $\beta$ occurs first then $c_j(\alpha) = 0$. In the first case, the next event must, therefore be $\beta$, while in the second case the next must be $\alpha$. (Recall that under (A1), no *new* clock is ever set to zero.) In either case, $\tau_{j+1} = \tau_j$. Observe, next, that regardless of the order in which $\alpha$ and $\beta$ occur, under (C2'),

$$\begin{aligned} Y_{j+1} &= \phi(\phi(Y_{j-1}, \alpha, U(\alpha, N(\alpha, j-1)+1)), \beta, U(\beta, N(\beta, j-1)+1))) \\ &= \phi(\phi(Y_{j-1}, \beta, U(\beta, N(\beta, j-1)+1)), \alpha, U(\alpha, N(\alpha, j-1)+1))). \end{aligned}$$

Furthermore, for any $\alpha' \in \mathcal{E}(Y_{j+1})$, either $\alpha'$ was in $\mathcal{E}(Y_{j-1})$ or $\alpha'$ was activated by the occurrence of $\alpha$ or $\beta$. In either case, inspection of (2.12) and (2.11) reveals that $c_{j+1}(\alpha')$ is independent of the order of $\alpha$ and $\beta$ since $\tau_{j+1} - \tau_j = 0$. For every $\alpha'' \in \mathbf{A}$, $N(\alpha'', j+1)$ is also independent of the order of $\alpha$ and $\beta$. But if $Y_{j+1}$, $c_{j+1}$ and $N(\cdot, j+1)$ are all independent of the order of $\alpha$ and $\beta$ (at $\theta$) then so is the rest of the sample path, since it is determined by recursion from these quantities. In particular, every $\tau_i(\theta)$ with $i > j$ is independent of the order of $\alpha$ and $\beta$ (at $\theta$). Thus, if $j'$ is the least index greater than $j+1$

for which $a_{j'}$ is discontinuous at $\theta$, then every $\tau_i$, $i < j'$ is continous at $\theta$. At $j'$, we may repeat the whole argument and proceed to the next discontinuous event (if any). Thus, we conclude that every $\tau_i$ is continuous.

The same argument shows that if $T(\alpha, k)$ is finite, it is continuous. For suppose that $(\alpha, k) = r_j$ so $T(\alpha, k) = \tau_j$ and $a_j = \alpha$. As argued above, in order that $a_j$ jump to, say, $\beta$, it is necessary that $c_{j-1}(\beta) = c_{j-1}(\alpha)$, in which case just after $a_j$ becomes $\beta$, $c_j(\alpha) = 0$. This implies that $\alpha$ is the next event, $a_{j+1}$, to occur, and it occurs just after $a_j$. This makes $(\alpha, k) = r_{j+1}$ and $T(\alpha, k) = \tau_{j+1} = \tau_j$. In short, changing the order of $\alpha$ and $\beta$ does not change $T(\alpha, k)$.

**Part (ii).** For $Y_i$ to be discontinuous there must be (at least) two events $\alpha$ and $\beta$ in $\mathcal{E}(Y_{i-1})$ with $c_{i-1}(\alpha) = c_{i-1}(\beta)$. As noted above, this implies that $\tau_{i+1} = \tau_i$.

**Part (iii).** Recall that $N(T)$ is defined by

$$N(T) = \sup\{n : \tau_n \leq T\}.$$

Since the $\tau_n$ are continuous in $\theta$, $N(T)$ jumps only when a transition occurs right at time $T$; i.e., when there is an $n$ such $\tau_n = T$. By definition, then, $\tau_{N(T)} = T$.

**Part (iv).** Since $T(\alpha, k)$ is continuous and every $\tau_n$ is continuous, a discontinuity of $N(T(\alpha, k))$ occurs only when two events occur at $T(\alpha, k)$. One of these is the $N(T(\alpha, k))$th event, the other is the $N(T(\alpha, k)) - 1$st. Since both occur at $T(\alpha, k)$, $T(\alpha, k) = \tau_{N(T(\alpha,k))-1}$. □

**Theorem 3.1.** Under the conditions of Lemma 3.1, $L_T$ and $L_{n_0}$ are almost surely continuous functions of $\theta$ throughout $\Theta$. The same is true of $L_{\alpha,k}$ if $T(\alpha, k) < \infty$ a.s. for all $\theta \in \Theta$.

**Proof.** Consider first $L_{n_0}$. Let $\{\theta_v\}$ be any sequence in $\Theta$ converging to $\theta$ and consider $L_{n_0}(\theta_v) - L_{n_0}(\theta)$ as $v \to \infty$. Invoking (2.25), we may treat separately each term

$$f(Y_i(\theta_v))[\tau_{i+1}(\theta_v) - \tau_i(\theta_v)] - f(Y_i(\theta))[\tau_{i+1}(\theta) - \tau_i(\theta)], \; i < n. \tag{3.5}$$

If $Y_i$ is continuous at $\theta$, then for all sufficiently large $v$, $Y_i(\theta_v) = Y_i(\theta)$ since $Y_i$ takes on only discrete values. In this case, continuity of the $\tau_i$'s implies that (3.5) goes to zero. Suppose, on the other hand, that $Y_i$ has a discontinuity at $\theta$. From Lemma 3.1(ii), this implies that $[\tau_{i+1}(\theta) - \tau_i(\theta)] = 0$; hence, part (i) implies that $[\tau_{i+1}(\theta_v) - \tau_i(\theta_v)] \to 0$ as $v \to \infty$. Boundedness of $f$ now implies that (3.5) converges to zero as $v \to \infty$.

For $L_T$, start with (2.24). If $N(T)$ is continuous at $\theta$, then the continuity of $L_T$ at $\theta$ works the same way as that of $L_{n_0}$. Suppose, therefore, that $N(T)$

jumps at $\theta$, say from $m$ to $m-1$. (A jump to $m+1$ works similarly.) Suppose that for all sufficiently small $h > 0$, one less event occurs in $[0, T]$ at $\theta + h$ than at $\theta$. Then, writing $m$ for $N(T)$ evaluated at $\theta$,

$$\begin{aligned} L_T(\theta+h) - L_T(\theta) = & \\ & \sum_{i=0}^{m-2} \{f(Y_i(\theta+h))[\tau_{i+1}(\theta+h) - \tau_i(\theta+h)] - f(Y_i(\theta))[\tau_{i+1}(\theta) - \tau_i(\theta)]\} \\ & +[T - \tau_{m-1}(\theta+h)]f(Y_{m-1}(\theta+h)) - [\tau_m(\theta) - \tau_{m-1}(\theta)]f(Y_{m-1}(\theta)) \\ & -[T - \tau_m(\theta)]f(Y_m(\theta)). \end{aligned}$$

Each of the first $m-1$ terms (the terms inside the summation) goes to zero by the argument used for $L_{n_0}$. For the remaining three terms, part (iii) of Lemma 3.1 implies that since $N(T)$ is discontinuous at $\theta$, $\tau_m(\theta) = T$. Substituting $\tau_m(\theta)$ for $T$, the last three terms simplify to

$$[\tau_m(\theta) - \tau_{m-1}(\theta+h)]f(Y_{m-1}(\theta+h)) - [\tau_m(\theta) - \tau_{m-1}(\theta)]f(Y_{m-1}(\theta)).$$

The argument used for $L_{n_0}$ shows that this, too, goes to zero. The case where $N(T)$ jumps *up* at $\theta$ works the same way.

Finally, for $L_{\alpha,k}$, if $N(T(\alpha,k))$ is continuous at $\theta$, then so is $L_{\alpha,k}$ by the argument used for $L_{n_0}$. Suppose, then, that as $\theta$ is increased, $N(T(\alpha,k))$ jumps from $m$ to $m-1$. From (2.26) and the fact that $T(\alpha,k) = \tau_{N(T(\alpha,k))}$ (always), we find that for small $h$,

$$\begin{aligned} L_{\alpha,k}(\theta+h) - L_{\alpha,k}(\theta) = & \\ & \sum_{i=1}^{m-2} \{f(Y_{i-1}(\theta+h))[\tau_i(\theta+h) - \tau_{i-1}(\theta+h)] - f(Y_{i-1}(\theta))[\tau_i(\theta) - \tau_{i-1}(\theta)]\} \\ & +[T_{\theta+h}(\alpha,k) - \tau_{m-2}(\theta+h)]f(Y_{m-2}(\theta+h)) \\ & -[\tau_{m-1}(\theta) - \tau_{m-2}(\theta)]f(Y_{m-2}(\theta)) \\ & -[T_\theta(\alpha,k) - \tau_{m-1}(\theta)]f(Y_{m-1}(\theta)). \end{aligned}$$

Part (iv) of Lemma 3.1 implies that $T_\theta(\alpha,k) = \tau_{m-1}(\theta)$. The argument used for $L_T$ now shows that this difference goes to zero. □

**Remark.** Though we will not show it here (see the notes at the end of the chapter), when (C2) holds and when every transition probability $p(s'; s, \alpha)$ is either zero or one, it is possible to write explicit recursions for event epochs purely in terms of $+$, min, and max. Specifically, for every $(\alpha, k)$, there is a $J < \infty$ and a set of indices $\{\mathbf{n}_\beta^j(\alpha,k), \beta \in \mathbf{A}, j = 1, \ldots, J\}$ such that

$$T_\theta(\alpha,k) = X_\theta(\alpha,k) + \min_j \max_\beta \{T_\theta(\beta, \mathbf{n}_\beta^j(\alpha,k))\}, \tag{3.6}$$

for all $\theta$. The indices $\{\mathbf{n}^j\}$ *do not* depend on $\theta$. It follows immediately from (3.6) that every $T_\theta(\alpha, k)$ is almost surely continuous in $\theta$ if every $X_\theta(\beta, j)$ is, $\beta \in \mathbf{A}$, $j = 1, 2, \ldots$. A generalization of (3.6) holds for GSMPs with nontrivial transition probabilities. This provides an alternative proof of part (i) of Lemma 3.1.

### Interchanging Expectation and Differentiation

To go from continuity to the interchange of expectation and differentiation we need an additional regularity condition. The following condition ensures that the clock samples $X_\theta(\alpha, k)$ do not change overly rapidly with $\theta$, and is quite broadly applicable. It leads to a simple proof of unbiasedness, because it allows us to bound derivatives in terms of quantities associated only with the "nominal" process $\{Z_t(\theta), t \geq 0\}$.

**(A3).** There is a constant $B > 0$ such that for all $\theta \in \Theta$ and all $\alpha$ and $k$,

$$\left| \frac{dX_\theta(\alpha, k)}{d\theta} \right| \leq B(X_\theta(\alpha, k) + 1).$$

We now come to our main results on estimating derivatives of expected performance measures for GSMPs. Part of the content of the following theorem and corollaries is that the derivatives of the expectations actually exist. For the remainder of this section, let $\Theta$ be a compact interval.

**Theorem 3.2.** Consider a GSMP for which (C2') holds (e.g., (C2) and (3.3) hold), and (A1)-(A3) are satisfied throughout $\Theta$.

**(i)** If $\mathbf{E}[\sup_{\theta\in\Theta} N(T)^2] < \infty$, then $d\mathbf{E}[L_T]/d\theta = \mathbf{E}[dL_T/d\theta]$ on $\Theta$.

**(ii)** If $\mathbf{E}[\sup_{\theta\in\Theta} \tau_n] < \infty$, then $d\mathbf{E}[L_{n_0}]/d\theta = \mathbf{E}[dL_{n_0}/d\theta]$ on $\Theta$.

**(iii)** If $\mathbf{E}[\sup_{\theta\in\Theta} T(\alpha, k)^2] < \infty$ and also $\mathbf{E}[\sup_{\theta\in\Theta} N(T(\alpha, k))^2] < \infty$, then $d\mathbf{E}[L_{\alpha,k}]/d\theta = \mathbf{E}[dL_{\alpha,k}/d\theta]$ on $\Theta$.

**Proof.** We proceed by finding bounds on the $d\tau_i/d\theta$'s, then on the $dL/d\theta$'s. From (2.15) we find that

$$\left| \frac{d\tau_i}{d\theta} \right| \leq \sum_{j=1}^{i} \left| \frac{dX(r_j)}{d\theta} \right| \eta(r_i; r_j).$$

Applying (A3), this is

$$\begin{aligned} &\leq \sum_{j=1}^{i} B(X(r_j) + 1)\eta(r_i; r_j) \\ &\leq B(\tau_i + i). \end{aligned} \tag{3.7}$$

Next, letting $\|f\| = \sup |f| < \infty$, from (2.21) we get

$$\begin{aligned} \left|\frac{dL_T}{d\theta}\right| &= \left|\sum_{i=1}^{N(T)} [f(Y_{i-1}) - f(Y_i)] \frac{d\tau_i}{d\theta}\right| \\ &\le 2\|f\| \sum_{i=1}^{N(T)} \left|\frac{d\tau_i}{d\theta}\right| \\ &\le 2\|f\| B \sum_{i=1}^{N(T)} [\tau_i + i]. \end{aligned}$$

And since $i \le N(T)$ implies $\tau_i \le T$, this is

$$\le 2\|f\|B \cdot N(T)[T + N(T)]. \tag{3.8}$$

Using (2.22), for $L_{n_0}$ we find, similarly, that

$$\left|\frac{dL_{n_0}}{d\theta}\right| \le 2\|f\|Bn_0[\tau_{n_0} + n_0]; \tag{3.9}$$

and for $L_{\alpha,k}$, using (2.23) we get

$$\left|\frac{dL_{\alpha,k}}{d\theta}\right| \le 2\|f\|B \cdot N(T(\alpha,k))[T(\alpha,k) + N(T(\alpha,k))]. \tag{3.10}$$

Next, for $L_T$, $L_{n_0}$ and $L_{\alpha,k}$, let $\mathcal{D}_T$, $\mathcal{D}_{n_0}$ and $\mathcal{D}_{\alpha,k}$ be the (random) set of $\theta \in \Theta$ at which the corresponding $L$ is differentiable. Each $L$ is a.s. continuous in $\theta$; moreover, each is differentiable except at discontinuities of some $a_i(\theta)$, so each is a.s. piecewise differentiable. Thus, we may apply the generalized mean value theorem (see Chapter 1, Section 1.3) to conclude that, whenever $\theta$ and $\theta + h$ are in $\Theta$,

$$\left|\frac{L(\theta+h) - L(\theta)}{h}\right| \le \sup_{\theta\in\mathcal{D}} \left|\frac{dL}{d\theta}\right|. \tag{3.11}$$

Using (3.8), (3.9) and (3.10), it is immediate that under the hypothesis in (i),

$$\mathbf{E}\left[\sup_{\theta\in\mathcal{D}_T} \left|\frac{dL_T}{d\theta}\right|\right] < \infty;$$

under the hypothesis in (ii),

$$\mathbf{E}\left[\sup_{\theta\in\mathcal{D}_{n_0}} \left|\frac{dL_{n_0}}{d\theta}\right|\right] < \infty;$$

and under the hypothesis in (iii),

$$\mathbf{E}\left[\sup_{\theta\in\mathcal{D}_{\alpha,k}}\left|\frac{dL_{\alpha,k}}{d\theta}\right|\right] < \infty;$$

Applying the dominated convergence theorem to (3.11), we find that for all $\theta \in \Theta$

$$\lim_{h\to 0}\mathbf{E}\left[\frac{L(\theta+h)-L(\theta)}{h}\right] = \mathbf{E}\left[\frac{dL(\theta)}{d\theta}\right].$$

That is, the limit on the left exists and is equal to the quantity on the right, which is what we needed to show. □

### Variations

The next two results follow immediately from the proof of Theorem 3.2.

**Corollary 3.1.** If all $dX_\theta(\alpha,k)/d\theta$ are bounded, i.e., (A3) can be replaced with $|dX_\theta(\alpha,k)/d\theta| \le B$ for all $\alpha$ and $k$, then in Theorem 3.2, (ii) is satisfied. A sufficient condition for the conclusion of (iii) is $\mathbf{E}[\sup_{\theta\in\Theta} N(T(\alpha,k))] < \infty$.

**Corollary 3.2.** If (A3) can be replaced with

$$\left|\frac{dX_\theta(\alpha,k)}{d\theta}\right| \le B \cdot X_\theta(\alpha,k)$$

for all $\alpha$ and $k$, then (i) holds if $\mathbf{E}[\sup_{\theta\in\Theta} N(T)] < \infty$.

It is useful to have conditions for Theorem 3.2 stated purely in terms of the clock samples. These are provided, in part, by the following result:

**Corollary 3.3.** In Theorem 3.2, sufficient conditions for (i) and (ii) are, respectively,
**(i')** For every $\alpha$, $P(\inf_{\theta\in\Theta} X_\theta(\alpha,k) = 0) < 1$;
**(ii')** For every $\alpha$, $\mathbf{E}[\sup_{\theta\in\Theta} X_\theta(\alpha,k)] < \infty$.

**Proof.** Let $N_\alpha$ be the renewal process derived from the i.i.d. sequence $X(\alpha,k)$, $k = 1,2,\ldots$; i.e.,

$$N_\alpha(t) = \max\{n : \sum_{k=1}^{n} X(\alpha,k) \le t\}.$$

Then $N(T) \le \sum_{\alpha\in\mathbf{A}} N_\alpha(T)$; and

$$\sup_{\theta\in\Theta} N(T)^2 \le \sum_\alpha \sup_{\theta\in\Theta} N_\alpha(T)^2 + \sum_\alpha\sum_{\beta\ne\alpha}(\sup_{\theta\in\Theta} N_\alpha(T))(\sup_{\theta\in\Theta} N_\beta(T)).$$

Since $N_\alpha$ and $N_\beta$ are independent for $\beta \neq \alpha$,

$$\mathbf{E}[\sup_{\theta\in\Theta} N(T)^2] \leq \mathbf{E}[\sum_\alpha \sup_{\theta\in\Theta} N_\alpha(T)^2] + \sum_\alpha \sum_{\beta\neq\alpha} \mathbf{E}[\sup_{\theta\in\Theta} N_\alpha(T)]\mathbf{E}[(\sup_{\theta\in\Theta} N_\beta(T)];$$

thus, it is enough to show that every $\mathbf{E}[\sup_{\theta\in\Theta} N_\alpha(T)^2]$ is finite. Note that

$$\sup_{\theta\in\Theta} N_\alpha(T) \leq \tilde{N}_\alpha(T) \equiv \max\{n : \sum_{k=1}^{n} \inf_{\theta\in\Theta} X_\theta(\alpha, k) \leq t\}.$$

Since $\{\inf_\theta X_\theta(\alpha, k)\}$ is an i.i.d. sequence, $\tilde{N}_\alpha$ is a renewal process. A standard result from renewal theory (Prabhu [65], p.155) states that $\tilde{N}_\alpha(T)$ has finite moments of all orders, provided its inter-renewal times are strictly positive with strictly positive probability, which is (i').

**Part (ii').** Since the clock for any event runs out at most $n$ times by $\tau_n$ (or directly from (2.14)) it is clear that

$$\tau_n(\theta) \leq \sum_\alpha \sum_{k=1}^{n} X_\theta(\alpha, k);$$

hence,

$$\sup_{\theta\in\Theta} \tau_n(\theta) \leq \sum_\alpha \sum_{k=1}^{n} \sup_{\theta\in\Theta} X_\theta(\alpha, k).$$

Thus, (ii') is sufficient for (ii) of Theorem 3.2. □

Condition (A1) excludes discrete clock setting distributions; but this condition is a bit stronger than what we really need for our results. The continuity of the $F_\alpha$'s is only used to ensure that multiple events never occur simultaneously—i.e., that for each $\theta$,

$$P(\tau_{n+1}(\theta) = \tau_n(\theta) \text{ for some } n) = 0. \tag{3.12}$$

If more than one of the distributions $F_\alpha$ has discontinuities, then (3.12) may be violated. But if we assume only

**(A1').** For every $\alpha \in \mathbf{A}$, $F_\alpha(0, \theta) = 0$,

and can verify (3.12) directly, then we do not need continuity of the $F_\alpha$'s.

**Corollary 3.4.** Theorem 3.2 and Corollaries 3.1-3.3 hold if (A1) is replaced by (A1') and (3.12).

Corollary 3.4 allows the use of families $\{F_\alpha(\cdot, \theta), \theta \in \Theta\}$ of discrete distributions in which the *location* of the atoms of $F_\alpha$ change with $\theta$ but their *probabilities* do not. Samples from distributions where the probabilities change violate (A2). See the discussion of discrete distributions in Chapter 1.

Finally, we consider a modification in which the function $f$ is allowed to depend on $\theta$—i.e., $f : \Theta \times \mathbf{S} \to \mathbf{R}$. Thus, $L_T$, for example, becomes

$$L_T = \int_0^T f(\theta, Z_t(\theta))dt;$$

the derivative $dL_T/d\theta$ is modified through addition of

$$\int_0^T \frac{\partial f}{\partial \theta}(\theta, Z_t(\theta))dt;$$

$L_{n_0}$ and $L_{\alpha,k}$ are modified analogously. (Here, $\partial f/\partial\theta$ is the partial derivative of $f$ with respect to its first argument.) This assumes that the derivative with respect to $\theta$ and the integral with respect to $t$ can be interchanged. The following gives a sufficient condition for this:

**Corollary 3.5.** Suppose $f$ is continuously differentiable in $\theta$ and $|\partial f/\partial\theta|$ is bounded on $\Theta \times \mathbf{S}$, then Theorem 3.2 and Corollaries 3.1-3.4 still hold.

### 3.1.3 Event Diagrams

The statement of (C2') might make it appear hard to verify, and even (C2) seems to require checking a huge number of cases. But for GSMPs with clear physical interpretations, such as networks of queues, the conditions are, in fact, quite readily checked. A useful aid in verifying the conditions is an *event diagram.* An event diagram for a GSMP is similar to the state transition diagrams often used to illustrate Markov chains. As in the Markov case, the nodes in an event diagram represent states; but the arcs, rather than being labeled with transition rates or probabilites, are labeled with both an event and the associated (GSMP) transition probability. (Figure 3.1 is a fragment of an abstract event diagram, in which the transition probabilities have been omitted.) Once we have an event diagram, we simply "follow the arrows" to check (C2).

Figure 3.3 depicts an event diagram for the GI/G/1 queue, in which $\alpha$ denotes arrival and $\beta$ denotes departure. Since all transition probabilities are unity, they are not shown. To verify (C2), start in any state from which arcs for two events emanate. In Figure 3.3, this means any state $s$ with $s > 0$. Follow the arcs to their destination nodes, $s-1$ and $s+1$. For (C2) to be satisfied, there must be a state reachable from $s-1$ through $\alpha$ (with the same probability as the transition $s \to s+1$) which is also reachable from $s+1$ through $\beta$ (with the same probability as $s \to s-1$). Clearly, $s$ itself is such a state.

Figure 3.4, which shows the event diagram for a GI/G/1/K finite capacity queue, illustrates one way (C2) can be violated. Starting in state $K$, if we

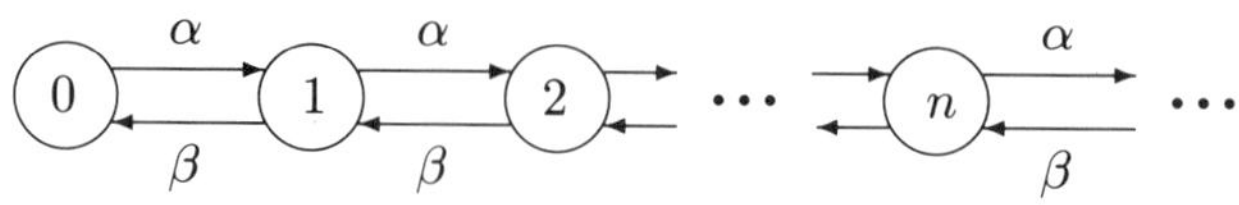


Figure 3.3: Event Diagram for GI/G/1 Queue

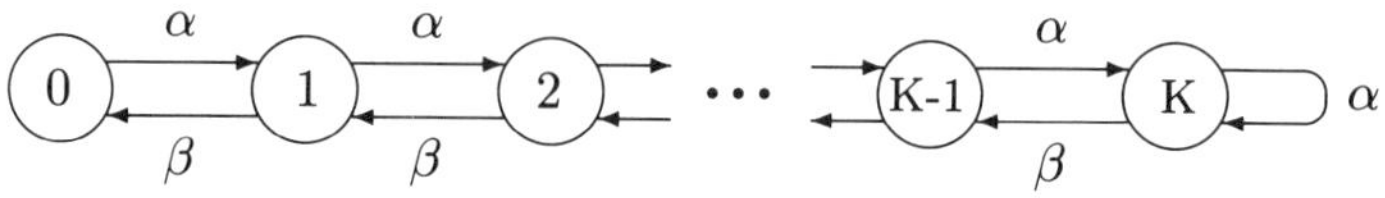


Figure 3.4: Event Diagram for GI/G/1/K Queue

follow first $\alpha$, then $\beta$, we end up in state $K-1$; but if we follow first $\beta$ then $\alpha$ we end up back in $K$, violating (C2) (and (C2') since there is only one choice for $\phi$). Algorithm 2.1 applied to the GI/G/1/K queue will, in fact, yield biased estimates. Intuitively, speeding up arrivals or slowing down service, might cause an initially accepted job to be blocked, substantially affecting the future evolution of the process. (Finite capacity queues are taken up more generally in Section 4.2.4 of Chapter 4.)

To illustrate an event diagram with transition probabilities, we consider a closed Jackson-like network; see Example 2 of Section 2.1, Chapter 2. We denote a typical state by a vector $\mathbf{n} = (n_1, \ldots, n_M)$ of queue lengths. The event $\beta_i$ denotes departure from queue $i$. State transitions are governed by a routing matrix $P$ via $p(\mathbf{n} - e_i + e_j; \mathbf{n}, \beta_i) = P_{ij}$. Figure 3.5 depicts a small portion of the complete event diagram for the network which, nevertheless, captures the essential features that cause (C2) to be satisfied: regardless of the order in which $\beta_i$ and $\beta_k$ occur, the same state is reached. The pattern in Figure 3.5 is repeated throughout the complete diagram for the network; and, as we will see later (Chapter 4, Section 4.2) in more detail, such networks do satisfy (C2). Figure 3.5 also shows that networks with state-dependent routing may violate (C2), since state-dependent routing may cause transition probabilities on opposing arcs to differ.

Figures 3.3-3.5 help to make another point about event diagrams. Drawing a complete diagram for a GSMP is rarely feasible; but it is often possible to identify patterns that make it sufficient to consider certain critcal regions that reveal whether or not (C2) is satisfied. To some extent, this begs the question, since one must know where the "critical" region lies. Our experience is that it becomes quite clear where to look after drawing just a few circles and arrows.

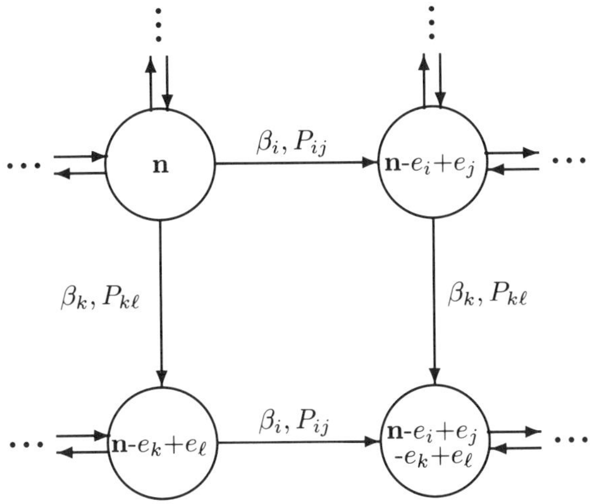


Figure 3.5: Partial Event Diagram for Jackson-like Network

## 3.2 GSMPs With Speeds

Sections 3.2.1-3.2.3 parallel the definition, construction and differentiation of parametric GSMPs developed in Chapter 2, but for GSMPs with speeds. Section 3.2.4 describes modifications to Section 3.1 needed to obtain unbiased derivative estimates in the presence of speeds.

### 3.2.1 Definition

*Speeds* are a mechanism for allowing clocks to run down at different rates in different states. (Thus far, we have only considered the case where all clocks always run down at unit rate.) Speeds offer additional modeling flexibility; so in this section we develop derivative estimates for GSMPs with speeds, stressing those points that differ from the case of constant rates.

From a modeling point of view (especially for queues) we see the following as the most useful applications of speeds:

(a) setting a clock speed temporarily to zero to "suspend" an event (e.g., pre-empting a job's service using a resume discipline),

and

(b) adjusting the rate at which an activity is performed (e.g., state-dependent service).

A third use is to set a rate to infinity to cause the instantaneous occurrence of an event. We consider the utility of such a mechanism limited, and will not consider it because of the technical difficulties it introduces. In fact, to obtain unbiased derivative estimates we will also need to exclude (a). This should not be surprising, given that setting a clock speed to zero amounts to (temporarily) deleting an event from the event list, which violates (C1). Thus, denoting by $\lambda_\alpha(s)$ the speed (clock rate) for event $\alpha$ in state $s$ (whenever $\alpha \in \mathcal{E}(s)$), let us require

$$\text{for all } s \in \mathbf{S} \text{ and all } \alpha \in \mathcal{E}(s),\ 0 < \lambda_\alpha(s) < \infty. \tag{3.13}$$

The definition of a GSMP with speeds via a discrete time, general state space Markov process is very similar to that given in Section 2.1 for GSMPs without speeds. If we assume (3.13), redefine $t^*$ to be

$$t^*(s,c) = \min\{c(\alpha)/\lambda_\alpha(s) : \alpha \in \mathcal{E}(s)\} \tag{3.14}$$

and let (for $\alpha \in \mathcal{E}(s)$)

$$c^*_\alpha(s,c) = c(\alpha) - \lambda_\alpha(s)t^*(s,c),$$

then the transition probability for $(Y_n, c_n)$ becomes

$$P((s,c),A) = p(s'; s, \alpha^*) \prod_{\alpha\in\mathcal{N}} F_\alpha(x(\alpha)) \prod_{\alpha\in\mathcal{O}} \mathbf{1}\{c^*_\alpha \le x(\alpha)\}.$$

(See Section 2.1 for notation.)

The sample path construction of Section 2.2 goes through as before, except that (2.11) is replaced with

$$c_{n+1}(\alpha) = c_n(\alpha) - \lambda_\alpha(Y_n)t^*(Y_n, c_n); \tag{3.15}$$

and in (2.7) and (3.15), $t^*$ is understood to mean (3.14).

To see why (3.14) and (3.15) are appropriate, note that it takes $c/\lambda$ time units to run down a clock with remaining time $c$ at rate $\lambda$. During this interval, a clock with $c'$ time remaining and rate $\lambda'$ would be run down to $c' - \lambda'(c/\lambda)$, if $c'/\lambda' > c/\lambda$.

### 3.2.2 Derivatives

Previously, we only allowed the clock samples $\{X_\theta(\alpha,k)\}$ to depend on a parameter $\theta$. We will now consider cases in which either the clock samples or the speeds depend on $\theta$. For ease of exposition, we consider these two cases separately. We first consider a situation where the speeds are differentiable funtions of a parameter $\theta$, and (3.13) holds for all $\theta \in \Theta$. We assume that the clock samples (and their distributions) *do not depend on* $\theta$. We will not explicitly include the argument $\theta$ in the speeds, but let the context determine whether $\lambda_\alpha(s)$ depends on $\theta$. Our first step is to find an analog of Lemma 2.2 to describe the change in the epochs of events under small changes in $\theta$.

In the presence of speeds, the interactions among the event epochs are more complex. Without speeds, a small delay in some $\tau_i$ propagates only to those $\tau_j$ for which $\eta(r_j; r_i) = 1$. With speeds, such a delay potentially affects *all* $\tau_j$ with $j \geq i$. For in delaying $\tau_i$, we not only delay the setting of new clocks (at $\tau_i$), we also increase the time that clocks for events in $\mathcal{E}(Y_{i-1})$ are run down at rate $\lambda_\alpha(Y_{i-1})$ rather than $\lambda_\alpha(Y_i)$. If $\lambda_\alpha(Y_{i-1}) > \lambda_\alpha(Y_i)$, then delaying $\tau_i$ could actually cause $\alpha$ to occur *sooner*, since the clock for $\alpha$ slows down after $\tau_i$.

The essential mechanism by which changes in the timing of one event propagate to other events can be described as follows. If $\alpha \in \mathcal{E}(Y_n)$, denote by $R_n(\alpha)$ the scheduled epoch of the next occurrence of $\alpha$—i.e.,

$$R_n(\alpha) = \tau_n + c_n(\alpha)/\lambda_\alpha(Y_n).$$

The event $\alpha$ would occur at $R_n(\alpha)$ if no other event occurred first and changed the speed $\lambda_\alpha$. Suppose that $a_{n+1} = \beta \neq \alpha$; then the occurrence of $\beta$ at $\tau_{n+1}$ may change the speed for $\alpha$ and therefore cause the next occurrence of $\alpha$ to be rescheduled according to

$$R_{n+1}(\alpha) = \tau_{n+1} + \frac{\lambda_\alpha(Y_n)}{\lambda_\alpha(Y_{n+1})}(R_n(\alpha) - \tau_{n+1}).$$

In words, the residual time $R_n(\alpha) - \tau_{n+1}$ until the scheduled occurrence of $\alpha$ is rescaled to the new clock rate. From this we get

$$\begin{aligned} \frac{dR_{n+1}(\alpha)}{d\theta} &= \left(\frac{\lambda_\alpha(Y_n)}{\lambda_\alpha(Y_{n+1})}\right)\frac{dR_n(\alpha)}{d\theta} + \left(1 - \frac{\lambda_\alpha(Y_n)}{\lambda_\alpha(Y_{n+1})}\right)\frac{d\tau_{n+1}}{d\theta} \\ &+ (R_n(\alpha) - \tau_{n+1})\frac{d}{d\theta}\left(\frac{\lambda_\alpha(Y_n)}{\lambda_\alpha(Y_{n+1})}\right), \end{aligned} \tag{3.16}$$

which shows how changes in the timing of one event introduce changes in the timing of another. At each transition, the updated "perturbation" $dR_{n+1}(\alpha)/d\theta$ is a combination of three terms: the previous perturbation $dR_n(\alpha)/d\theta$ in the scheduled occurrence of $\alpha$, the perturbation $d\tau_{n+1}/d\theta$ in the epoch of the current transition, and the derivative of the ratio of the $\alpha$-speeds before and after the transition. Note that if $\lambda_\alpha \equiv 1$, then $dR_{n+1}(\alpha)/d\theta = dR_n(\alpha)/d\theta$—a small change in the epoch of occurrence of $\beta$ will not alter the next occurrence of $\alpha$.

To keep track of these effects, we introduce

**Definition 3.1.** The *propagation process* $\psi = \{\psi_{\alpha\beta}(n), \alpha, \beta \in \mathbf{A}, n = 0, 1, 2, \ldots\}$ is a discrete time process that evolves as follows: For all $\alpha$ and $\beta$, $\psi_{\alpha\beta}(0) = 0$;

$\psi_{\alpha\beta}(1) = \mathbf{1}\{\alpha = \beta\}$; and for $n > 1$, and all $\alpha, \beta \in \mathbf{A}$,

$$\begin{aligned}
\beta \in \mathcal{O}(Y_n; Y_{n-1}, a_n) \quad &\Rightarrow \quad \psi_{\alpha\beta}(n) = (1 - \lambda_\beta(Y_{n-1})/\lambda_\beta(Y_n))\psi_{\alpha a_n}(n-1) \\
&\qquad +(\lambda_\beta(Y_{n-1})/\lambda_\beta(Y_n))\psi_{\alpha\beta}(n-1) \\
\beta \in \mathcal{N}(Y_n; Y_{n-1}, a_n) \quad &\Rightarrow \quad \psi_{\alpha\beta}(n) = \psi_{\alpha a_n}(n-1) \\
\beta \notin \mathcal{E}(Y_n) \quad &\Rightarrow \quad \psi_{\alpha\beta}(n) = \psi_{\alpha\beta}(n-1).
\end{aligned}$$

(Recall that $\mathcal{O}$ and $\mathcal{N}$ denote sets of old and new events.) From the basic process $\psi$ define a sequence of processes $\psi^{(i)}$, $i = 0, 1, 2, \ldots$ by setting $\psi^{(0)} = \psi$, and letting every $\psi^{(i)}$ evolve in the same way as $\psi$ but starting at the $i$th transition. Thus, $\psi^{(i)}(n) = 0$ for all $n \leq i$; $\psi^{(i)}_{\alpha\beta}(i+1) = \mathbf{1}\{\alpha = \beta\}$; and from $i+1$ on, $\psi^{(i)}$ follows the rules of Definition 3.1.

The following is the analog of Lemma 2.2 for derivatives with respect to speeds:

**Lemma 3.2.** Suppose (C1), (A1) and (3.13) hold, and that every $\lambda_\alpha(s)$ is differentiable in $\theta$. Then at each $\theta$, each $\tau_n$ is a.s. differentiable with

$$\frac{d\tau_n}{d\theta} = \sum_{i=0}^{n-1} \sum_{\alpha \in \mathbf{A}} \left( -\frac{\tau_{i+1} - \tau_i}{\lambda_\alpha(Y_i)} \right) \frac{d\lambda_\alpha(Y_i)}{d\theta} \psi^{(i)}_{\alpha a_n}(n). \tag{3.17}$$

One way to think of (3.17) is to imagine, instead of changing $\theta$, introducing a small perturbation, $\Delta\lambda_\alpha(Y_i)$, in the speed for $\alpha$ during the holding time $[\tau_i, \tau_{i+1})$ *only*. To first order, (3.17) says that the resulting change in $\tau_n$ is

$$\Delta\tau_n = \left( -\frac{\tau_{i+1} - \tau_i}{\lambda_\alpha(Y_i)} \right) \psi^{(i)}_{\alpha a_n}(n) \Delta\lambda_\alpha(Y_i).$$

Thus,

$$\left( -\frac{\tau_{i+1} - \tau_i}{\lambda_\alpha(Y_i)} \right) \frac{d\lambda_\alpha(Y_i)}{d\theta}$$

is the "perturbation generated" during $[\tau_i, \tau_{i+1})$ due to the change in $\lambda_\alpha(Y_i)$. This is consistent with the fact that $\lambda_\alpha(Y_i)$ acts as a scale parameter for the $\alpha$-clock during the interval $[\tau_i, \tau_{i+1})$. (See the discussion of scale parameters in Section 1.3.2 of Chapter 1.) The factor $\psi^{(i)}_{\alpha a_n}(n)$ is the extent to which the perturbation generated is propagated to $\tau_n$.

Verification that (3.17) correctly summarizes the effect of iterating (3.16) involves an intricate but conceptually straightforward proof by induction which does little to elucidate the role of the $\psi^{(i)}$'s. We therefore omit it. The proof of

Lemma 3.2 is similar to that of Proposition 5.2 of Section 5.3, which is spelled out in detail.

The propagation processes $\psi^{(i)}$ have a closer connection to the triggering indicators than even Lemma 3.2 may suggest. For suppose that every $\lambda_\alpha(s)$ is just $\mathbf{1}\{\alpha \in \mathcal{E}(s)\}$. Then every $\psi^{(i)}_{\alpha\beta}(n)$ is zero or one (like $\eta$). Moreover, the first case of Definition 3.1 reduces to $\psi_{\alpha\beta}(n) = \psi_{\alpha\beta}(n-1)$. Only the second case changes $\psi$: if a clock for $\beta$ is set at the occurrence of $a_n$ then $\psi_{\alpha\beta}(n)$ is set equal to zero or one as $\psi_{\alpha a_n}$ is equal to zero or one. In fact, when all speeds are unity,

$$\eta(r_n; r_i) = \psi^{(i-1)}_{a_i a_n}(n).$$

With this connection in mind, suppose, now, that it is the clock samples and not the speeds that depend on $\theta$. Suppose, further, that all events whose clock samples depend on $\theta$ have only unit speeds: for any $\alpha$ and $s$,

$$\lambda_\alpha(s) \neq 1 \Rightarrow \{X_\theta(\alpha, k), k = 1, 2, \ldots\} \text{ are constant in } \theta. \tag{3.18}$$

Then we have

**Lemma 3.3.** Under (C1), (A1), (A2), (3.13) and (3.18), at each $\theta$, each $\tau_n$ is a.s. differentiable with

$$\frac{d\tau_n}{d\theta} = \sum_{i=1}^{n} \frac{dX_\theta(r_i)}{d\theta} \psi^{(i-1)}_{\alpha a_n}(n). \tag{3.19}$$

Note that this specializes to Lemma 2.2. (Lemmas 3.1 and 3.2 apply to epochs $T(\alpha, k)$ as well; if $(\alpha, k) = r_n$, then $dT(\alpha, k)/d\theta = d\tau_n/d\theta$ in both cases.) Given expressions for the derivatives of the transition epochs, the derivatives of the performance measures $L_T$, $L_{n_0}$ and $L_{\alpha,k}$ are built up exactly as in Lemma 2.4, regardless of whether it is the clock samples or the speeds that depend on $\theta$. We will see in Section 3.2.4 that under conditions similar to those of Section 3.1.1 these expressions yield unbiased derivative estimators.

### 3.2.3 A Generic Algorithm for Speeds

We present a modification of the generic algorithm of Section 2.3.2 appropriate for GSMPs with speeds, when it is the speeds, not the clock samples, that depend on $\theta$. The case where the clock samples depend on $\theta$ is not fundamentally different.

As in Algorithm 2.1, $\delta(\alpha)$ is the current accumulated infinitesimal delay in the next occurrence of $\alpha$, and $d\tau_n/d\theta$ is given by the value of $\delta^*$ at the $n$th transition.

**Algorithm 3.1: Perturbation Propagation with Speeds**

0. Initialization.
    a. $\tau := 0$;
    b. Set $Y$ to initial state;
    c. For every $\alpha \in \mathcal{E}(Y)$
        Generate $X(\alpha)$ from $F_\alpha$;
        $c(\alpha) := X(\alpha)$;
        $\delta(\alpha) := 0$;
    d. For every $\alpha \notin \mathcal{E}(Y)$,
        $c(\alpha) := 0$;
        $\delta(\alpha) := 0$;

1. Next Transition.
    a. Find $\alpha^* \in \mathcal{E}(Y)$ with smallest $c(\alpha)/\lambda_\alpha(Y)$;
    b. $t^* := c(\alpha^*)/\lambda_\alpha(Y)$;
    c. $\tau := \tau + t^*$;
    d. For every $\alpha \in \mathcal{E}(Y)$
        $\delta(\alpha) := \delta(\alpha) - (t^*/\lambda_\alpha(Y))(d\lambda_\alpha(Y)/d\theta)$;
    e. $\delta^* := \delta(\alpha^*)$;
    f. Generate $Y'$ from $p(\cdot; Y, \alpha^*)$;
    g. For every $\alpha \in \mathcal{E}(Y') \backslash (\mathcal{E}(Y) - \{\alpha^*\})$
        Generate $X(\alpha)$ from $F_\alpha$;
        $c(\alpha) := X(\alpha)$;
        $\delta(\alpha) := \delta^*$;
    h. For every $\alpha \in \mathcal{E}(Y') \cap (\mathcal{E}(Y) - \{\alpha^*\})$
        $c(\alpha) := c(\alpha) - \lambda_\alpha(Y)t^*$;
        $\delta(\alpha) := (1 - \lambda_\alpha(Y)/\lambda_\alpha(Y'))\delta^* + (\lambda_\alpha(Y)/\lambda_\alpha(Y'))\delta(\alpha)$;
    i. $Y := Y'$;
2. Check Stopping Condition.
    If stopping condition is met, STOP; else go to 1.

### 3.2.4 Results for Speeds

If we assume (3.13) and also that either the speeds $\{\lambda_\alpha(s)\}$ or the clock samples $\{X(\alpha, k)\}$ are continuous in $\theta$ (depending on which we allow to depend on $\theta$), then under (C2') the continuity of the performance measures $L_T$, $L_{n_0}$ and $L_{\alpha,k}$ goes through just as in Theorem 3.1 of Section 3.1.2. However, verification that the $dL/d\theta$'s are unbiased estimators of the $d\mathbf{E}[L]/d\theta$'s requires further conditions on the speeds. The proof of Theorem 3.2 relied on the fact that in the absence of speeds,

$$\frac{d\tau_n}{d\theta} \le \sum_{i=1}^{n} \left| \frac{dX(r_i)}{d\theta} \right|,$$

since the triggering indicators are always zero or one. But in the general case, the propagation processes need not even be bounded. Repeated multiplication by ratios of rates (as in Definition 3.1) can cause the $\psi$'s (hence, the $d\tau_n/d\theta$'s) to grow quickly. This could, in principle, prevent the interchange of expectation and differentiation. We consider two sets of conditions on the speeds to control the growth of the $\psi$'s.

We begin with the following *increasing speeds condition:* For all $s \in \mathbf{S}$, if $\{\alpha, \beta\} \subseteq \mathcal{E}(s)$, $\beta \neq \alpha$, then

$$p(s'; s, \beta) > 0 \Rightarrow \lambda_\alpha(s) \le \lambda_\alpha(s'). \tag{3.20}$$

In words, at every state transition, the speed of every old event must increase. Consider

**Example 2.** *(continued) Jackson-like networks.* Suppose the service speed $\lambda_{\beta_i}$ at server $i$ depends only on the number of jobs at $i$, and is an increasing function of that number. At the occurrence of any $\beta_j$, $j \neq i$, the number of jobs at queue $i$ may increase or stay the same; hence, if a transition from $s$ to $s'$ is triggered by $\beta_j$, $\lambda_{\beta_i}(s') \ge \lambda_{\beta_i}(s)$, and (3.20) is satisfied.

Under (3.20), it is easy to see that the coefficients $\lambda_\alpha(Y_{n-1})/\lambda_\alpha(Y_n)$ and $1 - \lambda_\alpha(Y_{n-1})/\lambda_\alpha(Y_n)$ in Definition 3.1 are always between zero and one, since necessarily $\lambda_\alpha(Y_{n-1}) \le \lambda_\alpha(Y_n)$. This, in turn, implies that for all $i$, $\alpha$, $\beta$ and $n$,

$$|\psi_{\alpha\beta}^{(i)}(n)| \le 1. \tag{3.21}$$

This effectively reduces the proof of unbiasedness to the case of unit speeds.

**Theorem 3.3.** Consider a GSMP with speeds in which only the speeds depend on $\theta$. Suppose they are continuously differentiable in $\theta$, and that there are constants $B > 0$ and $\lambda_* > 0$ such that for all $\alpha$, $s$ and $\theta$

$$\left| \frac{d\lambda_\alpha(s)}{d\theta} \right| \le B,$$

and

$$\lambda_\alpha(s) > 0 \Rightarrow \lambda_\alpha(s) > \lambda_*.$$

Suppose that (A1), (3.13), (C2') and (3.20) all hold.

**(i)** If $\mathbf{E}[\sup_{\theta\in\Theta} N(T)] < \infty$, then $d\mathbf{E}[L_T]/d\theta = \mathbf{E}[dL_T/d\theta]$ on $\Theta$.

**(ii)** If $\mathbf{E}[\sup_{\theta\in\Theta} \tau_n] < \infty$, then $d\mathbf{E}[L_{n_0}]/d\theta = \mathbf{E}[dL_{n_0}/d\theta]$ on $\Theta$.

**(iii)** If $\mathbf{E}[\sup_{\theta\in\Theta} T(\alpha,k)^2] < \infty$ and $\mathbf{E}[\sup_{\theta\in\Theta} N(T(\alpha,k))^2] < \infty$, then $d\mathbf{E}[L_{\alpha,k}]/d\theta = \mathbf{E}[dL_{\alpha,k}/d\theta]$ on $\Theta$.

**Proof.** Under (3.20), the $\psi$'s are all between 0 and 1, so from (3.17) we find

$$\begin{aligned}\left|\frac{d\tau_n}{d\theta}\right| &\leq \sum_{i=0}\sum_{\alpha\in\mathbf{A}} \frac{\tau_{i+1}-\tau_i}{\lambda_\alpha(Y_i)}\left|\frac{d\lambda_\alpha(Y_i)}{d\theta}\right| \\ &\leq |\mathbf{A}|B\tau_n/\lambda_*.\end{aligned}$$

As in the proof of Theorem 3.2, we get

$$\left|\frac{dL_T}{d\theta}\right| \leq (2\|f\|B)\cdot N(T)\cdot T/\lambda_*.$$

The rest of the proof of (i) is the same as that of Theorem 3.2. Parts (ii) and (iii) work similarly. □

Because of (3.21), the proof of the following is the same as that of Theorem 3.2:

**Corollary 3.6.** Suppose the speeds do not depend on $\theta$, and only the clock samples of those events with unit speeds depend on $\theta$. Then if (C2'), (A1)-(A3) (3.13) and (3.20) hold, so do (i)-(iii) of Theorem 3.2.

We can relax the increasing speeds condition (3.20) at the expense of more stringent moment conditions on the event counting process $N$.

**Theorem 3.4.** Suppose the conditions of Theorem 3.3 are in effect, except that instead of (3.20) assume that all speeds are bounded above by some $\lambda^* < \infty$. Let $\rho = \lambda^*/\lambda_*$.

**(i)** If $\mathbf{E}[\sup_{\theta\in\Theta}(2\rho)^{2N(T)}] < \infty$, then $d\mathbf{E}[L_T]/d\theta = \mathbf{E}[dL_T/d\theta]$ on $\Theta$.
**(ii)** If $\mathbf{E}[\sup_{\theta\in\Theta}\tau_n] < \infty$, then $d\mathbf{E}[L_{n_0}]/d\theta = \mathbf{E}[dL_{n_0}/d\theta]$ on $\Theta$.
**(iii)** If $\mathbf{E}[\sup_{\theta\in\Theta}(N(T(\alpha,k))T(\alpha,k))^2] < \infty$, $\mathbf{E}[\sup_{\theta\in\Theta}(2\rho)^{2N(T(\alpha,k))}] < \infty$, then $d\mathbf{E}[L_{\alpha,k}]/d\theta = \mathbf{E}[dL_{\alpha,k}/d\theta]$ on $\Theta$.

**Proof.** Clearly, for $s$, $s'$ and $\alpha\in\mathcal{E}(s)\cap\mathcal{E}(s')$, $\lambda_\alpha(s')/\lambda_\alpha(s)\leq\rho$; moreover,

$$\left|1-\frac{\lambda_\alpha(s')}{\lambda_\alpha(s)}\right| \leq \max(1,\frac{\lambda_\alpha(s')}{\lambda_\alpha(s)}) \leq \rho.$$

Thus, for all $n$ and $i$,

$$\max_{\alpha,\beta}|\psi^{(i)}_{\alpha\beta}(n+1)| \leq (2\rho)\max_{\alpha,\beta}|\psi^{(i)}_{\alpha\beta}(n)|.$$

Since $|\psi^{(i)}_{\alpha\beta}(i+1)| \leq 1$, we conclude that $\max|\psi^{(i)}_{\alpha\beta}(n)| \leq (2\rho)^{n-i}$. From (3.17) we get, as in the proof of Theorem 3.3,

$$\left|\frac{d\tau_n}{d\theta}\right| \leq |\mathbf{A}|B(2\rho)^{n-1}\tau_n/\lambda_*.$$

As in Theorem 3.2, we now get

$$\left|\frac{dL_T}{d\theta}\right| \leq (2\|f\|B/\lambda_*)N(T) \cdot T \cdot (2\rho)^{N(T)-1},$$

$$\left|\frac{dL_{n_0}}{d\theta}\right| \leq (2\|f\|B/\lambda_*)n_0 \cdot \tau_{n_0} \cdot (2\rho)^{n_0-1},$$

and

$$\left|\frac{dL_{\alpha,k}}{d\theta}\right| \leq (2\|f\|B/\lambda_*)N(T(\alpha,k)) \cdot T(\alpha,k) \cdot (2\rho)^{N(T(\alpha,k))-1}.$$

The rest of the proof is the same as Theorem 3.2, using the moment conditions (i)-(iii). □

**Corollary 3.7.** In the situation of Corollary 3.6, drop (3.20), but suppose all non-zero speeds are bounded above by $\lambda^*$ and below by $\lambda_*$ as in Theorem 3.4. Drop (A3) but suppose all $|dX_\theta(\beta,j)/d\theta|$ are bounded by some constant $B$.

**(i)** If $\mathbf{E}[\sup_{\theta\in\Theta}(2\rho)^{2N(T)}] < \infty$, then $d\mathbf{E}[L_T]/d\theta = \mathbf{E}[dL_T/d\theta]$ on $\Theta$.

**(ii)** Without further conditions, $d\mathbf{E}[L_{n_0}]/d\theta = \mathbf{E}[dL_{n_0}/d\theta]$ on $\Theta$.

**(iii)** If $\mathbf{E}[\sup_{\theta\in\Theta}(2\rho)^{2N(T(\alpha,k))}] < \infty$, then $d\mathbf{E}[L_{\alpha,k}]/d\theta = \mathbf{E}[dL_{\alpha,k}/d\theta]$ on $\Theta$.

**Proof.** Arguing as in Theorem 3.4, but starting from (3.19), we get

$$\left|\frac{d\tau_n}{d\theta}\right| \leq |\mathbf{A}|Bn(2\rho)^{n-1}.$$

The rest of the proof is the same as the other cases. □

* * *

## Notes and Comments

Cao [5] discussed continuity as a critical precondition for the unbiasedness of perturbation analysis derivative estimates. Cao [7] verified continuity of the throughput function in closed Jackson networks, and passed from continuity to the interchange of differentiation and expectation. The commuting condition (C2) and Theorems 3.1 and 3.2 extract the essence of this argument and generalize it, making a case-by-case investigation unnecessary. The idea of the commuting conditions—that when two events change order, the state reached is unchanged—is from Glasserman [25], which considers a birth-death process. That simple example is surprisingly indicative of the general case. The GSMP formulation of Section 3.1 is from Glasserman [24,27]. At about the same time, Li and Ho [63] described a "homogeneity" property of sample paths—essentially the continuity that follows from (C2)—but without a characterization of the class of processes for which the property holds.

Speeds are sometimes included in the basic GSMP set-up; we postponed introducing them because they make stochastic derivative formulas substantially more complicated. The derivatives and the algorithm given here have their origins in Ho and Yang [54], where perturbation analysis estimates are derived for a load-dependent server. There turns out to be no significant difference between event speeds and state-dependent service rates, at least as far as derivative estimates are concerned. The formulation given here is from Glasserman [24]. The special case of load-dependent servers in a closed queueing network is investigated in Glasserman [29], where propagation processes are introduced.

Recently, Glasserman and Yao [35,36] have investigated GSMPs satisfying a condition slightly weaker than (C2) and shown that this structure has significant implications beyond the setting of derivative estimation. Briefly, (C2) requires that changing the order of events not change the state reached, while condition (M) in [35,36] only requires that the *event list* of the state reached not change. Condition (M) turns out to be essentially equivalent to the existence of recursions for events involving only +, min and max, as in (3.6). Hence, in addition to continuity, (M) implies that all event epochs are monotone increasing functions of the clock samples. One application of this monotonicity is guaranteed variance reduction using *common random numbers*.

Further structural properties are pursued in [36]. There, the set of possible (or "feasible") sequences of events in a GSMP are viewed as strings of a formal language. Condition (M) is equivalent to this language forming an *antimatroid with repetition*, and this connection gives rise to a rich combinatorial structure. It leads, in particular, to a simple condition—a strengthening of (M)—under which the event epochs are increasing and *convex* functions of the clock samples.

# Chapter 4

# Derivative Estimation in Networks of Queues

Networks of queues are generalized semi-Markov processes with special structure; and this structure plays an important role both in implementing derivative estimates and in verifying their applicability. Most of the results in this chapter are applications of those of Chapter 3. The emphasis here is on examples and on exploiting, wherever possible, special properties of queues that distinguish them from abstract GSMPs. In the first section, we look at two special properties which simplify the later discussion and are important in applications. Section 4.2 catalogs queueing systems that do and do not satisfy the commuting conditions of Chapter 3. Sections 4.3 and 4.4 discuss variants of the basic conditions which extend their applicability.

## 4.1 Special Structure of Queues

We begin our investigation of stochastic derivatives for queues by taking up two loosely related preliminary issues which involve properties of queues not shared by arbitrary GSMPs. The first issue is the choice of transition mapping $\phi$ to close the gap between conditions (C2) and (C2') of Chapter 3, Section 3.1.1. The second issue is incorporating waiting times into the class of performance measures we can handle.

### 4.1.1 The Structure of Transitions

In Section 3.1.1, in order to pass from condition (C2) to condition (C2') we required (3.3), which states that no two transitions out of the same state due to the same event can have exactly the same probability—i.e., for any event $\alpha$ and states $s, s', s''$, if $p(s'; s, \alpha) > 0$ and $p(s''; s, \alpha) > 0$ and $s' \neq s''$, then $p(s'; s, \alpha) \neq p(s''; s, \alpha)$. This extra condition is not much of a practical restriction (since the difference between the transition probabilities could be arbitrarily small), but is enough of a nuisance that it is worth getting around. We do so by taking advantage of the structure of state transitions in networks of queues.

Consider a Jackson-like queueing network, as described in Example 2 of Section 2.1. Denote by $\mathbf{n}$ and $\mathbf{n}'$ (rather than $s$ and $s'$) two states (queue-length vectors) of the network. At a departure, $\beta_i$, from queue $i$, a job may join any queue $j$ for which the transition probability $P_{ij}$ is greater than zero. Suppose that $n_i > 0$ and $n_i' > 0$. Of all possible transitions out of $\mathbf{n}$ and $\mathbf{n}'$ due to $\beta_i$, we may identify as "the same" transition

$$\mathbf{n} \rightarrow \mathbf{n} - e_i + e_j$$

and

$$\mathbf{n}' \rightarrow \mathbf{n}' - e_i + e_j.$$

In both cases, a job moves from $i$ to $j$. But in an abstract GSMP, even if $\beta \in \mathcal{E}(s)$ and $\beta \in \mathcal{E}(s')$ there may be no obvious way to make a correspondence between the possible transitions out of $s$ due to $\beta$ with those out of $s'$ due to $\beta$. In the proof of Proposition 3.1, we (implicitly) made a correspondence between transitions with the same probability, which is why we required that no two transitions have exactly the same probabilities! For queueing systems, we have a direct way to identify transitions, so we may drop this requirement.

A construction of $\phi$ for Jackson-like networks that uses the structure of transitions is obtained the same way $\phi$ was constructed in Proposition 3.1. For each $\mathbf{n}$ and $\beta_i$, and each $u \in [0, 1]$ define $\phi(\mathbf{n}, \beta_i, u)$ by *inversion:* Partition the unit interval into subintervals, an interval of length $P_{ij}$ for each $j$ for which $P_{ij} > 0$. If $u$ falls in the $j$th interval, set $\phi(\mathbf{n}, \beta_i, u)$ equal to $\mathbf{n} - e_i + e_j$. (As in Example 2 of Section 2.1, we can incorporate open networks by introducing a fictitious node 0 and letting $e_0$ be the vector of all zeros.) Using this choice of $\phi$, (C2') is satisfied if (C2) is; and we will verify in Section 4.2.1 that (C2) is, in fact, satisfied by Jackson-like networks.

For more complicated systems, $\phi$ can be similarly defined. In a network with multiple job classes or state-dependent routing, a different partition of $[0, 1]$ is used for each class or in each state, accordingly. We will generally not construct $\phi$ explicitly, since the analog of the construction above is usually obvious. Moreover, we will go back and forth between conditions (C2) and (C2') without the additional requirement (3.3).

With this choice of $\phi$ in mind, in verifying (C2) we sometimes work with *state transition mappings* rather than $\phi$. For example, we can define a mapping $\Phi_{ij}$ by $\Phi_{ij}\mathbf{n} = \mathbf{n} - e_i + e_j$, so that it yields the state reached upon the movement of a job from node $i$ to node $j$. Verification of (C2) for a Jackson-like network then entails showing that $\Phi_{ij}\Phi_{k\ell}\mathbf{n} = \Phi_{k\ell}\Phi_{ij}\mathbf{n}$ (if $n_i > 0$ and $n_k > 0$), and that the probabilities of the transitions $\Phi_{ij}$, $\Phi_{k\ell}$ do not depend on the order in which they occur. Working with this notation, rather than $\phi$, sometimes simplifies arguments, and allows us to drop the routing indicator argument $u$.

### 4.1.2 Waiting Times

Recall that the performance measures we have considered thus far have the form

$$L_T = \int_0^T f(Z_t)dt,$$

$$L_{n_0} = \int_0^{\tau_{n_0}} f(Z_t)dt,$$

or

$$L_{\alpha,k} = \int_0^{T(\alpha,k)} f(Z_t)dt,$$

where $f$ is a bounded real-valued function, $T$ is a fixed time, $\tau_{n_0}$ is the epoch of the $n_0th$ event, and $T(\alpha, k)$ is the epoch of the $k$th occurrence of $\alpha$. Among the most important performance measures for networks of queues are waiting times and sojourn times, which do not always fit into any of these forms. Indeed, waiting times have no obvious analog in abstract GSMPs, but depend on the special structure of queueing systems. We will exploit this structure to apply to waiting times our conditions for unbiased derivative estimation.

In a GI/G/1 queue, the waiting (or system) time of a job is the difference between its departure and arrival times. Thus, if $\alpha$ denotes arrival to the queue, $\beta$ departure from the queue, and if the queue is empty at time zero, then the waiting time of the $k$th job is $T(\beta, k) - T(\alpha, k)$. Each of $T(\alpha, k)$ and $T(\beta, k)$ has the general form $L_{\alpha,k}$ (with $f \equiv 1$), and unbiasedness of the stochastic derivative of each term implies unbiasedness of the difference. (To exclude the time in service, we could consider, instead, $T(\beta, k) - X(\beta, k) - T(\alpha, k)$.)

In a network of queues, the waiting time of a job at a particular queue is still the difference of its departure and arrival times. But while the departure of a job will generally correspond to the occurrence of a unique event, the arrival of a job will not. If, for example, jobs may join queue $m$ from either queue $i$ or queue $j$, then an arrival to $m$ may be due to either $\beta_i$ (departure from $i$) or $\beta_j$ (departure from $j$). In this case, the epoch of the $k$th arrival to $m$ cannot be expressed as $T(\beta, k)$ for any fixed $\beta$; and the waiting time at queue $m$ cannot be expressed in terms of $L_T$, $L_{n_0}$ and $L_{\alpha,k}$.

Nevertheless, a modification of the results of Chapter 3, allows us to consider waiting times. If, in the situation described above, $\beta_i$ and $\beta_j$ change order, the *identity* of the $k$th job to arrive at $m$ changes; but the *epoch* of the $k$th arrival does not. Thus, the arrival times at a queue are continuous in the clock samples, even though they do not have the form $T(\beta, k)$. We now formalize this idea in the GSMP framework.

Consider a function $\chi : \mathbf{A} \times [0,1] \to \{0,1\}$; $\chi(\alpha, u) = 1$ tells us that a particular type of transition occurs when a transition is triggered by $\alpha$ and routed by $u$. For $k = 1, 2, \ldots$, let

$$N_\chi(k) = \inf\{n \geq 0 : \sum_{i=1}^{n} \chi(a_i, U(r_i)) = k\},$$

recalling that $U(r_i)$ is the routing indicator for $a_i$. Define

$$T_\chi(k) = \tau_{N_\chi(k)};$$

then $T_\chi(k)$ is the epoch of the $k$th transition at which $\chi = 1$. If $\chi(\alpha, u) = \mathbf{1}\{\alpha = \beta\}$, then $T_\chi(k) = T(\beta, k)$, and if $\chi \equiv 1$ then $T_\chi(k) = \tau_k$. In a Jackson-like network, using the $\phi$ described in Section 4.1.1, we could let

$$\chi(\beta, u) = 1 \text{ if } u \text{ routes a job to queue } m \text{ when } \beta \text{ occurs.}$$

More precisely, with $\phi$ as above, there is a subinterval of $[0,1]$ (depending on $\beta$) such that at the occurrence of $\beta$ a job is routed to queue $m$ if and only if the routing indicator falls in that subinterval. Let $\chi(\beta, u)$ be the indicator that $u$ falls in that subinterval. Then $T_\chi(k)$ is the epoch of the $k$th arrival to $m$.

Suppose, now, that the clock samples depend smoothly on a parameter $\theta$ as in Section 2.3.1 of Chapter 3. Because a separate sequence of routing indicators $\{U(\alpha, k), k = 1, 2, \ldots\}$ is assigned to each event $\alpha$, and because the $U$'s do not depend on $\theta$, continuity (in $\theta$) of every $T(\alpha, j)$ implies continuity of every $T_\chi(k)$. When two events change order, they keep their routing indicators. Right at the point where an order change occurs, $N_\chi(k)$ may increase or decrease by one, but $T_\chi(k)$ is unchanged. If, therefore, we define

$$L_{\chi,k} = \int_0^{T_\chi(k)} f(Z_t)\,dt,$$

then the argument used for Theorem 3.2 proves

**Theorem 4.1.** Under the conditions of Theorem 3.1, if $\mathbf{E}[\sup_{\theta\in\Theta} N(T_\chi(k))^2] < \infty$ and $\mathbf{E}[\sup_{\theta\in\Theta} T_\chi(k)^2] < \infty$, then

$$\mathbf{E}\left[\frac{dL_{\chi,k}}{d\theta}\right] = \frac{d\mathbf{E}[L_{\chi,k}]}{d\theta}.$$

Finally, it is worth noting that there is a simple scheme for keeping track of derivatives of waiting times or sojourn times through arbitrary nodes or subnetworks. Let $\{\delta_n(\alpha), \alpha \in \mathbf{A}, n = 1, 2, \ldots\}$ be the sequence of scheduled event perturbations defined in Chapter 2, Section 2.3.1. With the $i$th job in a system associate the variable $d^i$, representing the derivative of that job's waiting or sojourn time through a designated node or subnetwork. If the $i$th job enters the node or subnetwork at the $n_1 th$ transition, then at this transition set $d^i = -\delta_{n_1-1}(a_{n_1})$. When the job leaves (the node or subnetwork), say at the $n_2 th$ transition, set $d^i = d^i + \delta_{n_2-1}(a_{n_2})$. This makes $d^i$ equal to $\tau'_{n_2} - \tau'_{n_1}$, which is, in fact, the derivative of the sojourn time, $\tau_{n_2} - \tau_{n_1}$.

## 4.2 Structural Conditions for Queues

This section applies the commuting conditions of Chapter 3 to a variety of standard queueing systems to find sufficient conditions for unbiased derivative estimation in terms of the queueing systems themselves. Keeping in mind the the choice of $\phi$ discussed in Section 4.1.1, we work with (C2) rather than (C2'). We begin with the simple Jackson-like network, then consider networks with multiple job classes, with state-dependent routing, and with finite buffer capacities, and finally consider a variety of queueing disciplines. In some cases, we find that (C2) imposes restrictive conditions on the structure of a queueing system; but the results of this section represent only the simplest means by which our derivative estimates can be demonstrated to be applicable. Further results are possible through the extensions of Section 4.3 and Section 4.4, and the methods of later chapters.

### 4.2.1 Jackson-like Networks

Jackson-like networks were defined in Section 2.1 of Chapter 2. Here we generalize the definition to allow multiple independent renewal arrival streams. Thus, let $\alpha_k$ denote arrival on the $k$th stream, and let such arrivals be routed according to $P^k_{0j}$:

$$p(\mathbf{n} + e_j; \mathbf{n}, \alpha_k) = P^k_{0j}.$$

As always, $\beta_i$ denotes service completion at node $i$, and

$$p(\mathbf{n} - e_i + e_j; \mathbf{n}, \beta_i) = P_{ij}.$$

We consider open and closed networks simultaneously by letting $e_0$ be the vector of all zeros and $P_{i0}$ be the probability that a job leaving $i$ leaves the network. In the closed case, for every $i$ and $k$, $P_{i0} = P^k_{0i} = 0$.

That Jackson-like networks satisfy the commuting conditions is illustrated in Figure 3.5 of Chapter 3. We now prove this.

**Theorem 4.2.** Every Jackson-like network satisfies (C2).

**Proof.** Let $\mathbf{n}$ be a state for which $\mathcal{E}(\mathbf{n})$ contains at least two events; say, $\beta_i, \beta_j \in \mathcal{E}(\mathbf{n})$. This is equivalent to $n_i > 0$ and $n_j > 0$. Let $\mathbf{n}'$ be such that $p(\mathbf{n}'; \mathbf{n}, \beta_i) > 0$. Then $\mathbf{n}' = \mathbf{n} - e_i + e_r$ for some $r$. Furthermore, $n'_j \geq n_j > 0$ so $\beta_j \in \mathcal{E}(\mathbf{n}')$. Let $\mathbf{n}''$ satisfy $p(\mathbf{n}''; \mathbf{n}', \beta_j) > 0$; then $\mathbf{n}'' = \mathbf{n}' - e_j + e_m$ for some $m$. To verify (C2), let $\mathbf{n}''' = \mathbf{n} - e_j + e_m$ and observe that

$$\begin{aligned}
\mathbf{n}'' &= \mathbf{n}' - e_j + e_m \\
&= (\mathbf{n} - e_i + e_r) - e_j + e_m \\
&= (\mathbf{n} - e_j + e_m) - e_i + e_r \\
&= \mathbf{n}''' - e_i + e_r.
\end{aligned}$$

Moreover,

$$p(\mathbf{n}'''; \mathbf{n}, \beta_j) = P_{jm} = p(\mathbf{n}''; \mathbf{n}', \beta_j)$$

and

$$p(\mathbf{n}''; \mathbf{n}''', \beta_i) = P_{ir} = p(\mathbf{n}'; \mathbf{n}, \beta_i).$$

Other cases—where the two events are two external arrivals or one arrival and one service completion—work the same way. □

For any pair of nodes $i$ and $r$, define $\Phi_{ir} : \mathbf{S} \to \mathbf{S}$ by $\Phi_{ir}\mathbf{n} = \mathbf{n} - e_i + e_r$ whenever $n_i > 0$ (or $i = 0$). Then the proof of Theorem 4.2 verifies that $\Phi_{ir}\Phi_{jm} = \Phi_{jm}\Phi_{ir}$ and that $p(\Phi_{ir}\mathbf{n}; \mathbf{n}, \beta_i) = P_{ir}$ is independent of $\mathbf{n}$.

### 4.2.2 Multiple Job Classes

Consider, now, a modification of the Jackson-like network in which we allow multiple job classes—call this a multi-class Jackson-like network. We allow different job classes to be distinguished by different service time and interarrival time distributions, and different routing probabilities. However, we do not consider priorities among classes, and we insist that the class of a job be preserved as it moves through the network.

As in an ordinary Jackson-like network, all nodes consist of single server, infinite capacity, first come first served queues. We denote by $\beta_i^c$ a service completion of a class $c$ job at queue $i$, and by $\alpha_k^c$ an arrival on the $k$th stream of class $c$ arrivals. This framework is actually sufficiently general to allow job classes to be assigned upon arrival, by incorporating class assignments into the (GSMP) transition probabilities. By augmenting the state, we could also allow the class of jobs arriving on a single (renewal) stream to alternate.

With multiple job classes, the vector of queue lengths is no longer an adequate state description; we must also keep track of the order of different jobs

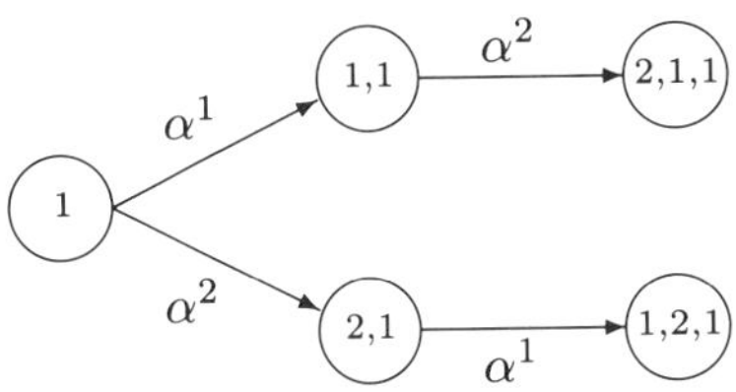


Figure 4.1: Partial event diagram for multi-class queue

in queue. Thus, denote a typical state of an $M$-node network by

$$s = ((n_1, \mathbf{c}_1), \ldots, (n_M, \mathbf{c}_M))$$

where

$$\mathbf{c}_i = (c_{in_i}, \ldots, c_{i1}),$$

and $c_{ik}$ is the class of the $k$th job at queue $i$. If the first job at $i$ is of class $c$, let $\Phi^c_{ij}s$ be the state reached from $s$ when that job leaves $i$ and joins queue $j$: $\Phi^c_{ij}s = s'$ where $n'_i = n_i - 1$, $n'_j = n_j + 1$, $n'_k = n_k$, $k \neq i, j$, and

$$\mathbf{c}'_i = (c_{in_i}, \ldots, c_{i2}),$$

$$\mathbf{c}'_j = (c, c_{jn_j}, \ldots, c_{j1}),$$

and $\mathbf{c}'_k = \mathbf{c}_k$, $k \neq i, j$. If $c_{i1} \neq c$, $\Phi^c_{ij}s$ is not defined.

To motivate our main result for multi-class Jackson-like networks, consider a single queue fed by two classes of jobs. There are two types of arrivals, $\alpha^1$ and $\alpha^2$, and two types of service completions, $\beta^1$ and $\beta^2$. The state records the order of different job classes in queue. Figure 4.1 shows part of the event diagram for this system, and illustrates why (C2) is violated: if the order in which jobs of different classes arrive is reversed, the state reached changes. In fact, if we let $\Phi^c_0 s$ denote the state reached from $s$ upon an arrival of class $c$, then for all $s$,

$$\Phi^1_0 \Phi^2_0 s \neq \Phi^2_0 \Phi^1_0 s;$$

arrivals of different classes do not "commute". This situation is typical of cases where (C2) fails in multi-class networks.

Say that node $i$ is *fed by a single source* if all arrivals to $i$ have the same origin, which may be another queue or one of the arrival streams. To state this precisely, suppose there is (at most) one event $\beta$ (service completion or arrival) for which there exist states $s$ and $s'$ such that $p(s'; s, \beta) > 0$ and $n'_i = n_i + 1$. Then $i$ is fed by a single source.

**Theorem 4.3** A multi-class Jackson-like network satisfies (C2) if and only if every node visited by more than one class of jobs is fed by a single source.

**Proof.** Let $s$ be a state for which $\mathcal{E}(s)$ contains at least two events; say, $\beta_i^c$ and $\beta_j^d$. The argument used in Theorem 4.2 goes through exactly as before, except in the case where $r = m$ (in the notation used there)—i.e., when the jobs leaving $i$ and $j$ both go to $r$. In this case, if $\beta_i^c$ is followed by $\beta_j^d$, the state reached is $s' \equiv \Phi_{jr}^d \Phi_{ir}^c s$ in which

$$\mathbf{c}'_r = (d, c, c_{rn_r}, \ldots, c_{r1}).$$

On the other hand, if $\beta_j^d$ occurs first, the state reached is $s'' \equiv \Phi_{ir}^c \Phi_{jr}^d s$, in which

$$\mathbf{c}''_r = (c, d, c_{rn_r}, \ldots, c_{r1}).$$

Clearly, for $k \neq r$, $\mathbf{c}''_k = \mathbf{c}'_k$. Thus, $\Phi_{ir}^c \Phi_{jr}^d s = \Phi_{jr}^d \Phi_{ir}^c s$ if and only if $d = c$. If the single-source condition holds, then $c$ must equal $d$ (since $i$ and $j$ feed $m$) and (C2) is satisfied (since the transition probabilities do not depend on the order of events). Conversely, if the condition does not hold, we can construct a situation like the one above (with $c \neq d$) that violates (C2). □

The simplest example satisfying the single source condition is the arrangement of $M$ nodes in a cycle with jobs moving from $i$ to $i+1$, $i = 1, \ldots, M-1$ and from $M$ back to 1. Various classes could be specified with different service time distributions at different nodes. Because jobs of different classes cannot overtake each other, (C2) is satisfied. A similar example is an open system of $M$ nodes in tandem where jobs of various classes arrive in a single stream. Weaker conditions than those of Theorem 4.3 will be developed in Section 4.4.

### 4.2.3 State-Dependent Routing

Consider a Jackson-like network in which the routing probabilities are allowed to depend on the state. For simplicity, we return to the case of a single class of jobs. Figure 3.5 of Section 3.1.3, Chapter 3 shows that arbitrary state-dependent routing is not compatible with (C2): if the occurrence of $\beta_i$ can change the routing for $\beta_k$ or vice-versa, then opposite sides of the event diagram will not have equal probabilities. In the extreme case, if $P_{k\ell} > 0$ in $\mathbf{n}$ but $P_{k\ell} = 0$ in $\mathbf{n} - e_i + e_j$, then the descending transition on the right would be deleted altogether. Nevertheless, a limited amount of state-dependent routing is compatible with (C2). We first provide a general result, then consider examples.

Let $\beta_0$ denote a generic event in a Jackson-like network—an external arrival or a service completion. Each $\beta_0$ determines a partition $\mathcal{S}(\beta_0)$ of the state space $\mathbf{S}$ into sets $(S_0(\beta_0), S_1(\beta_0), S_2(\beta_0), \ldots)$, with the property that the routing probabilities for $\beta_0$ are the same in states $s$ and $s'$ if and only if $s$ and $s'$ are in the same $S_i(\beta_0)$; and $S_0(\beta_0)$ (which may be empty) consists of those states that do not contain $\beta_0$ in their event list. If the routing of $\beta_0$ is not state-dependent, $\mathcal{S}(\beta_0) = (S_0(\beta_0), S_1(\beta_0))$, where $S_1(\beta_0) = \{s : \beta_0 \in \mathcal{E}(s)\}$; while if the routing of $\beta_0$ is different in every state, each $S_i(\beta_0)$, $i > 0$, consists of a single state. Let us say that $\beta_0$ has non-trivial state-dependent routing if $\mathcal{S}(\beta_0)$ contains at least two sets other than $S_0(\beta_0)$.

Denote by $\mathcal{E}_{\beta_0}(s)$ the subset of events in $\mathcal{E}(s)$ that can change the routing probabilities for $\beta_0$; that is, $\alpha \in \mathcal{E}_{\beta_0}(s)$ if $\alpha \in \mathcal{E}(s)$ and there exists a state $s'$ such that $p(s'; s, \alpha) > 0$ and $s'$ and $s$ are in different elements of $\mathcal{S}(\beta_0)$. In general, (C2) requires that a certain property hold for all states $s$ and all pairs of events $\alpha, \beta$ in $\mathcal{E}(s)$. Let us say that (C2) holds for $\beta_0$ if it holds for all states $s$ (with $\beta_0 \in \mathcal{E}(s)$) and all pairs of events $\alpha, \beta_0$ with $\alpha \in \mathcal{E}_{\beta_0}(s)$.

**Theorem 4.4.** In a Jackson-like network with state-dependent routing, (C2) holds if it holds for each $\beta_0$ with non-trivial state-dependent routing.

**Proof.** We only need to check that (C2) holds for states $s$ and generic events, $\alpha$ and $\beta$, with $\alpha, \beta \in \mathcal{E}(s)$, $\alpha \notin \mathcal{E}_\beta(s)$ and $\beta \notin \mathcal{E}_\alpha(s)$; by hypothesis, it holds in the other cases. But if $\alpha \notin \mathcal{E}_\beta(s)$ and $\beta \notin \mathcal{E}_\alpha(s)$, then the occurrence of either event will not alter the routing probabilities of the other; hence, the argument used for Theorem 4.2 (or simply Figure 3.5) shows that $\alpha$ and $\beta$ must satisfy (C2). □

The foregoing discussion is a bit abstract and may not give much insight into the types of routing mechanisms that do and do not satisfy (C2). We now present two examples; the first one violates (C2), the second one is consistent with Theorem 4.4.

Consider a bank of $M$ queues fed by a single stream of arrivals which are routed according to a "join shortest queue" policy. Let us suppose, quite arbitrarily, that the nodes are numbered, and that in case of a tie, an arrival is routed to the node with the smallest number among those that have the shortest queue. Jobs leaving any of the nodes leave the system.

Denote arrivals by $\alpha$ and departures from node $j$ by $\beta_j$. So long as there is a unique shortest queue, $i$, which is shortest by at least two, (C2) is satisfied, as the following diagram illustrates:

$$\begin{array}{ccccc} & & \alpha & & \\ & \mathbf{n} & \longrightarrow & \mathbf{n}+e_i & \\ \beta_j & \downarrow & & \downarrow & \beta_j \\ & \mathbf{n}-e_j & \longrightarrow & \mathbf{n}+e_i-e_j & \\ & & \alpha & & \end{array}$$

Provided $n_i < n_j - 1$, the occurrence of $\beta_j$ will not alter the routing of $\alpha$; $i$ still has the shortest queue following the occurrence of $\beta_j$. But if $j < i$, and $n_j = n_i + 1$, then the occurrence of $\beta_j$ will produce a tie between $i$ and $j$ that will be broken in favor of $j$. Thus, applying first $\alpha$ then $\beta_j$ makes the state $\mathbf{n} + e_i - e_j$; while taking these events in the opposite order returns the state to $\mathbf{n} - e_j + e_j = \mathbf{n}$. In the notation preceding Theorem 4.4, $\beta_j \in \mathcal{E}_\alpha(\mathbf{n})$, but $\{\beta_j, \alpha\}$ violates (C2) in $\mathbf{n}$. It may seem that this is a consequence of the way we have chosen to break ties, but other mechanisms for breaking ties run into essentially the same difficulty.

Now consider "round-robin" routing in which the queues are fed cyclicly. The first arrival is sent to node 1, the second to node 2, and so on, returning to node 1 after the last node has received an arrival. To make the example more interesting, consider two banks of queues, with $M_1$ and $M_2$ queues respectively. The first bank is fed by a stream ($\alpha$) of arrivals routed state-independently. Departures ($\alpha_i$, $i = 1, \ldots M_1$) from the first bank are routed round-robin to the second bank. Departures ($\beta_i$, $i = 0, \ldots, M_2 - 1$) from the second bank leave the system. If departures $\alpha_i$ and $\alpha_j$ change order, their destinations are reversed to preserve the round-robin policy.

Let us implement this mechanism by augmenting the state to include an indication of which queue in the second bank is to be fed next. Thus, a typical state looks like $(\mathbf{n}^1, \mathbf{n}^2, k)$ where $\mathbf{n}^1$ and $\mathbf{n}^2$ are the queue lengths in the two banks, and $k$ is the next queue in the second bank to be fed. At each occurrence of any $\alpha_i$, $k$ is incremented by one, modulo $M_2$. In the notation of Theorem 4.4, every $\mathcal{E}_{\alpha_i}(s)$ consists of all $\alpha_j$'s for which $n_j^1 > 0$ in $s$, since the occurrence of any $\alpha_j$ changes $k$. It is enough to consider pairs $\alpha_i$, $\alpha_j$ with $\alpha_j \in \mathcal{E}_{\alpha_i}(s)$. Let $\Phi_{\alpha_i} s$ be the state reached from $s$ upon the occurrence $\alpha_i$. Then if $n_i^1 > 0$ and $n_j^1 > 0$,

$$\begin{aligned}
\Phi_{\alpha_j}\Phi_{\alpha_i}(\mathbf{n}^1, \mathbf{n}^2, k) &= \Phi_{\alpha_j}(\mathbf{n}^1 - e_i, \mathbf{n}^2 + e_k, k+1) \\
&= (\mathbf{n}^1 - e_i - e_j, \mathbf{n}^2 + e_k + e_{k+1}, k+2) \\
&= \Phi_{\alpha_i}(\mathbf{n}^1 - e_j, \mathbf{n}^2 + e_k, k+1) \\
&= \Phi_{\alpha_i}\Phi_{\alpha_j}(\mathbf{n}^1, \mathbf{n}^2, k),
\end{aligned}$$

with addition modulo $M_2$ in the last component. Now (C2) is satisfied because the hypothesis of Theorem 4.4 is satisfied.

### 4.2.4 Finite Buffers

In all the systems we have considered so far, we have assumed that all queues have infinite capacity. We now consider Jackson-like networks of queues with limited capacity, restricting attention to the case of a single class of jobs with state-independent routing. There are several kinds of blocking that are commonly considered in the queueing theory literature. Here we will consider only two kinds: *internal* blocking of queues, in which a job waits after completing service for space to become available at its next destination, preventing initiation of service for the next job in queue; and *blocking with loss* in an open system, in which external arrivals that find a full buffer are simply lost. Other mechanisms can be analyzed similarly. We consider these two types of blocking separately. For ease of exposition, we consider internal blocking in closed networks only. Our results go through if, instead, we assumed only that no finite capacity queue is fed by external arrivals.

### Internal Blocking

In a closed network of finite capacity queues, we take as a typical state

$$s = ((n_1, b_1), \ldots, (n_M, b_M))$$

where $b_j = i$ if node $j$ is blocked by node $i$ and $b_j = 0$ if $j$ is not blocked. Let $N$ be the total number of jobs in the network and call a node a *blocking node* if its buffer (including room for a job in service) is strictly less than $N$. We denote a service completion at node $i$ by $\beta_i$. If node $i$ can be blocked, service completion at $i$ need not coincide with departure from $i$.

Since we are, for the time being, only considering closed networks, the single source condition used in Section 4.2.2 simplifies. Say that node $j$ is fed by node $i$ if $P_{ij} > 0$, and say that $j$ is fed by a single source if there is just one $i$ with $P_{ij} > 0$.

**Theorem 4.5.** If every blocking node is fed by a single source, then (C2) is satisfied.

**Proof.** The proof works essentially the same way as that of Theorem 4.2; the only case we describe separately is the order of $\beta_i$ and $\beta_j$ when $j$ is a blocking node and $i$ feeds only $j$. We use $\Phi_{\beta_i} s$ and $\Phi_{\beta_j} s$ to denote the states reached from $s$ upon the occurrence of $\beta_i$ and $\beta_j$, respectively. Suppose $n_j$ equals the capacity of node $j$ and $n_i > 0$. For simplicity, suppose from $j$ jobs go to $k$, a non-blocking node. Then

$$\begin{aligned}
&\Phi_{\beta_j} \Phi_{\beta_i} s \\
&= \Phi_{\beta_j}((n_1, b_1), \ldots, (n_i, j), \ldots, (n_j, 0), \ldots, (n_M, b_M)) \\
&= ((n_1, b_1), \ldots, (n_i - 1, 0), \ldots, (n_j, 0), \ldots, (n_k + 1, 0), \ldots, (n_M, b_M)) \\
&= \Phi_{\beta_i}((n_1, b_1), \ldots, (n_i, 0), \ldots, (n_j - 1, 0), \ldots, (n_k + 1, 0), \ldots, (n_M, b_M)) \\
&= \Phi_{\beta_i} \Phi_{\beta_j} s.
\end{aligned}$$

Other cases work the same way. □

It may not be immediately clear from the proof of Theorem 4.5 why we imposed the single-source condition. To see the importance of this condition, let $j$ be a blocking node with capacity $N - 2$ or less, and suppose that $j$ is fed by both $i$ and $k$. When the queue at $j$ is full, there are at least two jobs in the rest of the network, so it is potentially possible for $j$ to block both $i$

and $k$. A difficulty arises in unblocking these nodes, similar to the one we had in breaking ties for "join shortest queue" routing in Section 4.2.3. A GSMP model of the system must include a mechanism that determines the sequence in which $i$ and $k$ are to be unblocked upon the occurrence of $\beta_j$. The mechanism could be implemented through state augmentation or through the transition probabilities. We begin with two examples of the first type of mechanism.

**(i) First Blocked, First Unblocked.** Suppose whichever of $i$ and $k$ was blocked first is to be unblocked first. Augment the state to include an indication of which node was blocked first, when more than one is blocked by the same node. A change in the order of $\beta_i$ and $\beta_k$ can change this indication; therefore, the state reached through the occurrence of these events depends on their order.

**(ii) Priority Unblocking.** Suppose $k$ is always unblocked first, when both $i$ and $k$ are blocked. Suppose that, in some state, $i$ is blocked by $j$, but $k$ is not. Consider the order in which $\beta_j$ and $\beta_k$ occur. If $\beta_j$ occurs first, then $i$ becomes unblocked, and upon occurrence of $\beta_k$, $k$ becomes blocked. But if $\beta_k$ occurs first, then the occurrence of $\beta_j$ unblocks $k$ and leaves $i$ blocked.

Another alternative is to randomize between $i$ and $k$ in choosing which to unblock. This mechanism is most naturally described through the transition probabilities, with no augmentation of the state.

**(iii) Randomized Unblocking.** Suppose that when $i$ and $k$ are both blocked $i$ is unblocked with probability $q$, $0 < q < 1$, and $k$ with probability $1 - q$, at the occurrence of $\beta_j$. If only one of $i$ and $k$ is blocked, it is always unblocked at the occurrence of $\beta_j$. A change in the order of $\beta_j$ and $\beta_k$ can change the probability of a transition that unblocks $i$ from 1 to $q$, violating (C2).

While we have only considered three examples, there does not appear to be any mechanism for unblocking multiple queues compatible with (C2). However, the extensions of Section 4.4 and the methods of Chapters 5 and 6 allow a certain amount of simultaneous blocking.

It is worth noting that in spite of the similarity between the single source condition here and that given for multi-class networks in Section 4.2.2, the networks that satisfy these conditions are quite different. Start with any network with any number and arrangement of blocking nodes. In front of each multiple source blocking node place a new infinite capacity node through which all inputs to the blocking node must first pass. The resulting network satisfies the single source condition for blocking; no similar transformation satisfies the multi-class single source condition.

### Blocking With Loss

We turn, now, to blocking with loss in open systems. The simplest example is the GI/G/1/K finite capacity queue in which arrivals that find the buffer full are simply lost. Figure 3.4 of Section 3.1.3, Chapter 3, shows that (C2) is violated by this system. If we let $\alpha$ denote arrival and $\beta$ denote service completion, then the problem is that

$$\phi(\phi(K, \beta), \alpha) = \phi(K - 1, \alpha) = K,$$

whereas

$$\phi(\phi(K, \alpha), \beta) = \phi(K, \beta) = K - 1.$$

This situation is typical of blocking with loss. This should not be surprising since rejecting blocked jobs amounts to a form of state-dependent routing incompatible with the condition in Theorem 4.4.

There is an interesting class of exceptions which anticipate the discussion in Section 4.3 on the flexibility of the exponential distribution. We introduce them through a variant of blocking with loss. By a system with *buffer modulated arrivals* we mean one in which the external arrival process to a node is "shut off" whenever the buffer at that node is full. In other words, if an arrival, $\alpha$, is routed to node $i$ with positive probability, and if the buffer at $i$ is full in state $s$, then $\alpha \notin \mathcal{E}(s)$. With this mechanism, arrivals are never actually blocked. For (C1) to be satisfied, we require that a node with a finite buffer be fed by only one source. (Otherwise, the occurrence of one type of arrival may delete other pending arrivals from the event list.) In fact, we have the following analog of Theorem 4.5 for open networks:

**Theorem 4.6.** In an open Jackson-like network with buffer modulated arrivals, if every finite capacity queue is fed by a single source, then (C2) is satisfied.

Consider, once again, ordinary blocking with loss in a system in which all external arrivals form a single Poisson stream, as in an M/G/1/K queue. Because the arrivals are Poisson, they may be "shut off" when the buffer is full without altering the probabilistic evolution of the state. In other words, in systems with Poisson arrivals, the queue length processes under blocking with loss and buffer modulated arrivals coincide. Hence, we may as well use the latter. As an example, for the M/G/1/K queue an equally valid event diagram is obtained from Figure 3.4 if the loop from $K$ back to $K$ due to $\alpha$ is erased—i.e., if $\alpha$ is deleted from $\mathcal{E}(K)$. This alternative GSMP model of the M/G/1/K queue satisfies (C2). Changing the GSMP model in this way changes the resulting stochastic derivative (and the associated estimation algorithm) to one that works. We cannot, however, obtain a legitimate GSMP model of a GI/G/1/K queue in this way if the arrivals are not Poisson. For if we delete $\alpha$ from $\mathcal{E}(K)$, then at a transition from $K$ to $K-1$ the distribution of the residual arrival clock would be that of a new arrival clock, which is only correct if the interarrival times are exponential.

### 4.2.5 Queueing Disciplines

Thus far, we have only considered queues with a single server governed by a first come, first served discipline. Modeling more complex disciplines through the GSMP framework often requires the use of *speeds.* Recall from Section 3.2 of Chapter 3, that to ensure the continuity of $L_T$, $L_{n_0}$ and $L_{\alpha,k}$ in GSMPs with speeds we required, in addition to (C2), that, for all states $s$ and events $\alpha$,

$$\alpha \in \mathcal{E}(s) \Rightarrow 0 < \lambda_\alpha(s) < \infty. \tag{4.1}$$

Moreover, to show that the derivative estimates for these performance measures are unbiased, we considered alternative additional conditions on the speeds. The first was the *increasing speeds condition*

$$\{\alpha, \beta\} \subseteq \mathcal{E}(s) \text{ and } p(s'; s, \alpha) > 0 \Rightarrow \lambda_\beta(s) \leq \lambda_\beta(s'); \tag{4.2}$$

the second was the condition that there exist $\lambda_* > 0$ and $\lambda^* < \infty$ such that

$$\alpha \in \mathcal{E}(s) \Rightarrow \lambda_* < \lambda_\alpha(s) < \lambda^*. \tag{4.3}$$

Condition (4.3) led to significantly more stringent moment conditions on the event counting process $N(t)$. We now examine the applicability of (C2) to a variety of queueing disciplines, and the applicability of (4.1), (4.2) and (4.3) to disciplines that use speeds.

#### Multi-server queue

Consider a $\cdot/\mathrm{G}/m$ node consisting of $m < \infty$ identical servers sharing a single queue. In order to obtain a GSMP representation that makes the servers identical, we let service completion at any of the servers constitute the same event, $\beta$. This makes it possible for several clocks to be running simultaneously for the same event; thus, let $\mathcal{E}(s)$ contain as many copies of $\beta$ as there are busy servers. A departure from one server does not affect the others; i.e., it does not interrupt the other clocks.

With this variation on the usual GSMP set-up, (C2) is satisfied. We may take the state to be just the number of jobs; then $\phi(\phi(s,\alpha),\beta) = \phi(\phi(s,\beta),\alpha)$, whenever $s > 0$, and, of course, if $s > 1$ any two copies of $\beta$ "commute".

In a variant of the standard model a single server provides service at rate $\lambda(n)$ when there are $n$ jobs in its queue. This formulation uses speeds. It satisfies (C2), and also (4.1), if the server always works at a finite, strictly positive rate. If $\lambda$ is increasing in $n$, then (4.2) is satisfied. Another variant assigns service completion events to *jobs* rather than servers, and allows the $m$ servers to work at different (but state-independent) rates. Let the servers be ordered by increasing speed; when there is a departure from server $j$, any job at a server $i$, with $i < j$ moves to $i + 1$ (taking its residual service requirement) to receive faster service. This example satisfies (4.2), as well as (C2). However, if we assign events to jobs, we can only have a finite number of jobs, so the node must be part of a closed network.

### Infinite server queue

This type of node acts as a random delay with fixed distribution. Regardless of how we assign events, an infinite server queue in an open network requires that the event set $\mathbf{A}$ be infinite. But in a closed network, an "infinite" server queue can be taken to be one that has as many servers as there are jobs. Such a node does not require speeds; it is consistent with (C2).

### Processor sharing

In this discipline, each of $n$ jobs present receives service at rate $\lambda/n$ for some $\lambda$. We associate an event with each job in service; the speed for each of these events is the service rate $\lambda/n$. Since we associate events with jobs, we restrict attention to closed systems. Condition (C2) is satisfied; but (4.2) is always violated: the arrival of a job causes the service rate of every job to *decrease.* However, (4.3) is satisfied if $0 < \lambda < \infty$, which is, in any case, a necessary condition for (4.1).

### Pre-emptive last come, first served

There are two cases; neither satisfies both (C2) and (4.1). The *resume* variant uses the residual service time of a job when it returns to service after having been pre-empted. To obtain a GSMP formulation, associate an event with each job in the node. A job is pre-empted by having its service speed set to zero; it resumes service when its speed becomes positive again. The (necessary) use of speed zero violates (4.1). The second variant forces jobs to start a new service time when returning to service after being pre-empted. In this case, (C2) is violated because (C1) is: whenever there is a job in service, an arrival deletes an event—the current service completion event—from the event list.

### Kelly's Disciplines

Finally, we consider a general class of disciplines described in Kelly [59] (p.58), but without the exponential assummption made there. This class includes some of the examples above. A node consists of several *positions* which receive different amounts of service. A total service effort is provided at rate $\lambda(n)$ when there are $n$ jobs present; $\lambda(n) > 0$ whenever $n > 0$. A proportion $\gamma(k, n)$ of this effort is directed to the job in position $k$. When this job leaves the queue, the jobs in positions $k+1, k+2, \ldots, n$ move to positions $k, k+1, \ldots, n-1$. When a job arrives at the queue, it enters position $k$ ($k = 1, \ldots, n+1$) with probability $\delta(k, n)$; jobs previously in positions $k, k+1, \ldots, n$ move to $k+1, k+2, \ldots, n+1$. To satisfy (4.1) for arbitrary $\delta$, we need every $\lambda(n) < \infty$ and every $\gamma(k, n) > 0$. However, we can restrict the support of $\gamma(\cdot, n)$ to $\{1, \ldots, n_1\}$ for every $n \geq n_1$, provided we restrict $\delta(\cdot, n)$ to $\{n_1 + 1, \ldots, n + 1\}$ for every $n \geq n_1$. Taking $\gamma(1, \cdot) \equiv 1$ and $\delta(n+1, n) = 1$ we recover the first come, first served discipline. Condition (C2) is satisfied in all these cases.

## 4.3 Exponential Clocks

This section and the next introduce two techniques by which the general results of Section 4.2 can be extended in systems meeting additional conditions. The first technique exploits special properties of the exponential distribution, and is applicable when some of the clock distributions are exponential. The second takes advantage of situations where it is known in advance that only a subset of the clock distributions depend on the parameter $\theta$.

In modeling a system as a GSMP, there is often some flexibility in precisely what one takes to be the states, events, transition probabilities, etc. (though the non-interruptive condition (C1) significantly narrows the choices). This is especially true when some of the clock setting distributions are exponential, in which case we can expect to find many different GSMPs modeling the same system. Finding alternative GSMP models of a system is useful, since some may satisfy (C2) while others do not.

To show how exponential clock times lead to different GSMPs, we review basic properties of the exponential distribution. Let $X_1$ and $X_2$ be independent unit mean exponential random variables, and let $\lambda_1$ and $\lambda_2$ be positive real numbers. Then

$$P(X_i/\lambda_i \leq x) = 1 - e^{-\lambda_i x}, \ x > 0; \tag{4.4}$$

$$P(X_1/\lambda_1 < X_2/\lambda_2) = \frac{\lambda_1}{\lambda_1 + \lambda_2}; \tag{4.5}$$

and

$$P(\min(X_1/\lambda_1, X_2/\lambda_2) \leq x) = 1 - e^{-(\lambda_1+\lambda_2)x}, \ x > 0. \tag{4.6}$$

The familiar *memoryless* property is

$$P(X_1 - y \leq x | X_1 > y) = 1 - e^{-x}, \ x \geq 0. \tag{4.7}$$

Property (4.4) allows us to interpret an exponential, mean $1/\lambda$, unit speed clock, as a unit mean, speed $\lambda$ exponential clock, and vice-versa. Properties (4.5) and (4.6) allow us to split an event into multiple events in the following sense: Suppose a unit mean exponential clock $X$ is run down at rate $\lambda$. When the clock runs out, a transition to $s_1$ occurs with probability $p$, and a transition to $s_2$ occurs with probability $1-p$. We could replace $X$ with two independent, unit mean exponential clocks $X_1$ and $X_2$, run down at rates $p\lambda$ and $(1-p)\lambda$, respectively. If $X_i$ runs out first, then when it runs out we make a transition to $s_i$, $i = 1, 2$. Using $X_1$ and $X_2$ in this way, the probability that the transition is to $s_i$, and the distribution of the time until a transition occurs, are the same as those obtained using $X$ only. Reversing this procedure, we can merge multiple events into a single event. Finally, (4.7) allows us to substitute a new clock sample for an aged sample (or vice-versa) when convenient.

Precisely how (4.4)-(4.7) can be exploited is illustrated through the examples that follow. Our use of these properties here is fairly straightforward. Far more elaborate use will be made of them in Chapters 5 and 6.

### Jackson-like Networks

Since Jackson-like networks satisfy (C2) for all distributions, it may seem that no further benefit can be derived from the exponential distribution. But in some (albeit rare) cases, a process which could be viewed as a Jackson-like network is specified in some other way which violates our conditions. Consider a birth-death Markov process on $\{0, 1, 2, \ldots\}$. One might construct such a process by setting new clocks for transitions upon entry to each state. Viewing the process as a GSMP, this corresponds to making $n \to n+1$ and $n' \to n'+1$ different events and making $n \to n-1$ and $n' \to n'-1$ different events, when $n' \neq n$. This violates (C1): at the occurrence of the event $n \to n-1$, the event $n \to n+1$ is interrupted. But by viewing the process as an M/M/1 queue, one is led to use only two events for all states, which yields a GSMP satisfying (C1) and also (C2). Similarly, one may view a finite birth death process as the state of a closed cyclic network of two exponential queues.

### Multi-class Queues

This example is from Ho and Hu [52]. Consider the multi-class queue introduced in Section 4.2.2; there are two classes of arrivals, $\alpha^1$, $\alpha^2$ and two classes of departures, $\beta^1$, $\beta^2$. Figure 4.1 illustrates why (C2) is violated in general. But suppose the interarrival distributions $F_{\alpha^1}$ and $F_{\alpha^2}$ are exponential, with rates $\lambda_1$ and $\lambda_2$. We may merge the arrivals into a single Poisson stream with rate $\lambda = \lambda_1 + \lambda_2$, and assign an arrival (call it $\alpha$) class 1 with probability $p = \lambda_1/\lambda$ and class 2 with probability $1-p$. This in no way alters the probabilistic evolution of the process. It does, however, cause (C2) to be satisfied; Figure 4.1

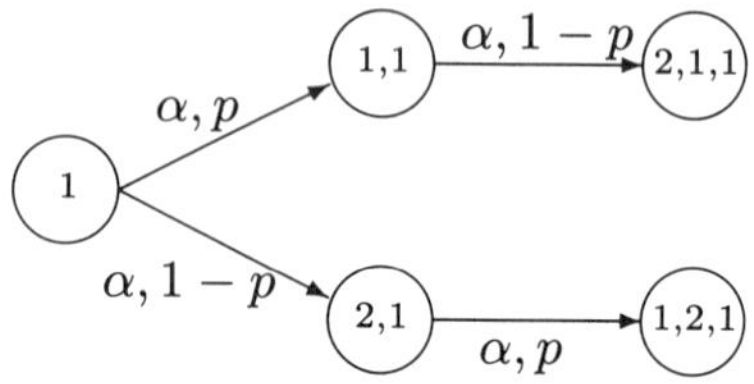


Figure 4.2: Multi-class queue with merged arrival streams

becomes Figure 4.2. The new event diagram satisfies (C2) because there is just one arrival event (adding more states and transitions to the diagram would not change this.) The price we pay for this transformation is that now $\lambda_1$ and $\lambda_2$ are parameters of the transition probabilities whereas previously they were parameters only of the interarrival times; hence, we could no longer take $\theta$ equal to $\lambda_1$ or $\lambda_2$.

### State-dependent Routing: Feedback

Consider a queue in which, upon completion of service ($\beta$), a job is fed back into the queue with probability $p_n$, and allowed to leave with probability $1-p_n$, where $n$ is the number in the queue. Unless $p_n$ is independent of $n$, this system violates (C2) because an arrival ($\alpha$) can change the probability of feedback. But suppose $F_\beta$ is exponential with rate $\lambda$. We can "split" $\beta$ into two events $\beta_f$ and $\beta_l$ with unit mean exponential clocks, and speeds

$$\lambda_{\beta_f}(n) = \lambda p_n,$$

and

$$\lambda_{\beta_l}(n) = \lambda(1 - p_n).$$

Upon occurrence of $\beta_f$, a job is fed back; upon occurrence of $\beta_l$, a job leaves. To preserve (C1), we include $\beta_f$ in $\mathcal{E}(0)$ and set $p(0; 0, \beta_l)$. It is easily checked that (C2) is now satisfied. This transformation has the opposite effect of the one we used in the multi-class queue in the following sense: in the transformed system, each $p_n$ is a parameter of the speeds *only* whereas previously it was a parameter of the routing probabilities. Thus, this transformation affords us an indirect way to get derivatives with respect to the $p_n$'s.

### Self-loops

An exponential event $\beta$ in a state $s$ with $p(s; s, \beta) = 1$ never affects the probabilistic evolution of a GSMP, so such an event may be added or deleted wherever convenient. Such a modification will generally affect (C1) and (C2). In Section 4.2.4 we saw that deleting such a "self-loop" made the M/G/1/K queue compatible with these conditions. In the feedback example above we introduced an event, $\beta_f$, which *only* made self-loops. We could, in fact, eliminate $\beta_f$ entirely from that example without altering the process or violating (C1)-(C2). Indeed, whenever an event leads only to self-loops it can be eliminated, even if it is not exponential.

## 4.4 Relevant Events

None of the results of Chapters 2 and 3 or Section 4.2 placed any restrictions on which clock distributions, $F_\alpha$, may depend on $\theta$. We have, in fact, allowed all the clock distributions to depend on the parameter. But our main condition, (C2), can be relaxed if we know in advance that only some of them do. This translates into relaxed conditions for the applicability of our derivative estimates to queueing systems when only some of the service time and interarrival time distributions depend on $\theta$.

To motivate this idea, consider a single-server queue fed by two classes of arrivals, $\alpha^1$ and $\alpha^2$. As discussed in Section 4.2.2, a change in the order of occurrence of $\alpha^1$ and $\alpha^2$ changes the resulting state, violating (C2). Let $\beta^1$ and $\beta^2$ denote departure of class 1 and class 2 jobs. It is easy to see that changes in the order of $\beta^1$ and $\beta^2$, or in the order of some $\beta^i$ and $\alpha^j$, do not change the resulting state and do not violate (C2). Furthermore, changes in *service times* may cause some $\alpha^i$ and some $\beta^j$ to change order, but *cannot* change the order of $\alpha^1$ and $\alpha^2$—the timing of arrivals is unaffected by the timing of departures. Thus, if the service time distributions $F_{\beta^1}$ and $F_{\beta^2}$ depend on $\theta$, but the interarrival time distributions $F_{\alpha^1}$ and $F_{\alpha^2}$ do not, no change in $\theta$ will cause $\alpha^1$ and $\alpha^2$ to change order. For this reason, estimates of derivatives with respect to $\theta$ should be unbiased.

We formalize this idea through the concept of *relevance.* We introduce it using the GSMP framework and notation of Chapters 2 and 3, then apply it to queueing systems.

### Relevance in GSMPs

**Definition 4.1.** For any fixed $\alpha_0 \in \mathbf{A}$, define the set of $\alpha_0$-*relevant* events recursively as follows:

**(i)** $\alpha_0$ is $\alpha_0$-relevant;

**(ii)** if $\alpha_1$ is $\alpha_0$-relevant, and there are states $s$ and $s'$ with $p(s'; s, \alpha_1) > 0$ and $\alpha_2 \in \mathcal{E}(s')\backslash\mathcal{E}(s)$, then $\alpha_2$ is $\alpha_0$-relevant.

Part (ii) states that if a clock for $\alpha_2$ is potentially set by the occurrence of an $\alpha_0$-relevant event, then $\alpha_2$ is $\alpha_0$-relevant. In fact, we have

**Lemma 4.1.** Unless $\alpha_1$ is $\alpha_0$-relevant, $\eta(\alpha_1, k_1; \alpha_0, k_0)$ is zero for all $k_1, k_0$.

**Proof.** From Definition 4.1, if $\eta(\alpha_1, k_1; \alpha_0, k_0) = 1$, then there is a sequence $(\beta_0, j_0), \ldots, (\beta_m, j_m)$ such that $(\beta_0, j_0) = (\alpha_0, k_0)$, $(\beta_m, j_m) = (\alpha_1, k_1)$, and the $j_i$th clock for $\beta_i$ is set at the $j_{i-1}$th occurrence of $\beta_{j-1}$, $i = 1, \ldots, m$. Part (ii) of Definition 4.1 applied repeatedly implies that every $\beta_i$ (in particular, $\alpha_1$) is $\alpha_0$-relevant. □

We now state, for fixed $\alpha_0 \in \mathbf{A}$, a relaxed version of (C2'):

**(R). Relevance Condition.** Condition (C2') holds whenever either $\alpha$ or $\beta$ is $\alpha_0$-relevant.

The multi-class queue with which we began this section satisfies (R) if $\alpha_0$ is taken to be either $\beta^1$ or $\beta^2$. The next theorem verifies unbiasedness when $\theta$ is a parameter of $F_{\alpha_0}$ only.

**Theorem 4.7.** If only $F_{\alpha_0}$ depends on $\theta$, then the conclusion of Theorem 3.2 holds with condition (C2') replaced by (R).

**Proof.** Theorem 3.2 relies on (C2') only via the continuity results of Theorem 3.1, which, in turn, rely on (C2') only via Lemma 3.1; thus, we only need to verify that Lemma 3.1 still holds. The proof of Lemma 3.1 checks that no discontinuities are introduced when changes in $\theta$ cause two events $\alpha$ and $\beta$ to occur at the same time. If either $\alpha$ or $\beta$ is $\alpha_0$-relevant then, by hypothesis, (C2') still holds for $\{\alpha, \beta\}$ so the original argument remains valid. We now argue that if neither $\alpha$ nor $\beta$ is $\alpha_0$-relevant, changes in $\theta$ will not cause them to change order. For any $k$, we have from Lemma 2.2 that

$$\frac{dT_\theta(\alpha, k)}{d\theta} = \sum_{\alpha', j'} \frac{dX_\theta(\alpha', j')}{d\theta} \eta(\alpha, k; \alpha', j').$$

Since only $\alpha_0$-clocks depend on $\theta$, this is

$$\frac{dT_\theta(\alpha, k)}{d\theta} = \sum_{j} \frac{dX_\theta(\alpha_0, j)}{d\theta} \eta(\alpha, k; \alpha_0, j).$$

But Lemma 4.1 implies that every $\eta(\alpha, k; \alpha_0, j) = 0$, since $\alpha$ is not $\alpha_0$-relevant. Thus, for every $k$, $dT(\alpha, k)/d\theta = 0$, and similarly for every $j$, $dT(\beta, j)/d\theta = 0$. Therefore, in the proof of Lemma 3.1, perturbations in $\theta$ cannot cause order changes among pairs of irrelevant events; at order changes involving a relevant event, the original argument of Lemma 3.1 applies. □

## Relevance in Queues

The abstract notion of relevance takes on a concrete interpretation in queueing networks. For ease of exposition, we only consider relevance with respect to service completion events. Relevance with respect to external arrivals is simpler but less interesting.

We begin with a simple characterization of relevance of one service completion event with respect to another in (infinite buffer, single class) Jackson-like networks. As before, $\beta_i$ denotes a service completion at node $i$. Say that node $j$ is reachable from node $i$ if there is a sequence of possible transitions that take a job from $i$ to $j$. More precisely, let $P$ be the routing matrix excluding the 0-row and column—i.e., excluding external arrivals and departures. Then $j$ is reachable from $i$ if the $ij$-entry of some power of $P$ is strictly positive.

**Lemma 4.2.** In a Jackson-like network, $\beta_j$ is $\beta_i$-relevant if and only if $j$ is reachable from $i$.

**Proof.** Suppose $\beta_j$ is $\beta_i$-relevant. Then there is a sequence of events

$$\beta_i, \beta_{k_1}, \ldots, \beta_{k_m}, \beta_j, \tag{4.8}$$

generated by Definition 4.1, each triggering the next. In a closed Jackson-like network, a service completion $\beta_k$ can only trigger another $\beta_{k'}$ if $P_{kk'} > 0$. Hence, (4.8) implies that

$$P_{ik_1} \cdots P_{k_m j} > 0, \tag{4.9}$$

which is precisely the condition for reachability. For the converse, given that $j$ is reachable from $i$, let $k_1, \ldots, k_m$ be a sequence of nodes satisfying (4.9). Starting from the state in which all jobs are at $i$, moving a job from $i$ to $j$ through nodes $k_1, \ldots, k_m$ generates the sequence of events (4.8), and shows that $\beta_j$ is $\beta_i$-relevant. □

It is easy to see that no external arrival is ever $\beta_i$-relevant, and that if $P$ is irreducible then every $\beta_i$ is relevant with respect to every external arrival.

The characterization of relevance is more complicated in networks with blocking or with multiple job classes. In the blocking case, for example, if $i$ is a blocking node, then even if $P_{ij} = 0$, $\beta_i$ may trigger $\beta_j$ if $P_{ji} > 0$. Indeed, in this case, $\beta_i$ potentially triggers $\beta_k$ for any $k$ reachable from $j$; see Figure 4.3.

To characterize relevance in networks with finite buffers, note that the reachability criterion in Lemma 4.2 depends only on which elements of the routing matrix are strictly positive, not on their values. Consider a matrix $\tilde{P}$ in which $\tilde{P}_{ij} > 0$ if and only if either $P_{ij} > 0$, or $P_{ji} > 0$ and $i$ is a blocking node. All other entries of $\tilde{P}$ are zero. Now call $j$ reachable from $i$ (under $\tilde{P}$) if some power of $\tilde{P}$ is strictly positive. The proof of Lemma 4.2 also proves

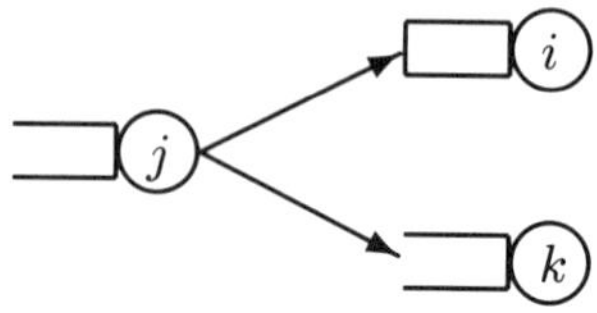


Figure 4.3: $\beta_i$ can trigger $\beta_k$ through $\beta_j$

**Lemma 4.3.** In a Jackson-like network with blocking, $\beta_j$ is $\beta_i$-relevant only if $j$ is reachable from $i$ under $\tilde{P}$.

The converse of Lemma 4.3 may fail if the number of jobs in the network is not sufficiently large: even if $j$ has a finite buffer and $P_{ij} > 0$, $\beta_j$ will not trigger $\beta_i$ unless $j$ actually blocks $i$.

For multi-class networks we can establish a similar result. As in Section 4.2.2, let $\beta_i^c$ denote service completion of a class $c$ job at node $i$, and let $P^c$ be the routing matrix for class $c$ jobs (excluding external arrivals and departures). Let $\tilde{P}((i,c),(j,c'))$ be positive if and only if either $c = c'$ and $P_{ij}^c > 0$, or $i = j$ and both class $c$ and $c'$ visit node $i$. Let all other entries be zero. Then

**Lemma 4.4.** In a multi-class Jackson-like network, $\beta_j^{c'}$ is $\beta_i^c$-relevant if and only if some power of $\tilde{P}((i,c),(j,c'))$ is strictly positive.

We now consider two applications of relevance. These results are immediate consequences of Theorems 4.3 and 4.4 and condition (R).

**Corollary 4.1.** A multi-class Jackson-like network satisfies (R) (for a fixed choice of $\alpha_0$) if no node visited by more than one class of jobs and fed by more than one source, is fed by any $\alpha_0$-relevant source.

In case every event is relevant, we recover Theorem 4.3. The example with which we began this section is a special case of Corollary 4.1. This corollary mainly covers cases of open networks where pairs of events that violate (C2) are "upstream" from the server whose service times depend on the parameter of differentiation.

For Jackson-like networks with blocking (and a single class of jobs), we consider separately internal blocking and blocking with loss.

**Corollary 4.2.** Consider a Jackson-like network with finite buffers in which no blocking node is fed by a external arrivals. Condition (R) is satisfied if whenever a blocking node $i$ is fed by distinct sources $j$ and $k$, none of $\beta_i$, $\beta_j$ and $\beta_k$ is relevant.

**Corollary 4.3.** Consider an open Jackson-like network with loss, in which the blocking nodes are fed only by external arrivals. Condition (R) is satisfied if no service completion at a blocking node is relevant, and no arrival that feeds a blocking node is relevant.

* * *

## Notes and Comments

Perturbation analysis derivative estimates were originally introduced for networks of queues. While GSMPs provide a convenient framework in which to state conditions and prove theorems, the ultimate test of the method is in the examples; and queueing systems remain the most interesting and important examples. This chapter is the payoff for hard work and heavy notation in the preceding two. A simple calculus emerges through which we can verify the interchange of expectation and differentiation without any math—in some cases, just by drawing pictures. We always have the option of retreating to the more rigorous setting of Chapter 3.

The structure of queueing systems is of interest in its own right. While some properties are easy to see in a GSMP formulation—most notably the structure of the state space—others are obscured. For example, it is sometimes hard to extract waiting times or routing topology from a GSMP. There is probably room for a class of models somewhere between specific queueing systems and generalized semi-Markov processes.

Most of the examples studied here are from Glasserman [28]. Earlier related work includes Cao [7] and Suri and Zazanis [79]. The "no simultaneous blocking" condition is proposed in Cao and Ho [9]. The shutting off of Poisson arrivals to full queues is used in Gong and Glasserman [46]. Relevance is introduced in Glasserman [24,27].

# Chapter 5

# Derivative Estimation in Markov Chains

We turn, in this chapter, to derivative estimation for Markov chains. Markov chains have substantial overlap with GSMPs and networks of queues, but their special properties and their importance as models make them worth considering separately. This chapter is not an application of previous results; instead, it develops another class of algorithms specific to the Markov case. Section 5.1 motivates and introduces a new condition, similar to (C2), appropriate for Markov chains. Section 5.2 exploits this condition and properties of the exponential distribution to construct parametric families Markov chains. Based on this construction, Section 5.3 develops a derivative estimation algorithm. Examples are considered in Section 5.4, and variance reduction is taken up in Section 5.5. Standard notation and properties of Markov chains are assumed.

## 5.1 Structure in Markov Chains

### 5.1.1 From GSMPs to Markov Chains

The setting of this chapter is the following: We are given a family $\{Q_\theta, \theta \in \Theta\}$ of generator matrices for Markov processes on a countable state space $\mathbf{S}$. For all $\theta \in \Theta$, some horizon $T$, and a fixed initial state $Z_0$, we are to construct $\{Z_t(\theta), 0 \leq t \leq T\}$, a Markov process on $[0, T]$ with generator $Q_\theta$. Letting

$$L(\theta) = \int_0^T f(Z_t(\theta))dt,$$

our goal is to obtain an unbiased estimator of $d\mathbf{E}[L(\theta)]/d\theta$. In particular, we seek conditions under which $dL(\theta)/d\theta$ is such an estimator, and an algorithm to compute it from a sample path of $\{Z_t(\theta), 0 \le t \le T\}$. This is the same problem we considered for generalized semi-Markov processes in Chapters 2 and 3, but here we make use of special properties of Markov chains and the exponential distribution. We first discuss connections between GSMPs and Markov chains. In referring to GSMPs, we use the notation of Chapters 2 and 3 without comment.

A GSMP in which all clock time distributions $\{F_\alpha, \alpha \in \mathbf{A}\}$ are exponential is a Markov chain. Suppose, for concreteness, that $F_\alpha(x) = 1 - \exp(-\mu_\alpha x)$, $\alpha \in \mathbf{A}$. Then the transition rate from $s$ to $s'$, $s' \neq s$, is given by

$$q(s, s') = \sum_{\alpha \in \mathcal{E}(s)} \mu_\alpha \lambda_\alpha(s) p(s'; s, \alpha).$$

If we allow the $\mu_\alpha$'s (hence, the $F_\alpha$'s) to depend on a parameter $\theta$, then we have a special case of the problem considered in Chapters 2 and 3. Thus, we know that if (C2) and some additional regularity conditions are satisfied, the stochastic derivative $dL/d\theta$ provides an unbiased derivative estimator.

If, instead, we start with an arbitrary Markov process on $\mathbf{S}$, there are many ways to represent it as a GSMP. In some cases—such as an ordinary Jackson network with exponential service times—the process has an obvious representation as a GSMP. If this representation satisfies (C2), then the results of Chapter 3 are applicable. In this case, Chapter 3 often provides the simplest way to proceed. But, in general, there maybe no obvious "physical" events in a Markov chain, and no clear choice of GSMP representation. In this case, two natural choices are the following:

**(i) Generator Representation.** With each pair of states $s, s'$ in $\mathbf{S}$ for which $q(s, s') > 0$ introduce an event $\alpha_{ss'}$; intuitively, $\alpha_{ss'}$ represents "transition from $s$ to $s'$". The clock times for $\alpha_{ss'}$ are exponential with mean $q^{-1}(s, s')$. All transition probabilities are "deterministic" in the sense that $p(s'; s, \alpha_{ss'}) \equiv 1$. This representation corresponds implicitly to the following simulation algorithm: Upon entry into $s$, generate an exponential, mean $q^{-1}(s, s')$ clock for every $s'$ with $q(s, s') > 0$. The smallest of these becomes the holding time in $s$, and the $s'$ associated with the smallest clock becomes the next state. After the transition to $s'$, all residual clocks are discarded and a new set of clocks is sampled.

**(ii) Semi-Markov Representation.** In this representation, a Markov chain is viewed as a semi-Markov process with exponential holding times. With each state $s$ associate an event $\alpha_s$ representing "exit from $s$". The clock distribution for $\alpha_s$ is exponential with mean $q^{-1}(s)$ where, as usual, $q(s) = \sum_{s' \neq s} q(s, s')$.

For Markov chains in general, upon exit from $s$ the next state is $s'$ with probability $q(s,s')/q(s)$; hence, this quantity is $p(s';s,\alpha_s)$. (We are implicitly assuming here that $0 < q(s) < \infty$.) From the perspective of simulation, this representation corresponds to generating (exponential) holding times directly, and generating transitions by sampling from the discrete distributions $p(\cdot;s,\alpha_s)$.

When, for every $s$, there is just one $s'$ with $q(s,s') > 0$, representation (i) reduces to (ii). Except in this case, under the generator representation (C2) is always violated. Indeed, since $\mathcal{E}(s) \cap \mathcal{E}(s') = \emptyset$ whenever $s' \neq s$, (C1) is violated. Using the semi-Markov representation, (C2) is always satisfied, but at a price: For the results of Chapter 3 to apply, we can only allow the clock distributions—not the transition probabilities—to depend on $\theta$. This translates to allowing $q_\theta(s)$ to depend on $\theta$ while insisting that $q_\theta(s,s')/q_\theta(s)$ be constant—in other words, $\partial_\theta[q_\theta(s,s')/q_\theta(s)] = 0$. Thus, only a special kind of dependence on $\theta$ can be handled using (ii), and we could not, for example, estimate derivatives with respect to some $q(s,s')$.

Neither of the representations is entirely satisfactory. If it is possible to find a different assignment of events to transitions which leads to a GSMP consistent with (C2), then that assignment (and the algorithm of Chapter 2) can, of course, be used. Thus, *here, we are primarily interested in processes for which either there is no obvious GSMP representation, or for which the obvious representation violates (C2), or for which the obvious representation does not permit differentiation with respect to the parameter of interest.*

For Markov chains with special structure, we present a construction that complements the GSMP construction of Chapter 2, and the resulting derivative estimation algorithm. We use the special structure to cut ties with the notion of "event". The role of events is taken, instead, by transition *arcs:* in a Markov chain, we say that there is an arc from state $s$ to state $s'$ if $q(s,s') > 0$. We assign clocks to arcs, and think of them as competing to determine the next transition, much as they do in a GSMP. The key difference is that in a Markov chain all clocks are essentially identical—we can take all from a single stream of unit mean, exponential random variables. Furthermore, *residual* clock times always have the same distribution as *new* clock times. These two properties give us considerable flexibility in how we assign and reassign clocks to transition arcs. This flexibility is the key to the new construction.

Because the construction we introduce is rather complicated, we should stress that it is only used to derive (and justify) a derivative estimation algorithm. In particular, there is no suggestion this construction be used as a simulation algorithm. Once we have derived the derivative estimator, we will show how it can be implemented with any simulation (or observation), without using the special construction.

### 5.1.2 The Common Successor Condition

To motivate a new variant of (C2) for Markov chains, we briefly review the significance of (C2). As in Chapters 2 and 3, we use $Y_0, Y_1, \ldots$ to denote the sequence of states, and $\tau_1, \tau_2, \ldots$ to denote the sequence of transition epochs. However, for the rest of this chapter we use $x, y$ and $z$, rather than $s$ and $s'$ to denote typical elements of the state space, **S**. Under (C2), we are able to construct GSMPs so that at a discontinuity of $Y_i(\theta)$, $\tau_{i+1}(\theta) - \tau_i(\theta) = 0$. This makes

$$\int_0^{\tau_n} f(Z_t(\theta))dt = \sum_{i=0}^{n} f(Y_i)[\tau_{i+1} - \tau_i]$$

continuous in $\theta$, even at jumps of $Y_i$. (See Lemma 3.1, Theorem 3.1, and Figure 3.2). Condition (C2) ensures that if the $i$th and $i+1st$ events change order, $Y_{i+1}$ remains unchanged. Thus, if $Y_i$ jumps from $y_i$ to $y_i^*$, then the state sequence $Y_0, \ldots Y_{i-1}, y_i, Y_{i+1}, \ldots$ jumps to $Y_0 \ldots, Y_{i-1}, y_i^*, Y_{i+1}, \ldots$.

Now consider a family of Markov processes $\{Z_t(\theta)\}$, with state sequences $\{Y_n(\theta)\}$ and transition epochs $\{\tau_n(\theta)\}$. We want to achieve the same effect as (C2) without reference to events. Suppose that, under some construction, under a change in $\theta$ the transition out of $Y_{i-1}$ is to $y_i^*$, rather than $y_i$. We would like to ensure that *(a)* from $y_i^*$ the process moves to the original $i+1st$ state $Y_{i+1}$, and *(b)* at the value of $\theta$ where the jump in the $i$th state occurs, $\tau_{i+1} - \tau_i = 0$. For this, the condition we require is

**(CM). Common Successor Condition.** For any pair of states $x, y$, if there is a state $z$ for which $q(z, x) > 0$ and $q(z, y) > 0$, then there must also be a state $z'$ for which $q(x, z') > 0$ and $q(y, z') > 0$.

If we think of the state space of a Markov chain as a directed graph with an arc from $x$ to $y$ whenever $q(x, y) > 0$, then the condition is that every pair of states with a common immediate predecessor must have a common immediate successor. In the case of a birth-death process, for example, the pairs of states with a common immediate predecessor are all of the form $(x-1, x+1)$ and have as common immediate successor $x$ (which is also the common immediate predecessor).

We now outline a construction that exploits (CM); details will be given in the next section. The construction starts from a sequence $\{X_i, i = 1, 2, \ldots\}$ of independent, unit mean, exponential random variables that do not depend on $\theta$. (This sequence implicitly determines the "common probability space" on which our Markov chains are defined. A point $\omega$ is a realization of this sequence.) These random variables play the role of "clock samples", and we refer to them as such. Transitions are generated much as in the generator representation above, except that after a transition residual clock times are re-used, rather than discarded.

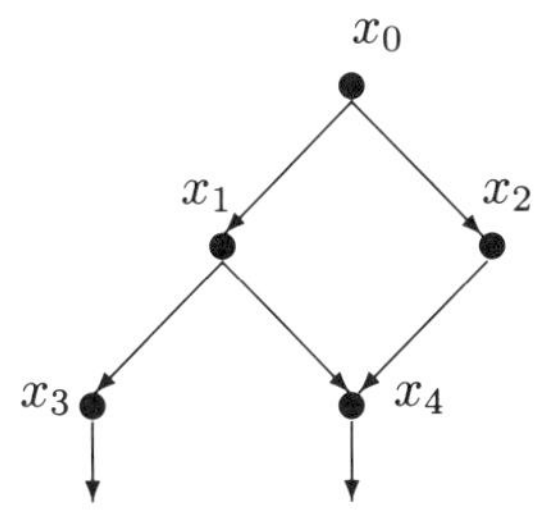


Figure 5.1: Fragment of a Markov chain

Consider the Markov chain partially illustrated in Figure 5.1. Start in state $x_0$ and assign clock samples $X_1$ and $X_2$ to transitions $(x_0, x_1)$ and $(x_0, x_2)$, respectively. Clock $X_i$ is run down at rate $q(x_0, x_i)$, $i = 1, 2$. Suppose that $X_1/q(x_0, x_1) < X_2/q(x_0, x_2)$. Then the holding time in $x_0$ is $X_1/q(x_0, x_1)$, and the next state is $x_1$. So far, this is the construction that follows from the generator representation; but now we introduce a twist. When $X_1$ runs out (and we make a transition to $x_1$), we reassign the residual time of $X_2$ to the transition $(x_1, x_4)$. Conditional on $X_1/q(x_0, x_1) < X_2/q(x_0, x_2)$, the residual time $[X_2/q(x_0, x_2)] - [X_1/q(x_0, x_1)]$ is exponential with mean $q^{-1}(x_0, x_2)$ (by the memoryless property); so

$$X_2 - \frac{q(x_0, x_2)}{q(x_0, x_1)} X_1 \tag{5.1}$$

is exponentially distributed with mean one. Thus, to generate the transition out of $x_1$, we may assign the residual time (5.1) to the transition $(x_1, x_4)$ (running it at rate $q(x_1, x_4)$), and assign a new clock sample $X_3$ to $(x_1, x_3)$ (running it at rate $q(x_1, x_3)$). If the transition out of $x_0$ had been to $x_2$, we would, instead, assign the residual time $X_1 - q(x_0, x_1)X_2/q(x_0, x_2)$ to the transition $(x_2, x_4)$. If we let $K(x, y)$ be a common immediate successor of $x$ and $y$ (whenever $x$ and $y$ have a common immediate predecessor), then the general principle is: upon a transition from $x$ to $x'$, if $q(x, y) > 0$ then assign the residual time on the clock for $(x, y)$ to the transition $(x', K(x', y))$. For any remaining transitions out of $x'$, draw a new clock sample from the sequence $\{X_i\}$.

To see that this construction has the desired properties, consider, again, the example of Figure 5.1. Suppose that initially the transition out of $x_0$ is to $x_1$, and that we increase $q(x_0, x_2)$ by $\delta > 0$. As we increase $\delta$, we eventually get to a point at which

$$X_1/q(x_0, x_1) > X_2/(q(x_0, x_2) + \delta),$$

in which case the transition out of $x_0$ is to $x_2$ rather than $x_1$. Just before the critical value of $\delta$ is reached, the residual time on $X_2$, following the transition to $x_1$, is zero—that is,

$$X_1/q(x_0,x_1) = X_2/(q(x_0,x_2)+\delta) \Rightarrow X_2 - \frac{q(x_0,x_2)+\delta}{q(x_0,x_1)}X_1 = 0.$$

Similary, just after the critical value of $\delta$, the residual time on $X_1$ (following, now, the transition to $x_2$) is zero. Thus, just before the critical value, the sequence of states is $x_0, x_1, x_4, \ldots$ and just after it is $x_0, x_2, x_4, \ldots$. Right at the critical value, the holding time in the second state ($x_1$ or $x_2$) is zero. This is precisely the behavior we wanted; see *(a)* and *(b)* above.

We glossed over one point in stating the general principle behind the construction. If, following a transition from $x$ to $x'$ we assign every clock previously assigned to $(x,y)$ to $(x', K(x',y))$, we may end up assigning multiple clocks to the same transition; nothing excludes the possibility that $K(x',y) = K(x',y')$ for $y \neq y'$. This does not present any difficulties. If $m$ clocks are assigned to the same transition $(x',y)$, each is run down at rate $q(x',y)/m$. The minimum of $m$ exponential random variables with rate $\lambda/m$ is exponential with rate $\lambda$, so this is equivalent to running a single clock at rate $q(x',y)$.

## 5.2 The Construction

We now carry out the construction in detail. As noted earlier, this construction is only used to derive an algorithm for derivative estimation. The algorithm itself (in the form to be given in Section 5.3.2) does not rely on the construction. Hence, a reader interested only in the algorithm and content with the intuitive description above may skip this section.

### Preliminaries

To simplify the construction, we impose a few reasonable conditions on the generators $\{Q_\theta, \theta \in \Theta\}$, in addition to condition (CM). We require that for each $\theta$, $Q_\theta$ be *conservative* and have no *absorbing* or *instantaneous* states, meaning that for every $x \in \mathbf{S}$,

$$0 < q_\theta(x) \equiv -q_\theta(x,x) = \sum_{y \neq x} q_\theta(x,y) < \infty.$$

We also assume that for every $x$ there are only finitely many $y$ for which $q_\theta(x,y) > 0$.

We need conditions on the dependence of $Q_\theta$ on $\theta$. We assume that the elements of $Q_\theta$ are continuously differentiable functions of $\theta$; and, furthermore, that transition arcs can be neither created nor eliminated through changes in

$\theta$. In other words, for all $x$ and $y$, the indicator function $\mathbf{1}\{q_\theta(x,y) > 0\}$ is constant in $\theta$.

With these conditions, we can undertake our construction of a Markov process with generator $Q_\theta$ (and fixed initial state) for a range of $\theta$ values. There are two parts: we first follow the construction outlined in the previous section at a "nominal" value $\theta_0$. This is the value at which derivatives are evaluated. We then "perturb" the evolution of the nominal process to obtain a Markov process with generator, say $Q_{\theta_0+h}$.

### Construction at Nominal Value

We begin at some $\theta_0$ but suppress this argument. Denote by $\{Z_t\}$ the Markov chain we construct. All processes will be constructed to be right continuous. As in Section 5.1.2, for chains satisfying (CM), whenever $x$ and $y$ have a common immediate predecessor, let $K(x,y)$ be a common immediate successor. In addition, let

$$\begin{aligned}
\tau_n &= \text{epoch of } n\text{th transition, } \tau_0 \equiv 0;\\
Y_n &= n\text{th state;}\\
J_n &= \text{index of clock that triggers } n\text{th transition;}\\
r_n(j) &= \text{arc to which clock } j \text{ is assigned at } \tau_n;\\
c_n(j) &= \text{time remaining on clock } j \text{ at } \tau_n;\\
C_n &= \text{set of clocks assigned at } \tau_n;\\
N_n &= \text{number of clocks assigned or used up to } \tau_n;\\
&= \max\{j \in C_n \cup \{J_1,\ldots,J_n\}\};\\
A(x) &= \text{set of arcs } (x,y) \text{ with } q(x,y) > 0;\\
d(a) &= \text{destination of arc } a\text{; e.g., } d((x,y)) = y.
\end{aligned}$$

Much of this notation is analogous to the GSMP notation. In particular, $Y_n$, $\tau_n$ and $c_n$ play essentially the same role as before. Since transition arcs take the place of events, $A(x)$ takes the place of $\mathcal{E}(s)$, and $r_n(j)$ can be thought of as the "event" to which clock $j$ is assigned at $\tau_n$. Since $J_n$ is the index of the clock that determines the $n$th transition, $r_{n-1}(J_n)$ plays the role of the $n$th event, $a_n$. Finally, $d(r_{n-1}(J_n))$ determines the $n+1st$ state in the same way that $\phi(Y_n, a_n)$ does for GSMPs without transition probabilities.

Assume for the moment that for every $x$,

$$K(x,y) = K(x,z) \Rightarrow y = z. \tag{5.2}$$

Then the trajectories of $\{Z_t\}$ are determined as follows. Initialize by setting $\tau_0 = 0$; fixing $Y_0$ to be the initial state; setting $N_0 = |A(Y_0)|$ and $C_0 =$

$\{1,\ldots,N_0\}$. If $j \in C_0$, then $c_0(j) = X_j$. For every $j \in C_0$, let $r_0(j)$ be an element of $A(Y_0)$, with $r_0(j') \neq r_0(j)$ if $j' \neq j$. Now repeat the following recursion:

$$\tau_{n+1} = \tau_n + \min\{c_n(j)/q(r_n(j)) : j \in C_n\}; \tag{5.3}$$

$$J_{n+1} = \min\{j \in C_n : c_n(j)/q(r_n(j)) = (\tau_{n+1} - \tau_n)\}. \tag{5.4}$$

$$Y_{n+1} = d(r_n(J_{n+1})). \tag{5.5}$$

If $|A(Y_{n+1})| = |C_n| - 1$, then

$$C_{n+1} = C_n - \{J_{n+1}\}, \tag{5.6}$$

and

$$N_{n+1} = N_n; \tag{5.7}$$

otherwise,

$$N_{n+1} = N_n + |A(Y_{n+1})| - |C_n| + 1, \tag{5.8}$$

and

$$C_{n+1} = C_n - \{J_{n+1}\} \cup \{N_n + 1, \ldots, N_{n+1}\}. \tag{5.9}$$

For $j \in C_{n+1} \cap C_n$ (i.e., for every "old" clock),

$$c_{n+1}(j) = c_n(j) - q(r_n(j))[\tau_{n+1} - \tau_n] \tag{5.10}$$

and

$$r_{n+1}(j) = (Y_{n+1}, K(Y_{n+1}, d(r_n(j))); \tag{5.11}$$

while every "new" clock $j \in C_{n+1} \backslash C_n$, is assigned a distinct $c_{n+1}(j)$ arbitrarily from $\{X_{N_n+1}, \ldots, X_{N_{n+1}}\}$ and an $r_{n+1}(j)$ from the unassigned elements of $A(Y_{n+1})$.

The critical step here is (5.11) which determines how residual clocks are reassigned following a transition. It is this step that exploits condition (CM), and ensures that a change in $Y_n$ will not change $Y_{n+1}, Y_{n+2}, \ldots$.

If we drop (5.2)—i.e., if we allow $K(x,y) = K(x,z)$ even if $y \neq z$—then it is possible in (5.11) for several clocks to be reassigned to the same arc. This requires only a minor modification, already touched on at the end of Section 5.1.2: When $m$ clocks are assigned to the same transition $(y_1, y_2)$, each is run at rate $q(y_1, y_2)/m$. With multiple clocks assigned to the same arc, it is also possible in (5.11) to have $d(r_n(j)) = Y_{n+1}$ (if $j$ and $J_{n+1}$ are assigned to the same arc), in which case $r_{n+1}(j) = K(Y_{n+1}, Y_{n+1})$. To cover this case, every $K(y,y)$ may be defined to be an arbitrary element of $A(y)$. In other words, any clocks other than $J_{n+1}$ previously assigned to $(Y_n, Y_{n+1})$ may be assigned arbitrarily to some transition out of $Y_{n+1}$.

Letting $Z_t = Y_n$ on $[\tau_n, \tau_{n+1})$, it is easy to see that

**Proposition 5.1.** $\{Z_t, t \geq 0\}$ is a Markov process with generator $Q$.

**Remark.** This construction coincides with the usual GSMP construction in certain special cases. For example, consider a series of $M$ exponential queues in tandem, with possibly state-dependent service rates, and Poisson arrivals. Represent the state by a vector $\mathbf{n} = (n_1, \ldots, n_M)$ of queue lengths. Assign a clock (service requirement) to every busy server. Suppose that nodes $i$ and $j$ are busy, $i+1 < j < M$ (other cases works the same way). The states $\mathbf{n} - e_i + e_{i+1}$ and $\mathbf{n} - e_j + e_{j+1}$ have $\mathbf{n}$ as common predecessor and $\mathbf{n} - e_i + e_{i+1} - e_j + e_{j+1}$ as common successor. Thus, according to the construction above, upon the occurrence of a transition from $\mathbf{n}$ to $\mathbf{n} - e_j + e_{j+1}$, the clock previously assigned to the transition $(\mathbf{n}, \mathbf{n} - e_i + e_{i+1})$ is reassigned to the transition $(\mathbf{n} - e_j + e_{j+1}, \mathbf{n} - e_j + e_{j+1} - e_i + e_{i+1})$. But this just corresponds to using the residual service requirement at server $i$ to schedule the next service completion at server $i$, which is also what happens when clocks are assigned to arrival and service-completion events in the GSMP construction.

### Construction at Perturbed Value

We now describe how the "perturbed" $\theta_0 + h$ process is obtained from the "nominal" $\theta_0$ process, denoting these by $\{Z_t(\theta_0)\}$ and $\{Z_t(\theta_0 + h)\}$. (We similarly append the parameter argument to other notation used above.) The evolution of $\{Z_t(\theta_0 + h)\}$ begins by following the construction above (but driven by $Q_{\theta_0 + h}$ rather than $Q_{\theta_0}$). As $h$ is increased (or decreased) it may happen that for some $n$, $J_n(\theta_0 + h) \neq J_n(\theta_0)$. At such a transition, it may happen that $Y_n(\theta_0 + h) \neq Y_n(\theta_0)$. (This would not *necessarily* happen since both $J_n(\theta_0 + h)$ and $J_n(\theta_0)$ could be assigned to the same arc.) Let, then, $n$ be the smallest index for which $J_n(\theta_0 + h)$ has a discontinuity at $\theta_0 + h$. At this point, we distinguish two cases.

**Case 1:** $Y_{n+1}(\theta_0 + h) = Y_{n+1}(\theta_0)$. Following (5.3)-(5.11), this is the case that arises when $J_n(\theta_0 + h) = J_{n+1}(\theta_0)$ and $J_{n+1}(\theta_0 + h) = J_n(\theta_0)$—i.e., when the *order* in which two consecutive clocks run out changes right at $\theta_0 + h$. In this case, $\tau_{n+1}(\theta_0 + h) = \tau_n(\theta_0 + h)$; zero time is spent in the intermediate state $Y_n$. Note that even though $Y_{n+1}(\theta_0 + h) = Y_{n+1}(\theta_0)$, it is possible that $C_{n+1}(\theta_0 + h) \neq C_{n+1}(\theta_0)$, and even for $j \in C_{n+1}(\theta_0 + h) \cap C_{n+1}(\theta_0)$, it is possible that $r_{n+1}(j, \theta_0 + h) \neq r_{n+1}(j, \theta_0)$. The set of clocks assigned after the $n + 1st$ transition and the arcs to which they are assigned may depend, in part, on $Y_n$, which is different for $\theta_0$ and $\theta_0 + h$. To ensure that $Z_t(\theta_0 + h)$ follows closely the evolution of $Z_t(\theta_0)$, we correct this discrepancy. We do so by reassigning clocks to arcs to match the evolution of $\{Z_t(\theta_0)\}$. (Since all new and residual clocks are exponential with mean one, how we assign them to arcs does not affect the probabilistic evolution of the process.)

The correction is as follows: Set $C_{n+1}(\theta_0 + h) = C_{n+1}(\theta_0)$, set $N_{n+1}(\theta_0 + h) = N_n(\theta_0)$, and for any $j \in C_{n+1}(\theta_0)$, set $r_{n+1}(j, \theta_0 + h) = r_{n+1}(j, \theta_0)$. For $j \in C_{n+1}(\theta_0 + h) \cap C_{n+1}(\theta_0)$, set $r_{n+1}(j, \theta_0 + h) = r_{n+1}(j, \theta_0)$. There is also the matter of the residual clock times at $\tau_{n+1}(\theta_0 + h)$. Since zero time is spent in the $n$th state $Y_n$ at $\theta_0 + h$, the residual clock times are unaffected by the discontinuity in $Y_n$; for $j \in C_{n+1}(\theta_0 + h)$, $c_{n+1}(j)$ is the same just after the discontinuity in $J_n$ as just before the discontinuity.

**Case 2:** $Y_{n+1}(\theta_0 + h) \neq Y_{n+1}(\theta_0)$. For this case to occur, more than two clocks must run out at the same time as $J_n$. As we will see, this case is actually negligible. In order, however, that $\{Z_t(\theta_0 + h)\}$ be defined even in this case, we simply allow it to continue to evolve according to (5.3)-(5.11), but driven by $Q_{\theta_0 + h}$.

The complication of possibly having to correct the assignment of clocks (in Case 1) is a consequence of not having "events" to make a correspondence between transitions out of different states. In the GSMP setting, clocks always stay with the same event. Hence, whether $Y_{n-1}, Y_n, Y_{n+1}$ is $y_{n-1}, y_n, y_{n+1}$ or $y_{n-1}, y_n^*, y_{n+1}$, there is no ambiguity in how to assign residual clocks upon entry to $y_{n+1}$. But under the construction above, the arc out of $y_{n+1}$ to which a particular clock is assigned may depend on the intermediate state $Y_n$. Case 1 above merely corrects this. We stress, once again, that this correction—and, indeed, the entire construction—is only needed to *derive the form* of the derivative estimator. In actually implementing the estimator it is neither necessary nor particularly desirable to use the construction above.

### Continuity

The significance of this construction is given by Lemmas 5.1 and 5.2, below. To state them, let $N_t(\theta)$ be the number of transitions made by $\{Z_t(\theta)\}$ in $(0, t]$; i.e.,

$$N_t(\theta) = \max\{n : \tau_n(\theta) \leq t\}.$$

We often suppress the argument $\theta$ and write $N_t$ when the context makes the parameter value clear. For the following, fix a real number $T > 0$ and an integer $n > 0$ and let $\tilde{N}$ denote either $n$ or $N_T(\theta_0)$.

**Lemma 5.1.** Suppose the elements of $Q_\theta$ are continuously differentiable functions of $\theta$. Suppose there are positive constants $B$, $q_*$ and $q^*$ independent of $\theta$ such that at every $\theta$ in $\Theta$, for all $x$ and $y$, $|q'_\theta(x, y)| \leq B$, and if $q_\theta(x, y) > 0$ then

$$0 < q_* < q_\theta(x, y) < q_\theta(x) < q^* < \infty.$$

Then for any $\theta_0 \in \Theta$, under the construction above, the following hold with probability $1 - O(h^2)$:

**(i)** Every $\tau_i$, $i \leq \tilde{N}$, is continuous in $\theta$ throughout $(\theta_0 - h, \theta_0 + h)$.
**(ii)** For every $i \leq \tilde{N}$, at any discontinuity of $Y_i$ in $(\theta_0 - h, \theta_0 + h)$, $\tau_{i+1} = \tau_i$.
**(iii)** At any discontinuity of $N_T$ in $(\theta_0 - h, \theta_0 + h)$, $\tau_{N_T} = T$.

Except for the $O(h^2)$ probability of a discontinuity, Lemma 5.1 is very similar to Lemma 3.1. A possible discontinuity arises when Case 2 is invoked or when Case 1 must be invoked twice. The proof that these events have probability $O(h^2)$ is somewhat intricate; it is relegated to Appenix B.

Let $f$ be a bounded, real-valued function on **S**, the state space of $Z$. For real $T > 0$ and integer $n_0 > 0$ define

$$L_T = \int_0^T f(Z_t)dt \tag{5.12}$$

$$L_{n_0} = \int_0^{\tau_{n_0}} f(Z_t)dt. \tag{5.13}$$

From Lemma 5.1 we get

**Lemma 5.2.** $L_T$ and $L_{n_0}$ are, with probability $1-O(h^2)$, continuous functions of $\theta$ throughout $(\theta_0 - h, \theta_0 + h)$.

**Proof.** Using Lemma 5.1, the proof is exactly the same as that of Theorem 3.1 of Section 3.1.2. □

Lemmas 5.1 and 5.2 differ from Lemma 3.1 and Theorem 3.1 in that they only ensure continuity throughout $(\theta_0 - h, \theta_0 + h)$ with probability $1 - O(h^2)$, rather than almost surely. This is a consequence of the possible need to correct the assignment of clocks, as in Case 1 and Case 2 above. For the purpose of interchanging derivatives and expectations, continuity with probability $1 - O(h^2)$ is nearly as good as almost sure continuity.

## 5.3 The Derivative Estimator

We now show how to calculate stochastic derivatives resulting from the construction of the previous section. The result, as we will show, is an unbiased derivative estimator. We first present a provisional form of the estimator based on keeping track of the evolution of the clocks in the construction of Section 5.2. We then simplify to obtain an estimator which depends only on $Z$, and makes no reference to clocks or the construction of Section 5.2.

### 5.3.1 Differentiation

Under the construction of the previous section, at any $\theta$, every $\tau_n$ is differentiable with probability one; the only points where $\tau_n$ may fail to be differentiable are the points of discontinuity of some $J_i$, $i \leq n$, where two clocks run out at the same time. Proceeding as in Lemma 2.4, we conclude that at each $\theta$, $L_T$ and $L_{n_0}$ are almost surely differentiable, and

$$\frac{dL_T}{d\theta} = \sum_{i=1}^{N_T} \frac{d\tau_i}{d\theta}[f(Y_{i-1}) - f(Y_i)], \tag{5.14}$$

$$\frac{dL_{n_0}}{d\theta} = \sum_{i=0}^{n_0-1} f(Y_i)[\frac{d\tau_{i+1}}{d\theta} - \frac{d\tau_i}{d\theta}]. \tag{5.15}$$

Hence, as with GSMPs, the key step is calculation of $\{d\tau_i/d\theta\}$.

For notational convenience, we use $q_n(j)$ for the rate at which clock $j$ is run down during $[\tau_n, \tau_{n+1})$; thus, $q_n(j)$ is $q(r_n(j))$ divided by the number of clocks in $C_n$ assigned to $(Y_n, r_n(j))$. With each clock $j$ associate an accumulator $D(j)$ which takes the value $D_n(j)$ at $\tau_n$. Roughly speaking, $D_n(j)$ is the infinitesimal delay in the time clock $j$ runs out if $j$ is the $n+1st$ clock to run out. The following algorithm updates the $D_n(j)$'s so that $D_n(J_n)$ is the derivative of the $n$th transition epoch.

**Algorithm 5.1: Perturbation Propagation in Clocks**

**Step 1.** Initialize every $D_0(j)$ to zero.
**Step 2.** At the $n+1st$ transition update according to

$$D_{n+1}(J_{n+1}) = D_n(J_{n+1}) - \frac{\tau_{n+1} - \tau_n}{q_n(J_{n+1})} q_n'(J_{n+1});$$

if $j \in C_{n+1} \cap C_n$, then

$$D_{n+1}(j) = (\frac{q_n(j)}{q_{n+1}(j)})(D_n(j) - \frac{\tau_{n+1} - \tau_n}{q_n(j)} q_n'(j)) + (1 - \frac{q_n(j)}{q_{n+1}(j)})D_{n+1}(J_{n+1});$$

and if $j \in C_{n+1} \backslash C_n$ then

$$D_{n+1}(j) = D_{n+1}(J_{n+1}).$$

We now show

**Proposition 5.2.** With probability one, for every $n \geq 0$, $d\tau_n/d\theta = D_n(J_n)$.

**Proof.** Define, for $j \in C_n$,

$$R_n(j) = \tau_n + c_n(j)/q_n(j);$$

then $R_n(j)$ is the time at which clock $j$ would run out if its rate remained $q_n(j)$ after $\tau_n$. Since $c_n(J_n) = 0$, $\tau_n = R_n(J_n) = R_{n-1}(J_n)$ so we need to show $R'_n(J_n) = D_n(J_n)$. In fact, we will show that for all $j$,

$$R'_n(j) = D_n(j) - \frac{c_n(j)}{q_n(j)^2} q'_n(j), \tag{5.16}$$

provided $j \in C_n$. This can be thought of as the derivative of the time at which clock $j$ is scheduled to run out at $\tau_n$. That (5.16) holds at $n = 0$ is immediate from $D_0(j) = 0$. Take as induction hypothesis that it holds up to $n$. If $j \in C_{n+1} \cap C_n$,

$$\begin{aligned} c_{n+1}(j) &= c_n(j) - q_n(j)[\tau_{n+1} - \tau_n] \\ &= q_n(j)[R_n(j) - \tau_n] - q_n(j)[\tau_{n+1} - \tau_n] \\ &= q_n(j)[R_n(j) - \tau_{n+1}] \end{aligned}$$

so

$$R_{n+1}(j) = \tau_{n+1} + q_n(j)[R_n(j) - \tau_{n+1}]/q_{n+1}(j).$$

Substituting $R_n(J_{n+1})$ for $\tau_{n+1}$ and differentiating:

$$\begin{aligned} R'_{n+1}(j) &= \left(1 - \frac{q_n(j)}{q_{n+1}(j)}\right) R'_n(J_{n+1}) + \left(\frac{q_n(j)}{q_{n+1}(j)}\right) R'_n(j) \\ &+ [R_n(j) - R_n(J_{n+1})] \left(\frac{q_n(j)}{q_{n+1}(j)}\right)'. \end{aligned}$$

Using the induction hypothesis (substituting according to (5.16)), the right side becomes

$$\left(1 - \frac{q_n(j)}{q_{n+1}(j)}\right)\left(D_n(J_{n+1}) - \frac{c_n(J_{n+1})}{q_n(J_{n+1})^2} q'_n(J_{n+1})\right) +$$
$$\left(\frac{q_n(j)}{q_{n+1}(j)}\right)\left(D_n(j) - \frac{c_n(j)}{q_n(j)^2} q'_n(j)\right) + [R_n(j) - R_n(J_{n+1})]\left(\frac{q_n(j)}{q_{n+1}(j)}\right)'.$$

This is also valid when $j \in C_{n+1} \backslash C_n$ provided we take $q_n(j)$ and $c_n(j)/q_n(j)^2$ to be identically zero in this case. Substituting according to

$$c_n(J_{n+1})/q_n(J_{n+1}) = \tau_{n+1} - \tau_n;$$

$$R_n(j) - R_n(J_{n+1}) = c_{n+1}(j)/q_n(j);$$

and

$$c_n(j) = q_n(j)[\tau_{n+1} - \tau_n] + c_{n+1}(j),$$

and using the definition of $D_{n+1}$, this simplifies to

$$R'_{n+1}(j) = D_{n+1}(j) - \left(\frac{q_n(j)}{q_{n+1}(j)}\right)\frac{c_{n+1}(j)}{q_n(j)^2}q'_n(j) + \frac{c_{n+1}(j)}{q_n(j)}\left(\frac{q_j(n)}{q_{n+1}(j)}\right)'.$$

Expanding the last derivative and cancelling terms we get

$$R'_{n+1}(j) = D_{n+1}(j) - \frac{c_{n+1}(j)}{q_{n+1}(j)^2}q'_{n+1}(j)$$

which is what we needed to show. □

**Remark.** The similarity between the evolution of the $D_n$'s and the expressions (3.16) and (3.17) of Section 3.2 for the derivatives of event epochs in GSMPs with speeds is no coincidence. The rates $q_n(j)$ act just like the clock speeds $\lambda_\alpha(Y_n)$, and the term

$$-\frac{\tau_{n+1} - \tau_n}{q_n(J_{n+1})}q'_n(J_{n+1}) \tag{5.17}$$

is just the "perturbation generated" during $[\tau_n, \tau_{n+1})$ by the change in $q_n(J_{n+1})$. See the discussion of scale parameters in Section 1.3.2 of Chapter 1.

We now come to the main result of this chapter. As discussed earlier, we always restrict attention to a set $\Theta$ throughout which every $\mathbf{1}\{q_\theta(x,y) > 0\}$ is a constant function of $\theta$—i.e., no transitions are created or eliminated through changes in the parameter. Let us also take $\Theta$ to be a compact interval.

**Theorem 5.1.** Let $\{Q_\theta, \theta \in \Theta\}$ satisfy (CM) and have continuously differentiable entries. Suppose there are constants $B$, $q_*$ and $q^*$ satisfying the conditions of Lemma 5.1 throughout $\Theta$. Then, at any $\theta \in \Theta$,

$$\mathbf{E}[\frac{dL}{d\theta}] = \frac{d\mathbf{E}[L]}{d\theta},$$

where $L$ is either $L_T$ or $L_{n_0}$.

**Proof.** Let $\Delta_h L = L(\theta + h) - L(\theta)$ for either $L = L_T$ or $L_{n_0}$. Let $\tilde{h}$ be the infimum over $h$ for which $L$ has a discontinuity in $(\theta - h, \theta + h)$ (which is strictly positive with probability one). Note that $\tilde{h}$ is random. We consider separately the two terms in

$$\mathbf{E}\left[\frac{\Delta_h L}{h}\right] = \mathbf{E}\left[\frac{\Delta_h L}{h}\mathbf{1}\{h < \tilde{h}\}\right] + \mathbf{E}\left[\frac{\Delta_h L}{h}\mathbf{1}\{h \geq \tilde{h}\}\right].$$

For the first term, we show that

$$\lim_{h\to 0}\mathbf{E}\left[\frac{\Delta_h L}{h}\mathbf{1}\{h < \tilde{h}\}\right] = \mathbf{E}\left[\frac{dL}{d\theta}\right]. \tag{5.18}$$

Since $L$ is continuous and piecewise differentiable on $(\theta - h, \theta + h)$ when $h < \tilde{h}$, the generalized mean value theorem implies

$$\left|\frac{\Delta_h L}{h}\mathbf{1}\{h < \tilde{h}\}\right| \leq \sup_{(\theta-h,\theta+h)} \left|\frac{dL}{d\theta}\right|.$$

Then (5.18) follows from the dominated convergence theorem if we can show that

$$\mathbf{E}\left[\sup_{(\theta-h,\theta+h)} \left|\frac{dL_i}{d\theta}\right|\right] < \infty.$$

Let $\|f\|$ be the supremum of the (bounded) function $|f|$, and let

$$\rho = 2(1 + \frac{q^*}{q_*}).$$

Note that

$$\max_{j\in C_1} |D_1(j)| \leq B(\frac{\tau_1 - \tau_0}{q_*})$$

and

$$\max_{j\in C_{n+1}} |D_{n+1}(j)| \leq \rho\left[\max_{j\in C_n} |D_n(j)| + B(\frac{\tau_{n+1} - \tau_n}{q_*})\right].$$

Thus,

$$\begin{aligned}\left|\frac{d\tau_n}{d\theta}\right| = |D_n(J_n)| &\leq \sum_{i=0}^{n-1}(\frac{\tau_{i+1} - \tau_i}{q_*})B\rho^{n-i-1} \\ &\leq B\tau_n\rho^n/q_*. \end{aligned} \tag{5.19}$$

Using (5.15), we find that

$$\begin{aligned}\left|\frac{dL_{n_0}}{d\theta}\right| &\leq 2\|f\|\sum_{i=1}^{n_0}\left|\frac{d\tau_i}{d\theta}\right| \\ &\leq 2\|f\|(B/q_*)\sum_{i=1}^{n_0}\tau_i\rho^i \\ &\leq 2\|f\|Bn_0\tau_{n_0}\rho^{n_0}/q_*.\end{aligned}$$

Since

$$\sup_{(\theta-h,\theta+h)} \tau_n \leq (X_1 + \cdots + X_n)/q_*,$$

and the right side is integrable, the result follows for $L_{n_0}$. For $L_T$, we get, in the same way, the bound

$$\left|\frac{dL_T}{d\theta}\right| \leq (2\|f\|B/q_*)\tau_{N_T}N_T\rho^{N_T} \leq (2\|f\|B/q_*)TN_T\rho^{N_T}.$$

At each $\theta$, $N_T$ is stochastically bounded by a Poisson random variable with parameter $q^*T$; hence, so is $\sup_{(\theta-h,\theta+h)}(N_T)$. For a Poisson random variable $\tilde{N}$, $\mathbf{E}[\tilde{N}\rho^{\tilde{N}}] < \infty$, which concludes verification of (5.18).

For the second term, we need to verify that as $h \to 0$, $\mathbf{E}[\Delta_h L/h \cdot \mathbf{1}\{h \geq \tilde{h}\}] \to 0$. Using the Cauchy-Schwarz inequality, we get

$$\mathbf{E}\left[\frac{\Delta_h L}{h}\mathbf{1}\{h \geq \tilde{h}\}\right]^2 \leq \mathbf{E}[\Delta_h L^2]\frac{P(\tilde{h} \leq h)}{h^2}.$$

Since $P(\tilde{h} \leq h) = O(h^2)$ (Lemma 5.2) we need only verify that $\mathbf{E}[\Delta_h L^2]$ goes to 0. With probability one, $\Delta_h L \to 0$ as $h \to 0$; and for all $\theta$, $\mathbf{E}[L_T^2(\theta)] \leq (\|f\|T)^2$ and $\mathbf{E}[L_{n_0}^2(\theta)] \leq 2n(\|f\|/q_*)^2$ so the dominated convergence theorem yields the result. □

### 5.3.2 An Improved Algorithm

One way to implement our derivative estimate is to generate sample paths of a Markov process using the construction in Section 5.2 and apply the recursion of Section 5.3.1. As we pointed out earlier, however, it is neither necessary nor particularly desirable to use the construction in applying the estimate. Indeed, if the algorithm is to be applied to observations of a real system, (as opposed to simulation) then using the construction of Section 5.2 is not even an option. Instead, we propose an algorithm that keeps track only implicitly of how the clocks *would be* evolving if they were driving the process. To put this another way, to get a derivative estimate from $Z = \{Z_t, t \geq 0\}$, we compute

$$\mathbf{E}\left[\frac{dL}{d\theta}\middle| Z\right],$$

where $L$ is either $L_T$ or $L_{n_0}$. This leads to considerable simplification of the estimator. Once conditioned on $Z$, the estimator can be applied to sample paths observed or simulated by any means.

If, for some $Q$, it is never necessary to assign multiple clocks to a single arc in the construction of Section 5.2 (i.e., if $y \neq y' \Rightarrow K(x,y) \neq K(x,y')$), then the evolution of the clocks is, in fact, completely determined by $Z$, and $\mathbf{E}[dL/d\theta|Z] = dL/d\theta$. In this case, it is possible to keep track of the $D_n(j)$'s from observation of $Z$. More generally, we get

$$\mathbf{E}\left[\frac{dL_T}{d\theta}\middle| Z\right] = \sum_{i=1}^{N_T}[f(Y_{i-1}) - f(Y_i)]\mathbf{E}[D_i(J_i)|Z], \tag{5.20}$$

and

$$\mathbf{E}\left[\frac{dL_{n_0}}{d\theta}\middle| Z\right] = \sum_{i=1}^{n_0-1} f(Y_i)\{\mathbf{E}[D_{i+1}(J_{i+1})|Z] - \mathbf{E}[D_i(J_i)|Z]\}. \tag{5.21}$$

Thus, we need to evaluate $\mathbf{E}[D_i(J_i)|Z]$.

Once a set of clocks are assigned to a common arc, they remain assigned to a common arc, and become indistinguishable. By construction, given $Z$, $J_i$ is equally likely to be any of the clocks assigned to $(Y_{i-1}, Y_i)$ at $\tau_i$; so, to get $\mathbf{E}[D_i(J_i)|Z]$ we *average*. More precisely, we first have

$$\mathbf{E}[D_1(J_1)|Z] = -\frac{\tau_1 - \tau_0}{q(Y_0, Y_1)} q'(Y_0, Y_1);$$

and then since the $D_n(j)$'s enter linearly in the recursion of the previous section, we can apply the recursion to their conditional expectations to get $\mathbf{E}[D_i(J_i)|Z]$. This is the key to the algorithm.

Given $Y_0, Y_1, \ldots, Y_n$, define,

$$S_n(x) = \{y : K(Y_n, y) = x, q(Y_{n-1}, y) > 0\};$$

then $S_n(x)$ is the set of states, $y$, such that a clock assigned to $(Y_{n-1}, y)$ at $\tau_{n-1}$ is reassigned to $(Y_n, x)$ at $\tau_n$. Note that $S_n(x)$ could be empty and could consist of just one element. Since the clocks cannot be "observed", we associate with each state $x$ an accumulator $\delta_n(x)$ corresponding to the *average* mentioned above. The following algorithm makes $\delta_n(Y_n) = \mathbf{E}[D_n(J_n)|Z]$, and implements our estimator without reference to clocks.

**Algorithm 5.2: Perturbation Propagation in Markov Chains**

**Step 1.** Initialize by setting every $\delta_0(x) = 0$, $dL_T = 0$ and $dL_{n_0} = 0$.
**Step 2.** At the $n$th transition, update by setting

$$\delta_n(Y_n) = \delta_{n-1}(Y_n) - \frac{\tau_n - \tau_{n-1}}{q(Y_{n-1}, Y_n)} q'(Y_{n-1}, Y_n);$$

for every $x$ with $q(Y_n, x) > 0$ and $|S_n(x)| > 0$,

$$\delta_n(x) =$$
$$\delta_n(Y_n) + \sum_{y \in S_n(x)} \frac{q(Y_{n-1}, y)}{q(Y_n, x)} \left( \delta_{n-1}(y) - \frac{\tau_n - \tau_{n-1}}{q(Y_{n-1}, y)} q'(Y_{n-1}, y) - \delta_n(Y_n) \right);$$

while if $q(Y_n, x) > 0$ and $|S_n(x)| = 0$, then

$$\delta_n(x) = \delta_n(Y_n).$$

Set

$$\begin{aligned} dL_T &= dL_T + \delta_n(Y_n)[f(Y_{n-1}) - f(Y_n)] \\ dL_{n_0} &= dL_{n_0} + f(Y_{n-1})[\delta_n(Y_n) - \delta_{n-1}(Y_{n-1})]. \end{aligned}$$

**Step 3.** After $n_0$ transitions, the value of $dL_{n_0}$ is $\mathbf{E}[dL_{n_0}/d\theta|Z]$; after $N_T$ transitions, the value of $dL_T$ is $\mathbf{E}[dL_T/d\theta|Z]$.

The three cases for updating $\delta_n$ in Step 2 correspond to the three cases for $D_n$ in Algorithm 5.1, Section 5.3.1. The second case implicitly averages the "perturbations" in all $|S_n(x)|$ clocks reassigned to $(Y_n, x)$ at $\tau_n$. (The $|S_n(x)|^{-1}$ that makes the sum an average is cancelled by the rate in the denominator, because if $r_n(j) = x$, then $q_n(j) = q(Y_n, x)/|S_n(x)|$.)

This algorithm makes explicit reference to states and to elements of $Q$. In practice, enumeration of the states of a Markov process can be difficult, and the generator $Q$ may only be known implicitly. (For example, in simulating a Jackson network, one would usually not list the states and transition rates.) For processes with special structure, the algorithm above can be further simplified to make explicit knowledge of the state unnecessary. An example of how this can be done is given in the next section.

## 5.4 Examples and Discussion

We illustrate the construction and results of Sections 5.2 and 5.3 through examples. These help clarify the significance and scope of condition (CM), on which our algorithms depend.

### Reversible Markov Chains

A continuous-time Markov chain is called *reversible* if there are strictly positive numbers $\{\nu(x), x \in \mathbf{S}\}$ satisyfing

$$q(x,y)\nu(x) = q(y,x)\nu(y)$$

whenever $q(x, y) > 0$ (cf. e.g., Kelly [59]). For such chains, any two immediate successors of $x$ have $x$ itself as a common immediate successor, so (CM) is always satisfied. Clearly, the condition of reversibility could be relaxed to $q(x, y) > 0 \Rightarrow q(y, x) > 0$ for all $x$ and $y$.

The special case where, in addition, every $y$ and $y'$ have *at most* one common immediate successor is worth considering separately. (A birth-death process is an example.) If $y$ and $y'$ have a common immediate successor, $x$, then $q(y, x)q(y', x) > 0 \Rightarrow q(x, y)q(x, y') > 0$, so $x$ is also their (unique) common immediate predecessor. For such chains, under the construction of Section 5.2, following a transition from $x$ to $y$, every clock previously assigned to some $(x, y')$ is reassigned to $(y, x)$, since $K(y, y')$ can only be $x$. For example, in a birth-death process, following a transition from $x$ to $x - 1$ (a death), the clock assigned to $(x, x + 1)$ (birth in $x$) is reassigned to $(x - 1, x)$ (birth in $x - 1$). In general, for these chains we have $S_n(y) = \emptyset$ if $y \neq Y_{n-1}$, and $S_n(Y_{n-1}) = \{y : q(Y_{n-1}, y) > 0, y \neq Y_n\}$ in Algorithm 5.2.

## Jackson Networks

We saw in Chapter 4 that Jackson networks, viewed as GSMPs, satisfy condition (C2), so it may seem unnecessary to consider (CM). But using (CM), and the estimator of Section 5.3, we are able to estimate certain derivatives which would be inaccessible through (C2) and the results of Chapter 3. In particular, using (CM), we can estimate derivatives with respect to service rates in the presence of state-dependent routing, and also derivatives with respect to the routing probabilities. Neither of these was generally possible in Chapter 4. This increased capability comes at the price of more restrictive network topologies.

When the process under consideration is a Jackson network, the restrictions forced by (CM) on the generator, $Q$, translate into analogous conditions on the routing matrix, $P$. This interplay between the connectivity of the state space and the connectivity of the queueing network plays an important role. It leads, for example, to some simplification of the algorithm in Section 5.3.2. In particular, we reformulate the algorithm in terms of *nodes* rather than *states*, which makes the necessary bookkeeping more convenient. We first consider the simplest setting—ordinary service and routing—then focus on state-dependent service and routing, and derivatives with respect to routing probabilities.

To be consistent with the rest of this chapter (rather than with Chapter 4) we use $x, y$, etc. (rather than $\mathbf{n}$, etc.) to represent the vector of queue lengths in a network. We denote by, e.g, $y(j)$ and $Y_n(j)$ the queue length at node $j$, and by $e_j$ the $j$th unit vector. Let $\mu_j$ be the service rate at $j$, and let $P$ be the matrix of routing probabilites. A fictitious node 0 (never idle) is the source of all external arrivals (if any) and the destination of all departures from the network (if any); cf. Chapter 2, Section 2.1, Example 2.

If we let $e_0$ be the vector of all zeros, all transition arcs take the form $(y, y - e_i + e_j)$ with transition rates $q(y, y - e_i + e_j) = \mathbf{1}\{y(i) > 0\}\mu_i P_{ij}$. (Here and below we take $y(0) > 0$ for all $y$.) For $i' \neq i$, pairs of states $y - e_i + e_j$ and $y - e_{i'} + e_{j'}$ have $y' = y - e_i + e_j - e_{i'} + e_{j'}$ as a common successor if they have $y$ as a common predecessor. (That's why (C2) is satisfied.) For if $y$ is a common predecessor, then $P_{ij} > 0$, $y(i) > 0$, $P_{i'j'} > 0$ and $y(i') > 0$, so $q(y - e_i + e_j, y') = \mu_{i'} P_{i'j'} > 0$ and $q(y - e_{i'} + e_{j'}, y') = \mu_i P_{ij} > 0$, and $y'$ is a common successor. But if $i' = i$, a problem may arise: $y - e_i + e_j$ and $y - e_i + e_{j'}$ may not have a common successor. In particular, $y' = y - 2e_i + e_j + e_{j'}$ fails to be a common successor if $y(i) = 1$. In this case, the movement of a job from $i$ to $j$ ($i$ to $j'$) leaves queue $i$ empty and makes a transition from $i$ to $j'$ ($i$ to $j$) impossible.

The additional condition we need for (CM) to hold is the following: For any pair of nodes $j$ and $k$, if there is an $i$ such that $P_{ij}P_{ik} > 0$, then there must also be a node $i^*$ such that $P_{ji^*}P_{ki^*} > 0$. In other words, every pair of *nodes* with a common immediate predecessor must have a common immediate successor, where, now, "predecessor" and "successor" refer to the topology of the network, *not* the state space. Let us call this condition ($CM_P$). If it is satisfied, then the *states* $y - e_i + e_j$ and $y - e_i + e_{j'}$ have $y - e_i + e_{i^*}$ as a common successor, where $i^*$ is the common successor (under $P$) of $j$ and $j'$.

Let us examine how clocks are assigned and reassigned using this structure and the construction of Section 5.2. When a job initiates service at node $i$, it is assigned a separate *potential* service time—a clock—for every $j$ for which $P_{ij} > 0$. Call this an $(i,j)$*-service time.* (Service *requirement* would be more accurate but is cumbersome.) Each such clock is run down at rate $\mu_i P_{ij}$. The $(i,j)$-service time that runs out first becomes the actual service time, and the corresponding $j$ is the destination of the job leaving $i$. All other (residual) clocks are reassigned. A clock previously assigned to $(y, y - e_{i'} + e_{j'})$, $i' \neq i$, is reassigned to $(y - e_i + e_j, y - e_i + e_j - e_{i'} + e_{j'})$; i.e., an $(i', j')$-service time before the transition continues to be an $(i', j')$-service time after the transition. For a clock previously assigned to $(y, y - e_i + e_{j'})$, $j' \neq j$, (another potential service time at $i$), there are two cases. If $y(i) > 1$, then it is reassigned to $(y - e_i + e_j, y - 2e_i + e_j - e_{j'})$; i.e., the residual $(i, j')$-service time becomes the new $(i, j')$-service time for the next job at $i$. But if $y(i) = 1$, then there is no next job, so it must be reassigned to

$$(y - e_i + e_j, K(y - e_i + e_j, y - e_i + e_{j'})) = (y - e_i + e_j, y - e_i + e_{i^*});$$

i.e., upon the movement of a job from $i$ to $j$, the $(i, j')$-service time becomes a $(j, i^*)$-service time, where $i^*$ is a common successor of nodes $j$ and $j'$ Thus, when we look at nodes rather than states we see that there are two ways a clock can be reassigned: it might be reassigned to a transition arc $(y, y')$ which corresponds to a job transition between the same pair of nodes $(i, j)$, or to a different pair $(j, i^*)$.

We come, now, to the actual algorithm. Suppose the service rates $\mu_i$ depend on $\theta$ (but are strictly positive for all $\theta$). We modify the algorithm of Section 5.3.2 to keep track of "perturbations" for pairs of nodes, $i, j$, in variables $\delta_n(i,j)$, rather than for states in variables $\delta_n(x)$. Thus, whereas previously $\delta_n(x)$ reflected the infinitesimal delay accumlated in a clock assigned to $(Y_{n-1}, x)$ at time $\tau_n$, now, $\delta_n(i,j)$ is the delay in an $(i,j)$-service time—that is, in a clock assigned to $(Y_{n-1}, Y_{n-1} - e_i + e_j)$—at time $\tau_n$. Let

$$\begin{aligned} M_n &= \text{node from which departure occurs at } n\text{th transition;} \\ A_n &= \text{node at which arrival occurs at } n\text{th transition;} \end{aligned}$$

then $\delta_n(M_n, A_n) = d\tau_n / d\theta$.

For nodes $i, j$ with a common predecessor (under $P$), let $K_P(i,j)$ be a common successor. From the discussion above, we find that if $Y_{n-1}(M_n) = 1$, then at the $n$th transition every $(M_n, k')$-service time, $k' \neq A_n$, is reassigned to $(A_n, K_P(A_n, k'))$. All other potential service times remain assigned to the same pair of nodes (though, in general, this means being reassigned to a different pair of states). Let $S_n(k)$ be the set of $(i, k')$ for which an $(i, k')$-service time is reassigned to the $(A_n, k)$-service time at the $n$th transition; that is, $(i, k') \in S_n(k)$ if either

$$i = M_n, \ P_{M_n k'} > 0, \ K_P(A_n, k') = k \text{ and } Y_{n-1}(M_n) = 1,$$

or

$$i = A_n, \ k' = k \text{ and } Y_{n-1}(A_n) > 0.$$

In the first case, the $(M_n, k')$-service time becomes an $(A_n, k)$-service time; in the second, the $(A_n, k)$-service time is reassigned to itself. For $i \neq M_n, A_n$, if $Y_{n-1}(i) > 0$, then an $(i, j)$-service time continues to be an $(i, j)$-service time, but this "reassignment" is not reflected in $S_n$.

Whenever $y(i) > 0$, $q_\theta(y, y - e_i + e_j) = \mu_i(\theta) P_{ij}$. With these substitutions Step 2 of Algorithm 5.2 becomes

**Step 2$'$.** At the $n$th transition, update by setting

$$\delta_n(M_n, A_n) = \delta_{n-1}(M_n, A_n) - \frac{\tau_n - \tau_{n-1}}{\mu_{M_n}} \mu'_{M_n};$$

for every $k$ with $P_{A_n k} > 0$ and $|S_n(k)| > 0$,

$$\begin{aligned}\delta_n(A_n, k) &= \delta_n(M_n, A_n) \\ &+ \sum_{(i,k') \in S_n(k)} \frac{\mu_i}{\mu_{A_n}} \left( \delta_{n-1}(i, k') - \frac{\tau_n - \tau_{n-1}}{\mu_i} \mu'_i - \delta_n(M_n, A_n) \right);\end{aligned}$$

if $Y_{n-1}(i) > 0$, $Y_n(i) > 0$, $i \neq A_n$, and $P_{ij} > 0$ then

$$\delta_n(i, j) = \delta_{n-1}(i, j) - \frac{\tau_n - \tau_{n-1}}{\mu_i} \mu'_i; \tag{5.22}$$

while if $P_{A_n k} > 0$ and $|S_n(k)| = 0$, then

$$\delta_n(A_n, k) = \delta_n(M_n, A_n).$$

Set

$$\begin{aligned} dL_T &= dL_T + \delta_n(M_n, A_n)[f(Y_{n-1}) - f(Y_n)] \\ dL_{n_0} &= dL_{n_0} + f(Y_{n-1})[\delta_n(M_n, A_n) - \delta_{n-1}(M_{n-1}, A_{n-1})]. \end{aligned}$$

The apparently new step (5.22) is actually a special case of $q(Y_n, x) > 0$ and $|S_n(x)| > 0$ in Algorithm 5.2. It corresponds to the case of $(i, j)$-service times that continue to be $(i, j)$-service times after the transition. We treat it separately here to distinguish reassignments of clocks to the same pair of nodes from reassignments to different pairs, reflected in $S_n(k)$.

**State-Dependent Routing and Service**

In the setting described above, every transition rate takes the form $q(y, y - e_i + e_j) = \mathbf{1}\{y(i) > 0\}\mu_i P_{ij}$ which depends on $y$ only through the obvious requirement that $y(i)$ be positive. In other words, the service rates and transition probabilities were assumed state-independent. But nothing in (CM) or in our construction requires this. We now examine state-dependent service and routing, indicating this by appending the argument $y$, as in $\mu_i(y)$ and $P_{ij}(y)$.

Let us first suppose that $\mu_i(y) > 0$ if (and only if) $y(i) > 0$. Let us also suppose that while the magnitudes of the routing probabilities may depend on $y$, the network topology, as determined by the set of $i, j$ for which $P_{ij} > 0$ does not. In other words, $\mathbf{1}\{P_{ij}(y) > 0\}$ does not depend on $y$. While this is not strictly necessary for (CM) to hold, it simplifies matters. In this case, (CM) is satisfied if $P$ satisfies the analogous condition (CM$_P$): $P_{ij}P_{ij'} > 0 \Rightarrow P_{ji^*}P_{j'i^*} > 0$ for some $i^*$. (Note that we may write $P_{ij} > 0$ unambiguously without the argument $y$.)

Suppose that the $\mu_i(y)$'s depend on $\theta$. Under the conditions above, the presence of state dependent service rates and routing probabilities does not significantly alter the calculation of $\delta_n(i,j)$. With $q(y, y-e_i+e_j) = \mu_i(y)P_{ij}(y)$, Step 2 of the Algorithm 5.2 becomes

**Step 2$''$.** At the $n$th transition, update by setting

$$\delta_n(M_n, A_n) = \delta_{n-1}(M_n, A_n) - \frac{\tau_n - \tau_{n-1}}{\mu_{M_n}(Y_n)}\mu'_{M_n}(Y_n);$$

for every $k$ with $P_{A_n k} > 0$ and $|S_n(k)| > 0$,

$$\begin{aligned}\delta_n(A_n, k) &= \delta_n(A_n, M_n) + \sum_{(i,k')\in S_n(k)} \frac{\mu_i(Y_{n-1})P_{ik'}(Y_{n-1})}{\mu_{A_n}(Y_n)P_{A_n k}(Y_n)}\Bigg(\delta_{n-1}(i,k') - \\ &\quad \frac{\tau_n - \tau_{n-1}}{\mu_i(Y_{n-1})}\mu'_i(Y_{n-1}) - \delta_n(M_n, A_n)\Bigg);\end{aligned}$$

if $Y_{n-1}(i) > 0$, $Y_n(i) > 0$, $i \neq A_n$, and $P_{ij} > 0$ then

$$\begin{aligned}\delta_n(i,j) &= \frac{\mu_i(Y_{n-1})P_{ij}(Y_{n-1})}{\mu_i(Y_n)P_{ij}(Y_n)}\left(\delta_{n-1}(i,j) - \frac{\tau_n - \tau_{n-1}}{\mu_i(Y_{n-1})}\mu'_i(Y_{n-1})\right) \\ &+ \left(1 - \frac{\mu_i(Y_{n-1})P_{ij}(Y_{n-1})}{\mu_i(Y_n)P_{ij}(Y_n)}\right)\delta_n(A_n, M_n);\end{aligned}$$

while if $P_{A_n k} > 0$ and $|S_n(k)| = 0$, then

$$\delta_n(A_n, k) = \delta_n(M_n, A_n).$$

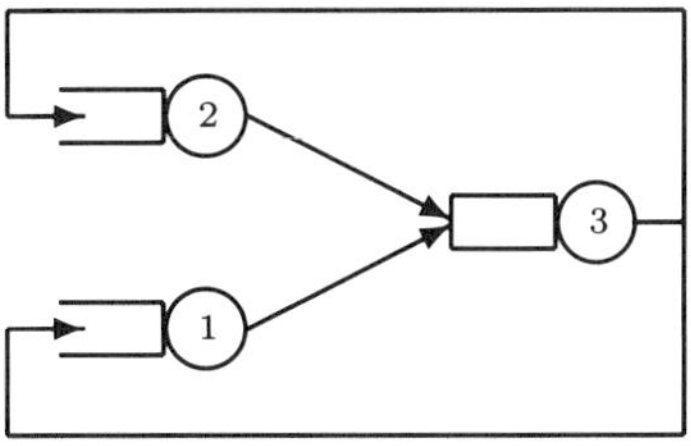


Figure 5.2: A network with blocking satisfying (CM), violating (C2)

Set

$$\begin{aligned} dL_T &= dL_T + \delta_n(M_n, A_n)[f(Y_{n-1}) - f(Y_n)] \\ dL_{n_0} &= dL_{n_0} + f(Y_{i-1})[\delta_n(M_n, A_n) - \delta_{n-1}(M_{n-1}, A_{n-1})]. \end{aligned}$$

Blocking is a special kind of state-dependent service—one that violates the condition $y(i) > 0 \Rightarrow \mu_i(y) > 0$. For closed networks in which $P$ satisfies ($\mathrm{CM}_P$), the single-source condition needed for (C2) (Chapter 4, Section 4.2.4) is sufficient for (CM), but it is not necessary. An alternative condition is that, if $j$ is a blocking node and $P_{ij} > 0$ then $P_{ji} > 0$. Consider, for example, the network illustrated in Figure 5.2. Only node 3 has a finite buffer, and that node can block both nodes 1 and 2. Letting $(y, i)$ denote the state in which the queue lengths are $y$ and node $i$ is blocked, states $(y, 1)$ and $(y, 2)$ have as common successor (and also common predecessor) state $y$, in which no node is blocked. If upon completing service node 1 becomes blocked, the residual service time at node 2 is therefore reassigned to the (3,1)-service time—i.e., to the potential service time at node 3 which takes a job back to node 1. Thus, when there is a change in which of nodes 1 and 2 completes service first, the sequence of states changes from $y, (y, 1), y, \ldots$ to $y, (y, 2), y, \ldots$. Algorithm 5.2 can be tailored to this setting as well, but we will not carry this out.

### Perturbing Routing Probabilities

One of the most interesting differences between the construction based on (CM) and that based on (C2), when applied to Jackson networks, is that the former allows the routing probabilities to depend on the parameter of differentiation, $\theta$. In viewing a Jackson network as a GSMP in which the clock times are the service and interarrival times, we implicity restrict the dependence on $\theta$ to the service and interarrival distributions. (For the results of Chapters 2 and 3 we required that only clock distributions depend on $\theta$.) But if we view a Jackson network as a Markov chain and use the construction of Section 5.2, then we can allow almost arbitrary dependence of the transition rates $q(y, y')$ on $\theta$. In particular, since $q(y, y-e_i+e_j) = \mathbf{1}\{y(i) > 0\}\mu_i P_{ij}$, we can allow $P$ to depend on $\theta$.

Suppose routing and service are state-independent, that $P$ satisfies ($\mathrm{CM}_P$), and, for simplicity, that $P$ but not $\mu$ depends on $\theta$. To ensure that $\mathbf{1}\{q_\theta(y, y') > 0\}$ is constant in $\theta$, we need $\mathbf{1}\{P_{ij}(\theta) > 0\}$ to be constant in $\theta$—that is, changes in $\theta$ do not change the network topology. To adapt Step 2, note that $q'(y, y - e_i + e_j)/q(y, y - e_i + e_j) = P'_{ij}/P_{ij}$. Thus, we get

**Step 2‴.** At the $n$th transition, update by setting

$$\delta_n(M_n, A_n) = \delta_{n-1}(M_n, A_n) - \frac{\tau_n - \tau_{n-1}}{P_{M_n A_n}} P'_{M_n A_n};$$

for every $k$ with $P_{A_n k} > 0$ and $|S_n(k)| > 0$,

$$\begin{aligned}\delta_n(A_n, k) &= \delta_n(M_n, A_n) \\ &+ \sum_{(i,k') \in S_n(k)} \frac{P_{ik'}}{P_{A_n k}} \left( \delta_{n-1}(i, k') - \frac{\tau_n - \tau_{n-1}}{P_{ik'}} P'_{ik'} - \delta_n(M_n, A_n) \right);\end{aligned}$$

if $Y_{n-1}(i) > 0$, $Y_n(i) > 0$, $i \neq A_n$, and $P_{ij} > 0$ then

$$\delta_n(i, j) = \delta_{n-1}(i, j) - \frac{\tau_n - \tau_{n-1}}{P_{ij}} P'_{ij};$$

while if $P_{A_n k} > 0$ and $|S_n(k)| = 0$, then

$$\delta_n(A_n, k) = \delta_n(M_n, A_n).$$

Set

$$\begin{aligned}dL_T &= dL_T + \delta_n(M_n, A_n)[f(Y_{n-1}) - f(Y_n)] \\ dL_{n_0} &= dL_{n_0} + f(Y_{i-1})[\delta_n(M_n, A_n) - \delta_{n-1}(M_{n-1}, A_{n-1})].\end{aligned}$$

### Markovian GSMPs

Jackson networks are GSMPs which are also Markov chains. As such, they illustrate the differences between conditions (CM) and (C2) and the constructions they entail. We now take up this topic more generally, considering arbitrary GSMPs with exponential clock times—that is, Markovian GSMPs. (Here, we continue to use $x$, $y$, etc. to denote states, even in referring to GSMPs.)

Despite the similarity between (CM) and (C2), neither implies the other in general. There is, however, a class of Markovian GSMPs for which (C2) is strictly stronger. These are GSMPs with deterministic transition probabilities—i.e., GSMPs in which for all states $y$ and events $\alpha \in \mathcal{E}(y)$ there is just one $y'$

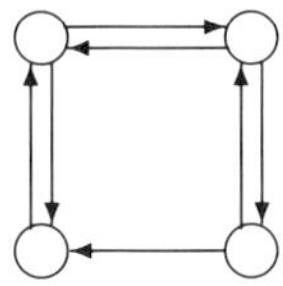

Figure 5.3: Transition diagram satisfying (CM) but not (C2)

for which $p(y'; y, \alpha) > 0$, necessarily with $p(y'; y, \alpha) = 1$. Using $\phi$ as defined in Chapter 2 and 3, let us denote this state $(y')$ by $\phi(y, \alpha)$.

**Proposition 5.3.** For a GSMP with deterministic transition probabilities (C2)⇒(CM).

**Proof.** Suppose $x$ and $y$ have a common immediate predecessor $z$. This means that there are events $\alpha$ and $\beta$ such that $\phi(z, \alpha) = x$ and $\phi(z, \beta) = y$. Since, by assumption, (C2) holds, we may define $z' = \phi(\phi(z, \alpha), \beta) = \phi(\phi(z, \beta), \alpha)$. Then $z' = \phi(x, \beta) = \phi(y, \alpha)$; that is, $z'$ is a common immediate successor of $x$ and $y$. □

The converse of Proposition 5.3 is false. Figure 5.3 shows the transition diagram of a chain that satisfies (CM) but does not satisfy (C2) under any labeling of the arcs, unless we allow transition probabilities (so that more than one arc emanating from a node can be labeled with the same event). Thus, there are Markovian GSMPs that satisfy (CM) but not (C2). If we allow transition probabilities, the opposite is also true, as can be seen from the example of Jackson networks. Ordinary Jackson networks—single class, infinite buffers—always satisfy (C2) (see Chapter 4, Section 4.2.1), but they do not satisfy (CM) without an additional condition on the routing matrix.

Let us examine these conditions more closely. Both place a condition on pairs of states reachable in a single transition from a common state. The difference lies in which pairs of states they consider, and what condition they impose. For (CM), *all* pairs of states $x, y$ with a common predecessor $z$ must have a common successor, regardless of the "physical" meaning of the transitions $z \to x$ and $z \to y$. For (C2) only pairs of states reachable through *different events* need have a "common successor": If $p(x; z, \alpha) > 0$ and $p(y; z, \alpha) > 0$, then $x$ and $y$ both have $z$ as a "common predecessor", but (C2) makes no requirement of them because they are reached from $z$ through the same event, $\alpha$. On the other hand, if $p(x; z, \alpha) > 0$ and $p(y; z, \beta) > 0$ then it is not enough for $x$ and $y$ simply to have a "common successor" $z'$. (C2) requires that $z'$ be reachable from $x$ *through* $\beta$ and from $y$ *through* $\alpha$. Moreover, it requires that $p(z'; x, \beta) = p(y; z, \beta)$ and $p(z'; y, \alpha) = p(x; z, \alpha)$, while (CM) imposes no conditions on the magnitudes of the transition rates between states. Thus, (C2) is weaker in that it puts a condition on fewer pairs of states, but it is stronger in that the condition it requires is more stringent.

### Scale Parameters

As we saw in Section 5.1.1, it is always possible to view a Markov chain as a GSMP satisfying (C2) via the semi-Markov representation. In this representation, there is always just one active clock so (C2) holds vacuously. In considering a parametric family of chains, however, this imposes a significant restriction on the possible dependence of $Q_\theta$ on $\theta$. We briefly consider this type of dependence.

Call $\theta$ a *scale parameter* for $Q_\theta$ if, for every $x$, whenever $q_\theta(x,y) > 0$ and $q_\theta(x,z) > 0$, $q'_\theta(x,y)/q_\theta(x,y) = q'_\theta(x,z)/q_\theta(x,z)$. This makes $[q_\theta(x,y)/q_\theta(x)]' = 0$ for every $x$ and $y$. In general, if $Y$ is the jump chain of a process with generator $Q$, then $P(Y_{n+1} = y|Y_n = x) = q(x,y)/q(x)$, $y \neq x$. Thus, a scale parameter is a parameter of the state holding times but not the embedded transition probabilities. In this case, the holding time in the $i$th state $Y_i$ is $\tau_{i+1} - \tau_i = X_{i+1}/q(Y_i)$, (where $X_{i+1}$ is the $i+1$st clock sample) so

$$\begin{aligned}\frac{d\tau_{i+1}}{d\theta} - \frac{d\tau_i}{d\theta} &= -\frac{X_{i+1}}{q(Y_i)^2}q'(Y_i)\\ &= -\frac{\tau_{i+1}-\tau_i}{q(Y_i)}q'(Y_i).\end{aligned}$$

This leads to the simple expressions

$$\frac{dL_{n_0}}{d\theta} = -\sum_{i=0}^{n_0-1} f(Y_i)\frac{\tau_{i+1}-\tau_i}{q(Y_i)}q'(Y_i) = -\int_0^{\tau_{n_0}} f(Z_t)\frac{q'(Z_t)}{q(Z_t)}dt,$$

and

$$\begin{aligned}\frac{dL_T}{d\theta} &= \sum_{i=1}^{N_T}\frac{d\tau_i}{d\theta}[f(Y_{i-1}) - f(Y_i)]\\ &= -\sum_{i=1}^{N_T}\int_0^{\tau_i}\frac{q'(Z_t)}{q(Z_t)}dt\,[f(Y_{i-1}) - f(Y_i)]\\ &= -\sum_{i=0}^{N_T-1} f(Y_i)\int_{\tau_i}^{\tau_{i+1}}\frac{q'(Z_t)}{q(Z_t)}dt + f(Y_{N_T})\int_0^{\tau_{N_T}}\frac{q'(Z_t)}{q(Z_t)}dt\\ &= -\int_0^{\tau_{N_T}} f(Z_t)\frac{q'(Z_t)}{q(Z_t)}dt + f(Z_T)\int_0^{\tau_{N_T}}\frac{q'(Z_t)}{q(Z_t)}dt.\end{aligned}$$

Here, Theorem 5.1 holds without the restriction to chains satisfying (CM). Furthermore, given $\{q'(x), x \in \mathbf{S}\}$, these expressions can be calculated directly from a sample path of $Z$ without Algorithm 5.2.

### The Limit of Small Transition Rates

Consider, again, the Markov chain illustrated in Figure 5.1. If, as we assume, $q(x_0, x_1) > 0$ and $q(x_0, x_2) > 0$, then $q(x_1, x_4) > 0$ and $q(x_2, x_4) > 0$ are necessary for (CM) to be satisfied. If we let $q(x_2, x_4) = \epsilon$, then so long as $\epsilon > 0$ we can apply Algorithm 5.2 and obtain an unbiased derivative estimate; but as soon as $\epsilon = 0$, the applicability of the algorithm is destroyed. This is paradoxical: we would typically expect $\mathbf{E}[L]$ and $d\mathbf{E}[L]/d\theta$ to be continuous in $\epsilon$, yet our algorithm appears to have a catastrophic discontinuity at $\epsilon = 0$. To look at this another way, if we start with a chain that violates (CM) it might be tempting to add transitions with small rates so that (CM) will hold, and then apply our algorithm, hoping that the added arcs will not substantially change the results.

The paradox above might be resolved by looking at the *variance* of the derivative estimator. Informally, if (CM) depends critically on the positivity of a particular $q(x, y)$, then we would expect that as $q(x, y) \to 0$ the variance of a derivative estimate based on Algorithm 5.2 will increase without bound. In this sense, the apparent discontinuity at $\epsilon = 0$ can be thought of as the limiting case of infinite variance. We do not see a viable way to make this heuristic argument rigorous. However, if it is valid, it suggests that adding arcs with small transition rates to satisfy (CM) is not likely to produce good derivative estimates. It also suggests that the methods of this chapter may not work well with chains that have some very small transition rates.

## 5.5 Discrete-Time Conversion

Algorithm 5.2 can be further improved—made more efficient—through a simple variance reduction technique known as *discrete-time conversion*, which is a special case of the more general method of conditional Monte Carlo; cf. Bratley, Fox, and Schrage [3],§2.6. The idea is to replace calculation of some estimate, $r$, from simulation (or observation) of $Z$ with calculation of another estimate, $\hat{r}$, from simulation (or observation) of the embedded chain $Y$. These estimates are related by

$$\hat{r} = \mathbf{E}[r|Y]. \tag{5.23}$$

Clearly, $\mathbf{E}[\hat{r}] = \mathbf{E}[r]$, so $\hat{r}$ is unbiased if $r$ is. Simulating $Y$ instead of $Z$ eliminates the need to generate exponential state holding times, and so reduces work. Moreover, taking conditional expectations as in (5.23) reduces variance (regardless of whether $Y$ is simulated or observed): Jensen's inequality for conditional expectations (Chung [12],p.302) implies $\mathbf{E}[\hat{r}^2] \leq \mathbf{E}[r^2]$, so $\mathbf{Var}[\hat{r}] \leq \mathbf{Var}[r]$. Thus, $\hat{r}$ is preferable to $r$, provided it is easily evaluated.

Fox and Glynn [21] apply this technique in the finite-horizon setting relevant here, and develop the theory in substantial generality. We review some of their results on estimation of $\mathbf{E}[L_{n_0}]$ and $\mathbf{E}[L_T]$ before considering discrete-time conversion for derivatives.

A well-known property of continuous-time Markov chains is that given the sequence, $Y$, of states, the holding times $\{\tau_{i+1}-\tau_i, i = 0, 1, \ldots\}$ are independent and exponentially distributed. Given $Y$, we also know that the $i$th holding time has mean $q^{-1}(Y_i)$. Hence,

$$\begin{aligned} \widehat{L_{n_0}} &= \mathbf{E}[L_{n_0}|Y] \\ &= \mathbf{E}\left[\sum_{i=0}^{n_0-1} f(Y_i)[\tau_{i+1} - \tau_i] \middle| Y\right] \\ &= \sum_{i=0}^{n_0-1} f(Y_i)/q(Y_i). \end{aligned}$$

Thus, calculation of $\widehat{L_{n_0}}$ simply requires replacing holding times by their conditional means.

The case of $L_T$ is more complicated because the number, $N_T$, of transitions made by $Z$ in $(0, T]$ is not determined by $Y$. Fox and Glynn [21] show how to get around this using *uniformization* (see Appendix 5.A). Suppose $q(x)$ is bounded in $x$; we uniformize at (finite) rate $q^* \geq \sup_x q(x)$. Let $N^*$ be the Poisson process of jumps of the uniformized chain, let $\tau_i^*$ be the epoch of its $i$th jump, and let $Y_i^*$ be the state just after $\tau_i^*$. Then

$$\begin{aligned} L_T &= \int_0^T f(Z_t)dt \\ &= \sum_{i=0}^{N_T^*-1} f(Y_i^*)[\tau_{i+1}^* - \tau_i^*] + f(Y_{N_T^*}^*)(T - \tau_{N_T^*}^*). \end{aligned}$$

Given $N_T^*$, the epochs of jumps of $N^*$ in $(0, T]$ are uniformly distributed over $(0, T]$, and have conditional mean $\mathbf{E}[\tau_{i+1}^* - \tau_i^* | N_T^*] = T/(N_T^* + 1)$ (see Appendix 5.A). The last holding time $T - \tau_{N_T^*}$ has the same conditional mean. Thus, if we let $M_i$ be the number of visits the uniformized chain makes to $Y_i$ before proceeding to $Y_{i+1}$ (or the number before $T$, in the case of $Y_{N_T}$), then

$$\begin{aligned} \mathbf{E}[L_T|Y^*, N_T^*] &= \mathbf{E}\left[\sum_{i=0}^{N_T^*-1} f(Y_i^*)[\tau_{i+1}^* - \tau_i^*] + f(Y_{N_T^*})(T - \tau_{N_T^*}) \middle| N_T^*, Y^*\right] \\ &= \sum_{i=0}^{N_T^*} f(Y_i^*)\frac{T}{N_T^* + 1} \\ &= \sum_{i=0}^{N_T} f(Y_i)\frac{TM_i}{N_T^* + 1}. \end{aligned}$$

As in the case of $L_{n_0}$, therefore, the holding times (including the last one, $T-\tau_N$) are replaced by conditional means, except that conditioning on $(Y^*, N_T^*)$ makes these $TM_i/(N_T^* + 1)$. Conditioning just on $Y$ would lead to lower variance, but $\mathbf{E}[L_T|Y]$ is hard to evaluate. Also, since the uniformized counting process, $N^*$, cannot be "observed" from $Z$, this method can only be implemented in simulation.

We now apply the same techniques to the derivatives of $L_{n_0}$ and $L_T$—actually, to their conditional expectations given $Z$. Thus, we go from conditioning on $Z$ to conditioning on $Y$ or $(N_T^*, Y^*)$.

In Algorithm 5.2, the holding time $\tau_n - \tau_{n-1}$ is introduced at the $n$th transition, then repeatedly added and multiplied by terms depending on $Y$ (through $q$ and $q'$). In other words, the holding times factor *linearly* into the algorithm, with coefficients determined by $Y$. There are, therefore, functions $g_0, g_1, g_2, \ldots$ such that

$$\delta_i(Y_i) = \sum_{j=0}^{i} g_{i-j}(Y_j, \ldots, Y_i)(\tau_{j+1} - \tau_j).$$

For example, $\delta_0 \equiv 0$ so $g_0 \equiv 0$; $\delta_1(Y_1) = -q'(Y_0, Y_1)(\tau_1 - \tau_0)/q(Y_0, Y_1)$ so $g_1(y_0, y_1) = -q'(y_0, y_1)/q(y_0, y_1)$. The higher order $g$'s are harder to evaluate but could, in principle, be written out. The key point is that the value of each $g_j$ is completely determined by $Y$ and is otherwise independent of $\tau_1, \tau_2, \ldots$. From (5.21) we get

$$\mathbf{E}[\frac{dL_{n_0}}{d\theta}|Z] = \sum_{i=0}^{n_0-1} f(Y_i)[\delta_{i+1}(Y_{i+1}) - \delta_i(Y_i)]$$

$$= \sum_{i=0}^{n_0-1} f(Y_i) \sum_{j=0}^{i} [g_{i+1-j}(Y_j, \ldots, Y_{i+1}) - g_{i-j}(Y_j, \ldots, Y_i)](\tau_{j+1} - \tau_j).$$

Therefore,

$$\mathbf{E}[\frac{dL_{n_0}}{d\theta}|Y] = \sum_{i=0}^{n_0-1} f(Y_i) \sum_{j=0}^{i} [g_{i+1-j}(Y_j, \ldots, Y_{i+1}) - g_{i-j}(Y_j, \ldots, Y_i)] q^{-1}(Y_j).$$

In other words, $\mathbf{E}[dL_{n_0}/d\theta|Z]$ is converted to discrete time simply by replacing holding times with their conditional means. The lower-variance estimator $\mathbf{E}[dL_{n_0}/d\theta|Y]$ can be evaluated using Algorithm 5.2, by substituting $q^{-1}(Y_n)$ for $\tau_n - \tau_{n-1}$ everywhere. Thus, we may estimate $\mathbf{E}[L_{n_0}]$ and $d\mathbf{E}[L_{n_0}]/d\theta$ simulating (or observing) only the embedded chain $Y$.

With $\mathbf{E}[dL_T/d\theta|Y]$ we face the same difficulty as with $\mathbf{E}[L_T|Y]$ so again we uniformize. Let $N^*$, $Y^*$ and $\tau^*$ be as above. Recall that $M_i$ is the number of visits the uniformized chain makes when the original chain makes a visit to $Y_i$. Let $m_0 = 0$ and $m_{j+1} = m_j + M_j$, $j = 1, 2, \ldots$; then $Y_i^* = Y_j$ for $m_j \leq i < m_{j+1}$. It is notationally convenient to let $\tau^*_{N_T^*+1} \equiv T$, even though $N^*$ may not have jump right at $T$. From (5.20) we get

$$\begin{aligned}
\mathbf{E}[\frac{dL_T}{d\theta}|Z] &= \sum_{i=1}^{N_T} \delta_i(Y_i)[f(Y_{i-1}) - f(Y_i)] \\
&= \sum_{i=1}^{N_T}[f(Y_{i-1}) - f(Y_i)] \sum_{j=0}^{i} g_{i-j}(Y_j, \ldots, Y_i)(\tau_{j+1} - \tau_j);
\end{aligned}$$

and therefore

$$\begin{aligned}
&\mathbf{E}[\frac{dL_T}{d\theta}|Y^*, N_T^*] \\
&= \sum_{i=1}^{N_T}[f(Y_{i-1}) - f(Y_i)] \sum_{j=0}^{i} g_{i-j}(Y_j, \ldots, Y_i)\mathbf{E}[\sum_{l=m_j}^{m_{j+1}-1} (\tau^*_{l+1} - \tau^*_l)|Y^*, N_T^*] \\
&= \sum_{i=1}^{N_T}[f(Y_{i-1}) - f(Y_i)] \sum_{j=0}^{i} g_{i-j}(Y_j, \ldots, Y_i)\frac{M_j T}{N_T^* + 1},
\end{aligned}$$

using the conditional expectation $\mathbf{E}[\tau^*_{l+1} - \tau^*_l|Y^*, N_T^*] = T/(N_T^* + 1)$. Thus, $\mathbf{E}[dL_T/d\theta|Y^*, N_T^*]$ can also be evaluated using Algorithm 5.2, except that now—in addition to generating $N_T^*$—we must use $M_n T/(N_T^* + 1)$ in place of $\tau_n - \tau_{n-1}$.

## 5.A Appendix: Uniformization

To *uniformize* a continuous-time Markov chain is to subordinate it to a Poisson process. Suppose the infinitesimal generator $Q$ satisfies $\sup_x -q(x, x) < q^*$ for some $q^* < \infty$; such a $Q$ is *uniformizable.* Uniformization constructs a Markov chain with generator $Q$ by generating transitions at a constant rate $q^*$—that is, according to a Poisson process with parameter $q^*$. At a jump of the Poisson process, the Markov chain makes a transition from its current state, $x$, to a different state, $y$, with probability $q(x, y)/q^*$; it makes a transition back to $x$ with probability $-q(x, x)/q^* = 1 - \sum_{y \neq x} q(x, y)/q^*$.

More precisely, let $P^* = Q/q^* + I$, where $I$ is the identity matrix; then $P^*$ is a Markov transition probability matrix. Let $Y^* = \{Y_n^*, n = 0, 1, 2\}$ be a discrete-time Markov chain with transition probabilies given by $P^*$. Let $N^*$ be a Poisson process with parameter $q^*$. Then $Z_t = Y^*_{N_t^*}$ is a continuous-time Markov chain with generator $Q$.

Let $\{\tau_n^*, n = 1, 2, \ldots\}$ be the epochs of jumps of $N^*$, and let $\tau_0^* = 0$. Transitions that take $Z$ back to the same state are called *null* jumps; the genuine transition epochs of $Z$—the non-null jumps—form a subsequence $\{\tau_n\}$ of $\{\tau_n^*\}$. The number of null jumps back to $x$ before a non-null jump out of $x$ is geometrically distributed with parameter $1 - q(x)/q^*$.

For all $n = 0, 1, \ldots$, $\tau_{n+1}^* - \tau_n^*$ is exponentially distributed with mean $1/q^*$. It is a standard property of the Poisson process that, given $N_T^*$, the number of jumps in $(0, T]$, the epochs of the jumps are uniformly distributed over $(0, T]$. If $N_T^* = m$, then for *each* subset of $i \leq m$ points, the probability that exactly those $i$ points occur in $(0, t]$, $0 < t \leq T$ is $(t/T)^i[(T-t)/T]^{m-i}$. Since there are $\binom{m}{i}$ ways of choosing those $i$ points,

$$P(N_t^* = i | N_T^* = m) = \binom{m}{i} \left(\frac{t}{T}\right)^i \left(\frac{T-t}{T}\right)^{m-i}.$$

Now, for $i < m$,

$$\begin{aligned}
\mathbf{E}[\tau_{i+1} - \tau_i | N_T^* = m] &= \mathbf{E}\left[\int_0^T \mathbf{1}\{N_t^* = i\} dt \,\middle|\, N_T^* = m\right] \\
&= \int_0^T P(N_t^* = i | N_T^* = m) dt \\
&= \int_0^T \binom{m}{i} \left(\frac{t}{T}\right)^i \left(\frac{T-t}{T}\right)^{m-i} dt.
\end{aligned}$$

(The second equality uses a generalization of Fubini's Theorem for conditional expectations; see Fox and Glynn [21] for the precise statement.) The last integral evaluates to $T/(m+1)$; thus, $\mathbf{E}[\tau_{i+1} - \tau_i | N_T^*] = T/(N_T^* + 1)$.

## 5.B Appendix: Proof of Lemma 5.1

As in Section 5.3.1, here we use $q_n(j)$ for the rate at which clock $j$ is run during the $n$th holding time $[\tau_n, \tau_{n+1})$.

There are two parts to the proof. The first part restricts attention to the case where at most one order change occurs; i.e., for at most one $i < \tilde{N}$, $J_i$ or $J_{i+1}$ has a discontinuity in $(\theta_0 - h, \theta_0 + h)$. We call this the *single order-change* case, and show that (i) and (ii) hold. This part is similar to the proof of Lemma 3.1. The second part of the proof verifies that this case has probability $1 - O(h^2)$.

By construction, at a point $\theta_0 + h$ where $J_1, \ldots, J_i$ are continuous, all $\tau_k$ and $Y_k$, $k \leq i$ are continuous as well. Increase $h$ from 0 and until some $J_i$ has a discontinuity, and let $\eta$ be the smallest index such that $J_\eta$ has a discontinuity in $(\theta_0 - h, \theta_0 + h)$. (Take $\eta = \infty$ when no such discontinuity arises.) If $\eta > \tilde{N}$, then (i) and (ii) hold automatically. Suppose, therefore, that $\eta \leq \tilde{N}$. In order that $J_\eta$ be discontinuous at $\theta_0 + h$, there must be two clocks $j$ and $k$ which run out at the same time; i.e.,

$$\tau_{\eta-1} + c_{\eta-1}(j)/q_{\eta-1}(j) = \tau_{\eta-1} + c_{\eta-1}(k)/q_{\eta-1}(k),$$

and, also,

$$c_{\eta-1}(j)/q_{\eta-1}(j) = c_{\eta-1}(k)/q_{\eta-1}(k) = \min\{c_{\eta-1}(i)/q_{\eta-1}(i) : i \in C_{\eta-1}\},$$

with all quantities evaluated at $\theta_0 + h$. From (5.3) we see that even if in $J_\eta$ changes at $\theta_0 + h$, right at $\theta_0 + h$

$$\begin{aligned} \tau_\eta &= \tau_{\eta-1} + c_{\eta-1}(j)/q_{\eta-1}(j) \\ &= \tau_{\eta-1} + c_{\eta-1}(k)/q_{\eta-1}(k) \end{aligned}$$

is continuous. We show below that this extends to every $\tau_i$, $i \leq \tilde{N}$.

A discontinuity in $J_\eta$ may introduce one in $Y_\eta$ — in particular, $Y_\eta$ would jump from $d(r_{\eta-1}(j))$ to $d(r_{\eta-1}(k))$ if $J_\eta$ jumped from $j$ to $k$ (see (5.5)). Regardless of this possibility, $c_{\eta-1}(j)/q_{\eta-1}(j) = c_{\eta-1}(k)/q_{\eta-1}(k)$ and (5.10) together imply that there will be at least one clock in $C_\eta$ with a residual time of zero. In light of (5.3), this means that $\tau_{\eta+1} = \tau_\eta$. We now extend this to every $\tau_i$, $i \leq \tilde{N}$.

Since we are, for the time being, restricting attention to the single order-change case, at most two clocks $j$ and $k$ run out at the same time (at $\tau_\eta = \tau_{\eta+1}$). Regardless of the order in which they occur, $Y_{\eta+1}$ is the same state—namely, $K(d(r_{\eta-1}(j), d(r_{\eta-1}(k)))$. Hence, Case 1 of the construction of $Z_t(\theta_0 + h)$ applies: $N_{\eta+1}$, $C_{\eta+1}$ and every $c_{\eta+1}(j')$ with $j' \in C_{\eta+1}$ remain unchanged at a discontinuity of $J_\eta$ and $Y_\eta$. Every $\tau_i$ and $Y_i$, $i > \eta + 1$, is continuous at $\theta_0 + h$ if $J_{\eta+2}, \ldots, J_i$ are. But in the single order-change case, every $J_i$, $\eta + 1 < i \leq \tilde{N}$ is, in fact, continuous. Thus, (i) and (ii) hold.

We now come to the second part of the proof. We need to verify that exceptions to the single order-change case occur with probability (at most) $O(h^2)$. There are two types of exceptions. It could happen that through an increase in $h$, more than two clocks run out at the same time; or, it could happen that there are two $Y_i$, $i \leq \tilde{N}$, with discontinuities in $(\theta_0 - h, \theta_0 + h)$. We refer to either of these cases as *multiple order-changes*.

To bound the probabilities of these events, we bound the change in the $\tau_i$ due to changes in $h$. Let $\tilde{h}$ be the infimum over $h$ for which a multiple order-change occurs in $(\theta_0 - h, \theta_0 + h)$ ($\tilde{h}$ is strictly positive with probability one). For $h < \tilde{h}$, the $\tau_i(\theta_0 + h)$ are continuous and have (at least) left and right hand derivatives. Taking the larger of these at each point, the bound (5.19) (which is valid throughout $(\theta_0 - \tilde{h}, \theta_0 + \tilde{h})$) yields

$$\left|\frac{d\tau_i}{d\theta}\right| \leq \tau_i \cdot B\rho^i/q_*,$$

where $\rho$ is a positive constant. Integrating, we get, for $h < \tilde{h}$,

$$e^{-hB\rho^i/q_*}\tau_i(\theta_0) \le \tau_i(\theta_0 + h) \le e^{hB\rho^i/q_*}\tau_i(\theta_0);$$

so,

$$|\tau_i(\theta_0 + h) - \tau_i(\theta_0)| \le (e^{hB\rho^i/q_*} - 1)\tau_i(\theta_0).$$

Using the mean value theorem, the right side can be replaced with $he^{hB\rho^i/q_*}\tau_i$. Since we are interested only in the limit as $h \to 0$, we may, without loss of generality, consider only $h < 1$. Thus, the bound becomes $he^{B\rho^i/q_*}\tau_i$, which we rewrite as $hb_i\tau_i$ by letting $b_i = e^{B\rho^i/q_*}$. For $J_i$ and $J_{i+1}$ to run out at the same time at $\theta_0 + h$, $h \le \tilde{h}$, we must have $\tau_{i+1}(\theta_0 + h) = \tau_i(\theta_0 + h)$; thus, the sum of the changes in $\tau_i$ and $\tau_{i+1}$ from $\theta_0$ to $\theta_0 + h$ must exceed $\tau_{i+1}(\theta_0) - \tau_i(\theta_0)$. That is,

$$\begin{aligned}
|\tau_{i+1}(\theta_0) - \tau_i(\theta_0)| &\le |\tau_{i+1}(\theta_0 + h) - \tau_{i+1}(\theta_0)| + |\tau_i(\theta_0 + h) - \tau_i(\theta_0)| \\
&\le b_ih\tau_i(\theta_0) + b_{i+1}h\tau_{i+1}(\theta_0) \\
&\le 2b_{i+1}h\tau_{i+1}(\theta_0).
\end{aligned}$$

Note that this last inequality is a statement about $\{Z_t(\theta_0)\}$ only, and is independent of the construction used. We may drop the argument $\theta_0$.

For fixed $i$,

$$P(\tau_{i+1} - \tau_i \le 2b_{i+1}h\tau_{i+1}) = O(h),$$

because $\tau_{i+1} - \tau_i$ is exponentially distributed. The probability that this holds for fixed $i$ and $j$, $(j \ne i)$ is $O(h^2)$, by the independence of the holding times. Hence, the probability that this holds for *some* $i$ and $j$ less than a fixed $n$ is $\binom{n}{2}$ times the probability that it holds for a particular $i$ and $j$, which is $O(h^2)$. The case $\tilde{N} = N_T$ is a bit more involved.

Uniformize the process at rate $q^*$. Let $N^*$ be the number of transitions in $(0, T]$ using uniformization, and denote the epochs of these transitions by $\tau_i^*$. We view the original (non-null) transition epochs as a subset of the $\tau_i^*$. Since all transitions are constrained to occur in $(0, T]$, for $i < N^*$ we may replace the bound $b_{i+1}h\tau_{i+1}^*$ with $b_{i+1}hT$, and consider

$$P(\text{for some } i, j < N^*,\ \tau_{i+1}^* - \tau_i^* \le 2b_{N^*}Th,\ \tau_{j+1}^* - \tau_j^* \le 2b_{N^*}Th). \tag{5.24}$$

Given $N^*$, the epochs of the $N^*$ transitions in $(0, T]$ are uniformly distributed. For fixed $i, j < N^*$ and any $t_i, t_j > 0$,

$$P(\tau_{i+1}^* - \tau_i^* \le t_i \text{ and } \tau_{j+1}^* - \tau_j^* \le t_j | N^*) \le (\frac{t_i}{T})(\frac{t_j}{T}).$$

Hence,

$$P(\tau_{i+1}^* - \tau_i^* \le 2b_{N^*}Th \text{ and } \tau_{j+1}^* - \tau_j^* \le 2b_{N^*}Th | N^*) \le (2b_{N^*}h)^2.$$

Since there are $\binom{N^*-1}{2} < N^{*2}$ ways of choosing $i$ and $j$,

$P(\text{some } i, j < N^*, \tau^*_{i+1} - \tau^*_i \leq 2b_{N^*}Th, \tau^*_{j+1} - \tau^*_j \leq 2b_{N^*}Th | N^*) \leq (N^* \cdot 2b_{N^*}h)^2.$

Unconditioning, we get $4h^2\mathbf{E}[(N^*b_{N^*})^2]$. Since $N^*$ has a Poisson distribution, the expectation is finite, and the product is $O(h^2)$.

Finally, for the *last* transition before $T$, a similar argument shows that

$$P(\tau_{N_T+1} - \tau_{N_T} \leq 2b_{N_T+1}\tau_{N_T+1}h) \leq$$
$$P(T - \tau_{N_T} \leq 2b_{N_T+1}\tau_{N_T+1}h, \tau_{N_T+1} - T \leq 2b_{N_T+1}\tau_{N_T+1}h) = O(h^2).\square$$

* * *

## Notes and Comments

At first glance, Markov chains appear to form a very restricted class of GSMPs—those for which all clocks are exponential. Even worse, they tend to obscure the structure of a system. Compare, for example, the (explicit) Markov chain description of a queueing system driven by phase-type distributions with a GSMP description. The Markov chain must introduce states for different phases of service, while the GSMP need only reflect the "physical" states of the system. Moreover, much of the theory of denumerable Markov chains assumes that the state space is a subset of the integers. From a practical viewpoint, this presupposes an enumeration of the states which, in addition to being burdensome, washes away all structure.

In light of these remarks, it may be surprising that we have considered Markov chains separately. But as the results of this chapter should indicate, the Markov property has interesting implications for derivative estimation. For the purpose of finding continuous constructions, Markov models trade structure for flexibility.

The construction and algorithm of this chapter are from Glasserman [30]. The treatment of discrete-time conversion and uniformization is based on Fox and Glynn [20,21]. It also benefitted from the comments of Bennett Fox on an early version of [30].

Markov chains, in either discrete or continuous time, form a convenient setting for derivative estimation via likelihood ratios; see Glynn [38,39]. Discrete-time conversion is applied to likelihood ratio estimators in [30].

Glasserman [31] considers derivative estimation for discrete-time Markov chains. The methods described there embed the discrete-time chain in continuous-time processes, take derivatives, then convert back to discrete time. One such procedure shows that derivative estimates based on likelihood ratios are also perturbation analysis estimates derived through a special construction.

When the time horizon is the time to hit a designated set (rather than a fixed time or a fixed transition), Fox and Glasserman [19] have a derivative estimator based on differentiating *Poisson's equation.* As shown in [19], this estimator can be derived from the likelihood ratio method by taking conditional expectations.

Methods for estimating differences due to genuine parameter changes (including integer-valued parameters) are considered in Ho and Li [53] and in Cassandras and Strickland [10].

# Chapter 6

# GSMPs Via Hazard Rates

In Chapter 5, we used the common successor condition (CM) to obtain a special construction of a parametric family of Markov chains, and showed that this construction leads to stochastic derivatives which are unbiased derivative estimators. In this chapter, we generalize this approach to a class of GSMPs satisfying an analog of (CM). As discussed in Section 5.4 of Chapter 5 (under "Markovian GSMPs") conditions (CM) and (C2) do not coincide when applied to GSMPs; hence, a derivative estimator for GSMPs based on (CM) complements the results of Chapters 2 and 3. Moreover, some of the queueing systems in Chapter 4 that do not satisfy (C2) fall within the scope of this chapter.

The construction we undertake here is essentially the same as that of Chapter 5. The new ideas—and most of the work—arise in extending the manipulation of clocks from the exponential to the general case. In this regard, the main tools of this chapter are the *hazard rate* of a positive random variable, and two attendant transformations. These results lead to a nice theory for the non-Markovian case; the price we pay is added complexity—notationally, algorithmically and computationally. The estimators we derive here are harder to implement than those of earlier chapters.

## 6.1 Hazard Transformations

For the results of Chapter 5, we made extensive use of special properties of the exponential distribution. Similar properties can be exploited with general clock time distributions via two simple but remarkable transformations from reliability theory. The first views a sample from a general clock distribution as the time it takes to run down an *exponential* sample at a time-varying rate—in other words, it *time-changes* the exponential. The second, for our purposes, shows how to "split" a single clock into multiple clocks, the minimum of which has the orginal clock distribution.

It is the *memoryless* property of the exponential distribution which allowed us, in Chapter 5, to reassign residual clock times to new transitions arbitrarily. We also used the following property: If $X_0$ and $X_1$ are independent and exponential with means $\mu_i^{-1}$, $i = 0, 1$, and $\xi$ is the index of the smaller of $X_0$ and $X_1$, then, for $i = 0, 1$ and $x \geq 0$,

$$\begin{aligned} P(\xi = i, X_\xi \leq x) &= \frac{\mu_i}{\mu_i + \mu_{1-i}}(1 - e^{-(\mu_0+\mu_1)x}) \\ &= P(\xi = i)P(X_\xi \leq x). \end{aligned} \tag{6.1}$$

In words, $X_\xi$ and $\xi$ are independent, and $X_\xi$ is exponential with mean $(\mu_0 + \mu_1)^{-1}$. It is this property that allowed us to assign different clocks to different transitions out of a state and use the smallest to determine both the holding time in that state and which transition actually occurs. We now show how to extend the memoryless property and (6.1) to general distributions.

Recall yet another property of the exponential distribution: the time it takes to run down a unit mean exponential clock at rate $\mu$ is exponentially distributed with mean $\mu^{-1}$. In other words, $X/\mu$ is exponential if $X$ is. Now imagine running down the (unit mean) exponential clock at a *time-varying* rate $\mu(t) \geq 0$. If we let $\widehat{X}$ be the time it takes for $X$ to be consumed, then

$$\widehat{X} = \inf\{x \geq 0 : \int_0^x \mu(t)dt = X\}; \tag{6.2}$$

this is always finite if $\int_0^\infty \mu(t)dt = \infty$. The distribution of $\widehat{X}$ is given by

$$\begin{aligned} P(\widehat{X} \leq x) &= P(X \leq \int_0^x \mu(t)dt) \\ &= 1 - \exp(-\int_0^x \mu(t)dt). \end{aligned}$$

Now let us reverse this transformation. Suppose we are given a distribution $F$ (with $F(0) = 0$) and we want to find a clock rate $\mu(t)$ that will give $\widehat{X}$ the distribution $F$. Thus, we want to solve

$$1 - \exp(-\int_0^x \mu(t)dt) = F(x)$$

for $\mu$. It is easy to see that if $F$ has a density, $f$, then a solution exists and it is given by the *hazard rate*

$$\mu(t) = -\frac{d}{dt}\ln(1 - F(t)) = \frac{f(t)}{1 - F(t)},$$

for all $t \geq 0$ such that $F(t) < 1$. We have proved the following well-known result:

**Lemma 6.1.** If $F$ has a hazard rate $\mu$ and $X$ has a unit mean exponential distribution, then $\widehat{X}$ defined in (6.2) has distribution $F$.

**Remark.** Ordinarily, $\mu(t)dt$ is interpreted as the "probability" that a sample, $Y$, from $F$ falls in $(t, t+dt)$, given that it exceeds $t$. To see this explicitly note that, for $h > 0$,

$$P(Y \leq t+h|Y > t) = \frac{F(t+h) - F(t)}{1 - F(t)}.$$

Dividing by $h$ and letting $h$ go to zero makes the right side $f(t)/(1 - F(t)) = \mu(t)$.

Anticipating the construction of Section 6.3, we interpret Lemma 6.1 as extending the memoryless property in the following sense: If the clock time for some event is to have the distribution $F$, it can be generated by running down an exponential sample at rate $\mu(t)$. If at any point the running of this clock is suspended, the *residual length of the exponential sample is still exponential.* Hence, it could be reassigned to some other event and henceforth run down at the hazard rate of the clock distribution for that event.

Next, we generalize (6.1). Our problem can be stated as follows. Given a distribution $F$ (with $F(0) = 0$) and positive numbers $p_1, \ldots, p_n$ summing to unity, find distributions $\widehat{F}_i$, $i = 1, \ldots, n$ such that if $\widehat{X}_i$, $i = 1, \ldots, n$, are independent with distributions $\widehat{F}_i$, and if

$$\xi = \min\{i : \widehat{X}_i \leq \widehat{X}_j, \text{ all } j \neq i\},$$

then

$$P(\xi = i, \widehat{X}_\xi \leq x) = p_i F(x). \tag{6.3}$$

In words, the probability that $\widehat{X}_i$ is the smallest is $p_i$, and the distribution of the minimum is $F(x)$, regardless of which $\widehat{X}_i$ is smallest.

**Lemma 6.2.** If $F$ is continuous, the distributions defined by

$$\widehat{F}_i(x) = 1 - [1 - F(x)]^{p_i}, \; i = 1, \ldots, n,$$

satisfy (6.3).

**Proof.** Continuity of $F$ makes $P(\widehat{X}_j \geq t) = 1 - \widehat{F}_j(t)$; thus,

$$P(\xi = i, \widehat{X}_\xi \leq x) = \int_0^x \prod_{j \neq i} [1 - \widehat{F}_j(t)] d\widehat{F}_i(t)$$

$$\begin{aligned}
&= \int_0^x \prod_{j\neq i}[1-F(t)]^{p_j}\cdot p_i[1-F(t)]^{p_i-1}dF(t)\\
&= p_i\int_0^x [1-F(t)]^{1-p_i}\cdot[1-F(t)]^{p_i-1}dF(t)\\
&= p_i\int_0^x dF(t)\\
&= p_iF(x).
\end{aligned}$$

□

Lemma 6.2 allows us to "split" events: If event $\alpha$ has clock distribution $F_\alpha$, and if for some state $s$ with $\alpha \in \mathcal{E}(s)$ there are exactly $n$ states $s_1, \ldots, s_n$ with $p_i \equiv p(s_i; s, \alpha) > 0$, $i = 1, \ldots, n$, then to each $s_i$ we can assign a *potential* clock time $\widehat{X}_{\alpha,i}$ with distribution

$$\widehat{F}_{\alpha,i} = 1 - [1 - F_\alpha]^{p_i}.$$

If $\widehat{X}_{\alpha,j}$ is the smallest of these, it becomes the *actual* clock time for $\alpha$, and the transition triggered by $\alpha$ is to $s_j$. Lemma 6.2 ensures that the actual clock time has distribution $F_\alpha$ and that the next state has distribution $p(\cdot; s, \alpha)$, independent of the actual clock time.

Combining the two lemmas, we can generate each $\widehat{X}_{\alpha,i}$ by running down an exponential $X_i$ at the hazard rate $\widehat{\mu}_{\alpha,i}$ of $\widehat{F}_{\alpha,i}$. If $\widehat{X}_{\alpha,j}$ runs out first then the *residual* $X_i$'s, $i \neq j$ are exponentially distributed and may be reassigned to other transitions. When $\widehat{X}_{\alpha,j}$ runs out, the residual life of $X_i$, $i \neq j$, is given by

$$X_i - \int_0^{\widehat{X}_{\alpha,j}} \widehat{\mu}_{\alpha,i}(t)dt.$$

**Remark.** In Lemma 6.2, if $F$ has a density then so does every $\widehat{F}_i$; thus, if $F$ has a hazard rate, $\mu$, then every $\widehat{F}_i$ has a hazard rate $\widehat{\mu}_i$. It is given by

$$\begin{aligned}
\widehat{\mu}_i(t) = \frac{\widehat{F}_i'(t)}{1-\widehat{F}_i(t)} &= \frac{p_i[1-F(t)]^{p_i-1}F'(t)}{[1-F(t)]^{p_i}}\\
&= p_i\frac{F'(t)}{[1-F(t)]} = p_i\mu(t).
\end{aligned}$$

Thus, for all $t$ and all $i$ and $j$, $\widehat{\mu}_i(t)/\widehat{\mu}_j(t) = p_i/p_j$; and the distributions defined by Lemma 6.2 are called a *proportional hazards* family.

With these results in hand, we can chart the way for a new construction of a GSMP. We impose a condition—different from (C2)—which, like (CM), requires that every pair of states with a common immediate predecessor have a common immediate successor; but "successor" and "predecessor" must now be understood in a GSMP sense. We generate transitions by splitting events and time-changing exponentials: With each possible transition out of a state, $s$, due to an event, $\alpha$, to another state, $s'$, we associate an exponential potential clock time, $X$. This clock is run at rate $p(s'; s, \alpha)\mu_\alpha(t)$. If it is the first to run out, we make a transition to $s'$; otherwise, following the transition to, say, $s''$ we reassign the residual time on $X$ to the common successor of $s'$ and $s''$. Under this construction, when the order in which two clocks run out changes, the state reached remains the same.

## 6.2 The Parametric Case

Before carrying out the construction, we consider the implications of Lemma 6.1 when the distribution $F$ depends on a parameter $\theta$. The relations we derive are fundamental building blocks for the derivative estimation algorithm that follows from the new construction.

It is convenient to think of $\mu(t)$ as the age-dependent rate of some service mechanism, and the unit mean exponential $X$ as the intrinsic *work requirement* brought by some job. (We call $\mu$ age-dependent rather than time-dependent because the argument, $t$, is not absolute time, but rather the time since the beginning of the current service interval.) In this setting, $\widehat{X}$ in (6.2) is the corresponding actual service time: it is the time it takes the server to complete $X$ units of work at rate $\mu(t)$. Suppose now that $F$ (hence $\mu$) depends on a parameter $\theta$, and consider

**Definition 6.1.** If $F(\cdot, \theta)$ has hazard rate $\mu(\cdot, \theta)$, define the *shifted cumulative hazard function*

$$M(c, t, \theta) = \int_0^c \mu(t + u, \theta) du. \tag{6.4}$$

If $M(\infty, t, \theta) = \infty$, define the *inverse* shifted cumulative hazard function by

$$M^{-1}(x, t, \theta) = \inf\{y \geq 0 : \int_0^y \mu(t + u, \theta) du = x\}. \tag{6.5}$$

We interpret $M(c, t, \theta)$ as the work done in $c$ time units as the service time ages from $t$ to $t + c$, and $M^{-1}(x, t, \theta)$ as the time it takes to complete $x$ units of work if the server starts working at rate $\mu(t, \theta)$. The function $M(\cdot, 0, \theta)$ is the ordinary cumulative hazard function for $F(\cdot, \theta)$. Note that (dropping the dependence on $\theta$) (6.2) implies that $X = M(\widehat{X}, 0)$; $X$ is the hazard accumulated over $[0, \widehat{X}]$.

### 6.2.1 Derivatives

Consider the change in the work done over an interval $[0, c]$ (starting at rate $\mu(0, \theta)$) under a perturbation in $\theta$. Assuming regularity conditions (spelled out below) the work done changes at rate

$$\frac{\partial}{\partial \theta} M(c, 0, \theta) = \int_0^c \frac{\partial}{\partial \theta} \mu(u, \theta) du.$$

Similarly, the derivative of the time it takes to complete a specified amount, $x$, of work (again, starting at rate $\mu(0, \theta)$) is $\partial_\theta M^{-1}(x, 0, \theta)$.

Suppose, now, that at time zero the server begins work on a requirement $X_1$; but at time $t > 0$, before the server completes this work, it is given a different work requirement $X_2$. This could be, for example, the *residual* work from a different server, which is reassigned upon a transition; it would still have a unit mean exponential distribution. When the work assignment changes, the server continues to work at the same age-dependent rate, in effect unaware of the change. If no other changes occur, the server completes work at time $\tau$ given by $\tau = t + M^{-1}(X_2, t, \theta)$. If only $\mu$ and $F$ depend on $\theta$, then

$$\frac{d\tau}{d\theta} = \frac{\partial}{\partial \theta} M^{-1}(X_2, t, \theta).$$

If $\mu(s, \theta) \equiv \mu(\theta)$ is age-independent, then $M^{-1}(x, t, \theta) = x/\mu(\theta)$, so this becomes

$$\frac{d\tau}{d\theta} = -\frac{X_2}{\mu^2(\theta)} \mu'(\theta) = -\frac{\tau - t}{\mu(\theta)} \mu'(\theta),$$

which is familiar as the perturbation generated via a small change in the rate of an exponential random variable; see equation (5.17) of Chapter 5 and (1.14) of Chapter 1.

More generally, let $\mu$ be age-dependent and suppose that the switch point $t$ and the work requirement $X_2$ also depend on $\theta$. For example, $t$ could be a service completion epoch at another server whose rate also depends on $\theta$; and if $X_2$ is the residual requirement at yet another server, then increasing $t$ decreases $X_2$. In this case, the chain rule yields

$$\frac{d\tau}{d\theta} = \frac{\partial}{\partial x} M^{-1}(X_2, t, \theta) \frac{dX_2}{d\theta} + \frac{\partial}{\partial t} M^{-1}(X_2, t, \theta) \frac{dt}{d\theta} + \frac{\partial}{\partial \theta} M^{-1}(X_2, t, \theta). \quad (6.6)$$

In the same way, if we look at $M(c, t, \theta)$ and allow $c$ and $t$ to depend on $\theta$ we get

$$\frac{d}{d\theta} M(c, t, \theta) = \frac{\partial}{\partial c} M(c, t, \theta) \frac{dc}{d\theta} + \frac{\partial}{\partial t} M(c, t, \theta) \frac{dt}{d\theta} + \frac{\partial}{\partial \theta} M(c, t, \theta). \quad (6.7)$$

We will justify (6.6) and (6.7) and also give more explicit expressions for the partial derivatives.

It is convenient to have expressions for the shifted cumulative hazard and its inverse in terms of $F$. Since $\mu(t,\theta) = -\partial_t \ln(1-F(t,\theta))$, it is easy to verify that

$$M(c,t,\theta) = -\ln\left(\frac{1-F(t+c,\theta)}{1-F(t,\theta)}\right). \tag{6.8}$$

Inverting this, we find that

$$M^{-1}(x,t,\theta) = F^{-1}\left(1-e^{-x}(1-F(t,\theta)),\theta\right) - t, \tag{6.9}$$

where

$$F^{-1}(y,\theta) = \inf\{x \geq 0 : F(x,\theta) \geq y\}.$$

To ensure differentiability, we need some conditions on the family $\{F(\cdot,\theta), \theta \in \Theta\}$ of distributions. The following set of hypotheses is particularly convenient:

**(H).**(i) $F$ is $C^1$ in $(x,\theta)$ on $(0,\infty)\times\Theta$;
(ii) $F(0,\theta)=0$ and the density $f(x,\theta)$ is strictly positive for all $0<x<\infty$ and all $\theta\in\Theta$;
(iii) for all $x$, $f(x,\theta)$ is $C^1$ in $\theta$.

These conditions are sufficient to ensure that $\mu$ exists, is strictly positive, and is continuously differentiable throughout $(0,\infty)\times\Theta$. If we merely assume that $F$ is differentiable in $\theta$ we have

**Definition 6.2.** The *parametric hazard rate* of $F$ is the function

$$\mu^*(t,\theta) = \frac{\partial_\theta F(t,\theta)}{1-F(t,\theta)}.$$

The parametric hazard rate, $\mu^*$, is "dual" to the ordinary hazard rate $\mu$: it is the ratio of a partial derivative of $F$ to $(1-F)$, but the derivative in the numerator is with respect to $\theta$ rather than $t$. Intuitively, $\mu^*(t,\theta)$ can be thought of as the rate at which $X(\theta)$ crosses level $t$ (downward) under an increase in $\theta$. (Compare this with the interpretation of $\mu$ in the remark following Lemma 6.1.) This intuition is exact when $X(\theta)$ is generated by inversion from a uniform $U$ and $F$ is increasing in $\theta$:

$$\begin{aligned}\lim_{h\downarrow 0}\frac{1}{h}P(X(\theta+h)\leq t|X(\theta)>t) &= \lim_{h\downarrow 0}\frac{1}{h}P(F^{-1}(U,\theta+h)\leq t|F^{-1}(U,\theta)>t)\\ &= \lim_{h\downarrow 0}\frac{1}{h}P(U\leq F(t,\theta+h)|U>F(t,\theta))\\ &= \lim_{h\downarrow 0}\frac{1}{h}\frac{F(t,\theta+h)-F(t,\theta)}{1-F(t,\theta)} = \mu^*(t,\theta).\end{aligned}$$

If the order of differentiation (of $F$ with respect to $t$ and $\theta$) can be interchanged, then $\partial_\theta \mu = \partial_t \mu^*$.

**Proposition 6.1.** Under condition (H), if $x$, $t$, and $c$ are differentiable in $\theta$ then so are $M$ and $M^{-1}$. Their derivatives are given by

$$[M(c,t,\theta)]' = \mu(t+c,\theta)\frac{dc}{d\theta} + [\mu(t+c,\theta) - \mu(t,\theta)]\frac{dt}{d\theta} + [\mu^*(t+c,\theta) - \mu^*(t,\theta)],$$

and, with $y = M^{-1}(x,t,\theta)$,

$$[M^{-1}(x,t,\theta)]' = \frac{1}{\mu(t+y,\theta)}\left\{\frac{dx}{d\theta} - [\mu(t+y,\theta) - \mu(t,\theta)]\frac{dt}{d\theta} - [\mu^*(t+y,\theta) - \mu^*(t,\theta)]\right\}.$$

**Proof.** We begin by noting that

$$\begin{aligned}\int_0^c \partial_\theta \mu(t+u,\theta)du &= \int_0^c \partial_u \mu^*(t+u,\theta)du \\ &= \mu^*(t+c,\theta) - \mu^*(t,\theta). \qquad (6.10)\end{aligned}$$

The interchange of derivatives is permitted because (H) implies that the function $\ln(1 - F(s,\theta))$ is continuously differentiable in $(s,\theta)$.

We now consider $M$. From (6.8) it is clear that the conditions on $F$ make $M$ continuously differentiable. The three terms in the expression for $M'$ correspond to the three partial derivatives in (6.7). The first two follow readily from differentiation of (6.8). For the third term,

$$\begin{aligned}\frac{\partial}{\partial\theta}M(c,t,\theta) &= \frac{\partial}{\partial\theta}\int_0^c \mu(t+u,\theta)du \\ &= \int_0^c \partial_\theta \mu(t+u,\theta)du \\ &= \mu^*(t+c,\theta) - \mu^*(t,\theta).\end{aligned}$$

Bringing the derivative inside the integral is permissible because, under (H), $\mu(s,\theta)$ is $C^1$ in $\theta$ for all $s$.

Turning to $M^{-1}$, (H) and (6.9) imply differentiability. To evaluate the derivative, we use implicit differentiation. Let $y = M^{-1}(x,t,\theta)$; then

$$\int_0^{y(\theta+h)} \mu(t(\theta+h)+u,\theta+h)du - \int_0^{y(\theta)} \mu(t(\theta)+u,\theta)du = x(\theta+h) - x(\theta).$$

Dividing both sides by $h$, letting $h$ go to zero and applying the chain rule, we get

$$\mu(t+y,\theta)y' + [\mu(t+y,\theta) - \mu(t,\theta)]t' + \int_0^y \partial_\theta \mu(t+u,\theta)du = x'.$$

Solving for $y'$ and substituting according to (6.10) we get (6.6). □

**Remark.** An interesting special case of Proposition 6.1 occurs when $t$ is identically zero and $x$ is independent of $\theta$. This is essentially the setting of Lemma 6.1, but with the parameter $\theta$ included. Let $X$ have a unit mean exponential distribution and let $\widehat{X}(\theta) = M^{-1}(X, 0, \theta)$; $\widehat{X}$ depends on $\theta$ only through $\mu(t,\theta)$. As before, $\widehat{X}(\theta)$ is the time it takes the server to complete $X$ units of work, working at rate $\mu(t,\theta)$. From Proposition 6.1 we get

$$\begin{aligned}\frac{d\widehat{X}}{d\theta} &= -\frac{1}{\mu(\widehat{X},\theta)}[\mu^*(\widehat{X},\theta) - \mu^*(0,\theta)] \\ &= -\frac{1}{\mu(\widehat{X},\theta)}\left[\frac{\partial_\theta F(\widehat{X},\theta)}{1-F(\widehat{X},\theta)} - \frac{\partial_\theta F(0,\theta)}{1-F(0,\theta)}\right].\end{aligned}$$

Since $F(0,\theta) \equiv 0$, $\partial_\theta F(0,\theta) = 0$. Upon substitution for $\mu$ this simplifies to

$$\begin{aligned} &= -\frac{1-F(\widehat{X},\theta)}{\partial_x F(\widehat{X},\theta)} \cdot \frac{\partial_\theta F(\widehat{X},\theta)}{1-F(\widehat{X},\theta)} \\ &= -\frac{\partial_\theta F(\widehat{X},\theta)}{\partial_x F(\widehat{X},\theta)},\end{aligned}$$

which corresponds to the conclusion of Theorem 1.3 of Chapter 1 for the sensitivity to $\theta$ of a sample from $F(x,\theta)$, based on *inversion*. Here, we have derived the same expression without reference to inversion.

## 6.2.2 Example Distributions

Because the functions $\mu$, $\mu^*$, $M$ and $M^{-1}$ are central to the derivative estimation algorithm of Section 6.4, it is useful to have a few examples of just what these functions look like in a few simple cases. Four of the simplest distributions are the exponential, the Erlang, the hyperexponential, and the Weibull. We now examine these.

### The Exponential Distribution

The results of Section 6.2.1 are most familiar—if trivial—in the exponential case. Let us take $F(t,\theta) = 1 - \exp(-\mu(\theta)t)$, noting that $\mu(\theta)$ is indeed the (age-independent) hazard rate of $F(\cdot,\theta)$. Since an exponential distribution is completely characterized by its mean $(1/\mu(\theta))$ this is the only way to parameterize a family of exponentials.

Differentiation shows that

$$\mu^*(t,\theta) = \frac{\partial_\theta F(t,\theta)}{1-F(t,\theta)} = \frac{\mu'(\theta)te^{-\mu(\theta)t}}{e^{-\mu(\theta)t}} = \mu'(\theta)t,$$

so $\mu^*$ is linear in $t$ for each $\theta$. Also,

$$M(c,t,\theta) = \mu(\theta)c \text{ and } M^{-1}(x,t,\theta) = \frac{x}{\mu(\theta)}.$$

If $c$ and $x$ are differentiable in $\theta$, $[M(c,t,\theta)]' = \mu(\theta)c' + \mu'(\theta)c = \mu(\theta)c' + [\mu^*(t+c,\theta) - \mu^*(t,\theta)]$, and, with $y = M^{-1}(x,t,\theta)$,

$$\begin{aligned}
[M^{-1}(x,t,\theta)]' &= \frac{1}{\mu(\theta)}\frac{dx}{d\theta} - \frac{x}{\mu^2(\theta)}\mu'(\theta) \\
&= \frac{1}{\mu(\theta)}\left\{\frac{dx}{d\theta} - y\mu'(\theta)\right\} \\
&= \frac{1}{\mu(\theta)}\left\{\frac{dx}{d\theta} - [\mu^*(t+y,\theta) - \mu^*(t,\theta)]\right\}.
\end{aligned}$$

These, of course, coincide with Proposition 6.1. The absence of $dt/d\theta$ in these expressions results from the cancellation of the coefficients in Proposition 6.1 in the age-independent case: $\mu(t+c,\theta) = \mu(t,\theta)$, $\mu(t+y,\theta) = \mu(t,\theta)$.

### The Erlang Distribution

An Erlang is the distribution of a convolution of exponential random variables with the same mean. In the case of a convolution of two exponentials with mean $\lambda^{-1}(\theta)$, it becomes

$$F(t,\theta) = 1 - e^{-\lambda(\theta)t} - \lambda(\theta)te^{-\lambda(\theta)t},$$

with density $f(t,\theta) = \lambda^2(\theta)t\exp(-\lambda(\theta)t)$. This makes the hazard rate

$$\mu(t,\theta) = \frac{\lambda^2(\theta)t}{1+\lambda(\theta)t},$$

and

$$\mu^*(t,\theta) = \frac{\lambda(\theta)\lambda'(\theta)t^2}{1+\lambda(\theta)t}.$$

Straightforward differentiation verifies that $\partial_\theta\mu = \partial_t\mu^*$. As functions of $t$, $\mu(\cdot,\theta)$ increases monotonically from zero to $\lambda(\theta)$ and $\mu^*(\cdot,\theta)$ increases from zero to infinity.

Using (6.8), we find that

$$\begin{aligned} M(c,t,\theta) &= -\ln\left(\frac{1-F(t+c,\theta)}{1-F(t,\theta)}\right) \\ &= -\ln\left(\frac{e^{-(t+c)\lambda(\theta)}[1+(t+c)\lambda(\theta)]}{e^{-t\lambda(\theta)}[1+t\lambda(\theta)]}\right) \\ &= c\lambda(\theta) - \ln\left(\frac{1+(t+c)\lambda(\theta)}{1+t\lambda(\theta)}\right). \end{aligned}$$

However, (6.9) does not yield an explicit expression for $M^{-1}(x,t,\theta)$ because we have no expression for $F^{-1}$. This, unfortunately, is a common problem. It is the same problem faced by the inversion method of generating random variables, where it is sometimes resolved by approximating $F^{-1}$ with a known function. An approximation to $F^{-1}$ yields an approximation to $M^{-1}$ via (6.9).

### The Hyperexponential Distribution

A random mixture of exponentials has a hyperexponential distribution. When $n$ exponentials with means $\lambda_i^{-1}(\theta)$, $i=1,\ldots,n$, are mixed with probabilities $p_i(\theta)$, $i=1,\ldots,n$, $(p_1(\theta)+\cdots+p_n(\theta)=1)$ the distribution is

$$F(t,\theta) = 1 - \sum_{i=1}^{n} p_i(\theta)e^{-\lambda_i(\theta)t}.$$

The corresponding density is

$$f(t,\theta) = \sum_{i=1}^{n} p_i(\theta)\lambda_i(\theta)e^{-\lambda_i(\theta)t}.$$

The functions $\mu$ and $\mu^*$ are given by

$$\mu(t,\theta) = \frac{\sum_{i=1}^{n} p_i(\theta)\lambda_i(\theta)e^{-\lambda_i(\theta)t}}{\sum_{i=1}^{n} p_i(\theta)e^{-\lambda_i(\theta)t}}$$

and

$$\mu^*(t,\theta) = -\frac{\sum_{i=1}^{n}[p_i'(\theta) - p_i(\theta)\lambda_i'(\theta)t]e^{-\lambda_i(\theta)t}}{\sum_{i=1}^{n} p_i(\theta)e^{-\lambda_i(\theta)t}}.$$

For each $\theta$, $\mu(\cdot,\theta)$ decreases monotonically from $\sum_i p_i(\theta)\lambda_i(\theta)$ to $\lambda_{[1]}(\theta)$, where $\lambda_{[1]}(\theta)$ is the smallest of the $\lambda_i(\theta)$'s. In contrast, $\mu^*(0,\theta)=0$ and $\mu^*(\infty,\theta) = \lambda_{[1]}'(\theta)$, but $\mu^*(\cdot,\theta)$ need not be monotonic.

Directly from (6.8) we get

$$M(c,t,\theta) = -\ln\left(\frac{\sum_{i=1}^{n} p_i(\theta)e^{-\lambda_i(\theta)(t+c)}}{\sum_{i=1}^{n} p_i(\theta)e^{-\lambda_i(\theta)t}}\right).$$

However, since we have no closed-form expression for $F^{-1}$, we do not have one for $M^{-1}$ either.

**The Weibull Distribution**

The Weibull distribution, parameterized by $\gamma(\theta)$, is given by

$$F(t,\theta) = 1 - \exp(-t^{\gamma(\theta)}).$$

This distribution arises frequently in reliability and in the theory of order statistics. Its density is

$$f(t,\theta) = \gamma(\theta)t^{\gamma(\theta)-1}\exp(-t^{\gamma(\theta)}),$$

so

$$\mu(t,\theta) = \gamma(\theta)t^{\gamma(\theta)-1} \quad \text{and} \quad \mu^*(t,\theta) = \gamma'(\theta)t^{\gamma(\theta)} \ln t.$$

Using (6.8) it is easy to see that

$$M(c,t,\theta) = (t+c)^{\gamma(\theta)} - t^{\gamma(\theta)}.$$

We also have an explicit expression for $M^{-1}$. Since

$$F^{-1}(x,\theta) = [-\ln(1-x)]^{1/\gamma},$$

substituting in (6.9) we get

$$M^{-1}(x,t,\theta) = (x+t^{\gamma})^{1/\gamma} - t.$$

When $\gamma \equiv 1$, the Weibull reduces to an exponential and $M^{-1}(x,t,\theta) \equiv x$.

## 6.3 The Hazard Rate Construction

We now apply the results of Section 6.1 to obtain a construction of a GSMP. We start from a sequence of exponential, unit mean random variables representing potential clock times. The construction is quite similar to that of Chapter 5 so we will not spell it out in as much detail. Instead, we give an outline of the construction, stressing the differences between the current GSMP setting and the ordinary Markov case of Chapter 5.

The results of Chapter 5 relied on the common successor condition (CM) for Markov chains, so we first need an analog for GSMPs. This is given by

**(CG). Common Successor Condition for GSMPs.** For all states $s_1, s_2 \in$ **S**, if there is a state $s$ and events $\alpha, \beta \in \mathcal{E}(s)$ (possibly with $\alpha = \beta$) such that $p(s_1; s, \alpha) > 0$ and $p(s_2; s, \beta) > 0$, then there must also be a state $s'$ and events

$\alpha' \in \mathcal{E}(s_2)$ and $\beta' \in \mathcal{E}(s_1)$ (possibly with $\alpha' = \beta'$) such that $p(s'; s_2, \alpha') > 0$ and $p(s'; s_1, \beta') > 0$.

If, in general, we say that $s'$ is an immediate successor of $s$ (and $s$ is an immediate predecessor of $s'$) when there exists an $\alpha \in \mathcal{E}(s)$ such that $p(s'; s, \alpha) > 0$, then (CG) says that any pair of states with a common immediate predecessor have a common immediate successor. As discussed in Chapter 5, this condition neither implies nor is implied by condition (C2). See the discussion following Proposition 5.3.

To simplify notation for the construction, we impose an additional condition which is not strictly necessary. We require that

$$\text{for all } s \text{ and } s' \text{ there is at most one } \alpha \text{ such that } p(s'; s, \alpha) > 0. \tag{6.11}$$

This allows us to define (when $s'$ is indeed an immediate sucessor of $s$)

$$\varepsilon_s(s') = \text{the unique } \alpha \text{ for which } p(s'; s, \alpha) > 0.$$

We can also refer directly to a transition $(s, s')$ with no ambiguity in the event that triggers it. Furthermore, if $s_1$ and $s_2$ have a common immediate predecessor, we may define

$$K(s_1, s_2) = K(s_2, s_1) = \text{a common immediate successor of } s_1, s_2.$$

without having to specify which events are involved.

To carry out the construction we need some notation. As always, $Y_n$ denotes the $n$th state, $a_n$ the $n$th event, and $\tau_n$ the epoch of the $n$th transition. However, rather than associate clock times with events, we associate *potential* clock times with individual *transitions*. Define $A(s)$ to be the immediate successors of $s$; i.e., the set of $s'$ for which there is an $\alpha \in \mathcal{E}(s)$ such that $p(s'; s, \alpha) > 0$. Then for every $s' \in A(Y_n)$ we let

$$c_n(s') = \text{potential clock time for transition } (Y_n, s').$$

We also let

$$R_n(s') = \text{scheduled occurrence of transition to } s',$$

and if $\alpha \in \mathcal{E}(Y_n)$,

$$t_n(\alpha) = \text{age of } \alpha \text{ at } n\text{th transition}.$$

The age of an event is just the time elapsed since (potential) clocks were set for it.

For each $\alpha \in \mathbf{A}$, $F_\alpha$ is (as before) the distribution of *actual* clock times for $\alpha$. If $s' \in A(s)$ and $\varepsilon_s(s') = \alpha$, let

$$F_{s,s'}(x) = 1 - [1 - F_\alpha(x)]^{p(s';s,\alpha)}.$$

As in Lemma 6.2, this "splits" the event $\alpha$ into multiple possible transitions, each of which has an associated clock time. Define $\mu_\alpha$ to be the hazard rate of $F_\alpha$ and note that $\mu_{s,s'}(t) = p(s'; s, \alpha)\mu_\alpha(t)$ is the hazard rate of $F_{s,s'}$. Let $M_{s,s'}$ and $M^{-1}_{s,s'}$ be defined from $\mu_{s,s'}$ via (6.4) and (6.5). Then if $s' \in A(Y_n)$,

$$R_n(s') = \tau_n + M^{-1}_{Y_n,s'}(c_n(s'), t_n(\varepsilon_{Y_n}(s'))).$$

In words, the clock associated with the transition $(Y_n, s')$ is scheduled to run out in the time it takes to consume $c_n(s')$ units of "work" at rate $\mu_{Y_n,s'}$ starting with an age of $t_n(\varepsilon_{Y_n}(s'))$.

We now come to the construction itself. Take as given a sequence $\{X_i, i = 1, 2, \ldots\}$ of independent, unit mean exponential random variables. Initialize by setting $Y_0$ to the initial state, setting $\tau_0 = 0$, and setting clocks for the possible transitions: for each $s' \in A(Y_0)$, draw an $X_i$ and set $c_0(s') = X_i$. Also, set $R_0(s') = M^{-1}_{Y_0,s'}(c_0(s'), 0)$ and set every $t_0(\alpha)$ to zero. Suppose for the moment that, for all $s$,

$$K(s, s_1) = K(s, s_2) \Leftrightarrow s_1 = s_2; \tag{6.12}$$

and now generate transitions as follows: The next state is the one associated with the shortest clock,

$$Y_{n+1} = \arg\min\{c_n(s') : s' \in A(Y_n)\};$$

the next event is the one associated with the transition $(Y_n, Y_{n+1})$,

$$a_{n+1} = \varepsilon_{Y_n}(Y_{n+1});$$

and the epoch of the next transition is just the scheduled time of the next transition to $Y_{n+1}$,

$$\tau_{n+1} = R_n(Y_{n+1}).$$

For *old* events $\alpha \in \mathcal{O}(Y_{n+1}; Y_n, a_{n+1})$, increase the current age by the time elapsed since the last transition—i.e., set

$$t_{n+1}(\alpha) = t_n(\alpha) + [\tau_{n+1} - \tau_n];$$

and for *new* events $\alpha \in \mathcal{N}(Y_{n+1}; Y_n, a_{n+1})$ set the age to zero, $t_{n+1}(\alpha) = 0$. For $s \in A(Y_n)$, $s \neq Y_{n+1}$ reassign the residual clock time for $(Y_n, s)$ to $(Y_{n+1}, s')$ with $s' = K(Y_{n+1}, s)$; i.e., set

$$c_{n+1}(s') = c_n(s) - M_{Y_n,s}(\tau_{n+1} - \tau_n, t_n(\varepsilon_{Y_n}(s))). \tag{6.13}$$

For any remaining $s' \in A(Y_{n+1})$, set a new clock by drawing an $X_i$ and setting

$$c_{n+1}(s') = X_i.$$

For *all* $s' \in A(Y_{n+1})$, set

$$R_{n+1}(s') = \tau_{n+1} + M^{-1}_{Y_n,s'}(c_{n+1}(s'), t_{n+1}(\varepsilon_{Y_{n+1}}(s')));$$

this is the scheduled time of the next transition to $s'$. The second term on the right is the time it takes to consume the residual clock time $c_{n+1}(s')$ working at rate $\mu_{Y_n,s}$, starting at an age of $t_{n+1}(\varepsilon_{Y_{n+1}}(s'))$.

If we drop (6.12), then it is possible in (6.13) to have multiple clocks previously assigned to transitions $(Y_n, s)$ reassigned to a single transition $(Y_{n+1}, s')$. If $m$ clocks would be so reassigned, replace (6.13) with

$$c_{n+1}(s') = \frac{1}{m}\min\{c_n(s) - M_{Y_n,s}(\tau_{n+1} - \tau_n, t_n(\varepsilon_{Y_n}(s))) : K(Y_{n+1}, s) = s', s \in A(Y_n)\}. \tag{6.14}$$

Equation (6.13) is the only one that requires comment. In reassigning the clock for $(Y_n, s)$ to $(Y_{n+1}, s')$, we want $c_{n+1}(s')$ to be the *residual* time on that clock. At the previous transition, the time on that clock was $c_n(s)$. At time $\tau_n + t$, $0 \le t < \tau_{n+1} - \tau_n$, the clock was running at rate $\mu_{Y_n,s}(t_n(\varepsilon_{Y_n}(s)) + t)$. Thus, over the interval $[\tau_n, \tau_{n+1})$ it was run down an amount

$$\int_0^{\tau_{n+1}-\tau_n} \mu_{Y_n,s}(t_n(\varepsilon_{Y_n}(s)) + t)dt = M_{Y_n,s}(\tau_{n+1} - \tau_n, t_n(\varepsilon_{Y_n}(s))).$$

Subtracting this amount from $c_n(s)$ therefore yields the residual clock time. In (6.14) we are simply taking the shortest residual clock time; we could equally well reassign all of them to $(Y_{n+1}, s')$, but this would complicate the notation.

If we set $Z_t$ equal to $Y_n$ on $[\tau_n, \tau_{n+1})$, we have

**Proposition 6.2.** $\{Z_t\}$ is the GSMP determined by $(\mathbf{S}, \mathbf{A}, \mathcal{E}, p, \{F_\alpha, \alpha \in \mathbf{A}\})$.

Proving Proposition 6.2 is a matter of verifying that our assignment of clocks produces the correct stochastic behavior. This, in turn, is a consequence of Lemmas 6.1 and 6.2, which ensure that our mechanisms for "splitting" events and consuming exponential clocks at age-dependent rates are legitimate.

Now suppose we allow the distributions $F_\alpha$, $\alpha \in \mathbf{A}$ to depend on a parameter $\theta \in \Theta$. This makes all $M_{Y_n,s}$ and $M^{-1}_{Y_n,s}$ functions of $\theta$, so changes in $\theta$ can change the timing of events. In particular, a change in $\theta$ can change the order in which clocks run out. But under our construction, right at the point at which changing $\theta$ changes the identity of, say, the $n$th clock, the $n$th and $n+1$st clocks simply change order. Moreover, right at the point at which they change order, the $n + 1$st state $Y_{n+1}$ is unchanged, and the holding time $\tau_{n+1} - \tau_n$ in the

$n$th state is zero. In other words, we have the same type of dependence on $\theta$ that we saw in Chapters 3 and 5. But as in Chapter 5, it is possible that when the $n$th and $n+1$st clocks change order, the assignment of clocks in $Y_{n+1}$ changes. Hence, we must apply the same correction we used in Section 5.2 of Chapter 5; see Cases 1 and 2 under "Construction at Perturbed Value". We will not elaborate on this correction here since it is no different from that of Chapter 5. Furthermore, this reassignment of clocks is needed only to complete the construction. It plays no role in the resulting derivative calculations.

With this construction (and some additional technical conditions) the transitions epochs $\tau_n$, $n = 1, 2, \ldots$, become continuous in the sense of Lemma 5.1; the precise statement comes in Lemma 6.3 of Section 6.4.2. From this, we obtain continuity of performance measures, just as in Lemma 5.2. However, in contrast, the epochs $T(\alpha, k)$, $k = 1, 2, \ldots$, of occurrences of a fixed event $\alpha$ may not be continuous, as they are under the construction of Chapter 2. The reason for this is that, in the current construction, when events change order residual clocks may be reassigned to different events. For example, if the $n$th and $n+1$st clocks change order, continuity of the event epochs may be preserved only by changing the identity of the $n$th and $n+1$st events. This point will be illustrated and discussed further in the case of the GI/G/1/K queue in Section 6.5.1.

## 6.4 Derivative Estimation

In this section, we translate the construction of Section 6.3 into a derivative estimation algorithm. We begin by re-introducing performance measures studied in previous chapters and evaluating their derivatives. We then give a generic algorithm for keeping track of the derivatives $d\tau_n/d\theta$ of the transition epochs. Finally, in Section 6.4.2 we show that derivative estimates obtained via this algorithm are unbiased.

### 6.4.1 A Generic Algorithm

Let $n_0$ be a positive integer and $T$ a positive real number. Let

$$L_{n_0} = \int_0^{\tau_n} f(Z_t(\theta))dt$$

and

$$L_T = \int_0^T f(Z_t(\theta))dt.$$

These performance measures inherit continuity from the $\tau_n$'s; see Lemma 6.3 and Lemma 5.2. In contrast, $L_{\alpha,k}$ (as defined in Chapter 2, Section 2.3.1) may be discontinuous because $T(\alpha, k)$ may be discontinuous.

If every clock distribution $F_\alpha$ satisfies (H), then $L_{n_0}$ and $L_T$ are almost surely differentiable at every $\theta \in \Theta$. As in Chapters 2 and 5, by first rewriting $L_{n_0}$ and $L_T$ as sums, we obtain

$$\frac{dL_{n_0}}{d\theta} = \sum_{i=0}^{n-1} f(Y_i)[\frac{d\tau_{i+1}}{d\theta} - \frac{d\tau_i}{d\theta}],$$

and

$$\frac{dL_T}{d\theta} = \sum_{i=1}^{N_T} \frac{d\tau_i}{d\theta}[f(Y_{i-1}) - f(Y_i)],$$

where $N_t$ is the number of transitions made by $Z_t$ in $(0,t]$. Thus, the problem of calculating these derivatives essentially reduces to that of calculating $\{d\tau_i/d\theta, i = 1, 2, \ldots\}$.

We now present an algorithm for simulating a GSMP, following the construction of Section 6.3, that simultaneously keeps track of the derivatives of the transition epochs. Much as in Section 5.3.2 of Chapter 5, for $s \in A(Y_n)$ let

$$S_n(s) = \{s' : K(Y_n, s') = s,\ s' \in A(Y_{n-1})\};$$

then, for $s' \in S_n(s)$, a clock previously assigned to the transition $(Y_{n-1}, s')$ is reassigned to $(Y_n, s)$ upon entry to $Y_n$. With this notation, we have

**Algorithm 6.1: Perturbation Propagation via Hazard Rates**

0. Initialization.
   a. $\tau_0 := 0$;
   b. Set $Y_0$ to initial state;
   c. For every $\alpha \in \mathcal{E}(Y_0)$
      $t_0(\alpha) := 0$;
      $t'_0(\alpha) := 0$;
   d. For every $s \in A(Y_0)$
      Generate exponential $X$, $c_0(s) := X$;
      $c'_0(s) := 0$;
      $R_0(s) := M^{-1}_{Y_0,s}(c_0(s), 0, \theta)$;
      $R'_0(s) := [M^{-1}_{Y_0,s}(c_0(s), 0, \theta)]'$;

1. Next Transition.
   a. Find $s \in A(Y_n)$ with smallest $c_n(s)$, $Y_{n+1} := s$;
   b. $a_{n+1} := \varepsilon_{Y_n}(Y_{n+1})$;
   c. $\tau_{n+1} := R_n(Y_{n+1})$;
   d. $\tau'_{n+1} := R'_n(Y_{n+1})$;

e. For every $\alpha \in \mathcal{E}(Y_{n+1}) \cap (\mathcal{E}(Y_n) - \{a_{n+1}\})$
$t_{n+1}(\alpha) := t_n(\alpha) + (\tau_{n+1} - \tau_n);$
$t'_{n+1}(\alpha) := t'_n(\alpha) + (\tau'_{n+1} - \tau'_n);$
f. For every $\alpha \in \mathcal{E}(Y_{n+1}) \backslash (\mathcal{E}(Y_n) - \{a_{n+1}\})$
$t_{n+1}(\alpha) := 0;$
$t'_{n+1}(\alpha) := 0;$
g. For every $s \in A(Y_{n+1})$
If $|S_{n+1}(s)| > 0$
$s^* := \arg\min\{c_n(s') - M_{Y_n,s'}(\tau_{n+1} - \tau_n, t_n(\varepsilon_{Y_n}(s')), \theta) : s' \in S_{n+1}(s)\};$
$c_{n+1}(s) := |S_{n+1}(s)|^{-1}[c_n(s^*) - M_{Y_n,s^*}(\tau_{n+1} - \tau_n, t_n(\varepsilon_{Y_n}(s^*)), \theta)];$
$c'_{n+1}(s) := |S_{n+1}(s)|^{-1}[c'_n(s^*) - [M_{Y_n,s^*}(\tau_{n+1} - \tau_n, t_n(\varepsilon_{Y_n}(s^*)), \theta)]'];$
If $|S_{n+1}(s)| = 0$
Generate exponential $X$, $c_{n+1}(s) := X;$
$c'_{n+1}(s) := 0;$
$R_{n+1}(s) := \tau_{n+1} + M^{-1}_{Y_{n+1},s}(c_{n+1}(s), t_{n+1}(\varepsilon_{Y_{n+1}}(s)), \theta);$
$R'_{n+1}(s) := \tau'_{n+1} + [M^{-1}_{Y_{n+1},s}(c_{n+1}(s), t_{n+1}(\varepsilon_{Y_{n+1}}(s)), \theta)]';$

2. Check Stopping Condition.
If stopping condition is met, STOP; else go to 1.

This algorithm is substantially more work to implement and run than that of Section 2.3.2 of Chapter 2. Most of the steps simply carry out the construction of Section 6.3, but even these are more complicated than corresponding steps in Algorithm 2.1. Evaluation of the shifted cumulative hazard function, its inverse, and their derivatives can be computationally intensive; and Algorithm 6.1 uses many evaluations at each transition. In cases where expressions for $M$ or $M^{-1}$ are not available, implementation is not even an option. Hence, this algorithm is not as immediately applicable as those of other chapters. Some simplifications in special cases are considered in Section 6.5.

### 6.4.2 Conditions for Unbiasedness

We now verify that the construction of Section 6.3 and Algorithm 6.1 lead to unbiased derivative estimators. Rather than seek the most general conditions that permit the interchange of derivative and expectation, we strive instead for simplicity. In particular, we work with conditions close to those used in Chapter 5.

We first state the continuity result that follows from our construction. As above, fix a positive integer $n_0$ and a positive real number $T$. Let $\tilde{N}$ denote either $n_0$ or $N_T(\theta)$.

**Lemma 6.3.** Suppose $(\mathbf{S}, \mathbf{A}, \mathcal{E}, p)$ satisfies (CG) and that $\{F_\alpha, \alpha \in \mathbf{A}\}$ satisfy (H). Suppose there are constants $q_*, q^*$ and $B$, independent of $\theta$, such that for all $\alpha \in \mathbf{A}$, $\theta \in \Theta$, and $t \geq 0$

$$0 < q_* < \mu_\alpha(t, \theta) < q^* < \infty, \tag{6.15}$$

and for all $t_1, t_2 \geq 0$

$$|\mu^*_\alpha(t_2, \theta) - \mu^*_\alpha(t_1, \theta)| \leq B|t_2 - t_1|. \tag{6.16}$$

Then for any $\theta_0 \in \Theta$ the following hold with probability $1 - O(h^2)$:

**(i)** Every $\tau_i$, $i \leq \tilde{N}$, is continuous in $\theta$ throughout $(\theta_0 - h, \theta_0 + h)$.

**(ii)** For every $i \leq \tilde{N}$, at any discontinuity of $Y_i$ in $(\theta_0 - h, \theta_0 + h)$, $\tau_{i+1} = \tau_i$.

**(iii)** At any discontinuity of $N_T$ in $(\theta_0 - h, \theta_0 + h)$, $\tau_{N_T} = T$.

The conditions in (6.15) and (6.16) effectively reduce Lemma 6.3 to the setting of Lemma 5.1. The proof is essentially the same (though more complicated) so we do not repeat it. Requiring (6.15) bounds all clock distributions above and below by exponentials with rates $q_*$ and $q^*$. It ensures, in particular, that the transitions of $Z$ can be dominated by a Poisson process. Moreover, it implies that for all $\alpha$, $x$, $t$, and $\theta$,

$$q_* x \leq M_\alpha(x, t, \theta) \leq q^* x,$$

and

$$\frac{x}{q^*} \leq M_\alpha^{-1}(x, t, \theta) \leq \frac{x}{q_*}.$$

Condition (6.16) bounds the dependence of every $M_\alpha$ on $\theta$ in the sense that it makes

$$\begin{aligned} |\partial_\theta M_\alpha(x, t, \theta)| &= \left| \int_0^x \partial_\theta \mu_\alpha(t + u, \theta) du \right| \\ &= |\mu^*_\alpha(t + x, \theta) - \mu^*_\alpha(t, \theta)| \\ &\leq Bx. \end{aligned}$$

In the exponential case, $\mu_\alpha(t, \theta) \equiv \mu_\alpha(\theta)$, we get $\partial_\theta M_\alpha(x, t, \theta) = \partial_\theta \mu_\alpha(\theta) x$, so (6.16) reduces to $|\mu'_\alpha| \leq B$. This is analogous to the condition $|q'(x, y)| \leq B$ in Lemma 5.1.

From Lemma 6.3 we get the following analog of Theorem 5.1:

**Theorem 6.1.** Under the conditions of Lemma 6.3, at any $\theta \in \Theta$

$$\mathbf{E}[\frac{dL}{d\theta}] = \frac{d\mathbf{E}[L]}{d\theta},$$

where $L$ is either $L_T$ or $L_{n_0}$.

## 6.5 Examples

The construction of Section 6.3 and the algorithm of Section 6.4.1 obtain generality at the price of notational complexity, so in this section we consider concrete examples. The systems we consider are of interest in their own right, and also help clarify the main ideas of this chapter.

### 6.5.1 The GI/G/1/K Queue

As discussed in Section 3.1.3 of Chapter 3 and Section 4.2.4 of Chapter 4, the finite capacity GI/G/1/K queue violates condition (C2), so the results of Chapters 2 and 3 cannot be applied to this system. However, it does satisfy (CG), so the methods of this chapter are applicable. Furthermore, the GI/G/1/K queue has two special properties which make it a particularly simple and interesting example: it involves no transition probabilities (every $p(s'; s, \alpha)$ is zero or one), and it permits a strengthening of Lemma 6.3 to almost sure continuity.

Let us first consider (CG). To avoid confusion with the common successor function $K(\cdot, \cdot)$, for the rest of this section we use $k$ to denote the buffer capacity. Let $\mathbf{S}$ be the set of possible queue lengths $\{0, 1, \ldots, k\}$; let $\alpha$ denote arrival and $\beta$ denote service completion. If $0 < s < k$, then $s$ is an immediate predecessor of both $s-1$ and $s+1$, and also a common immediate successor: $s = K(s-1, s+1)$. We know from Section 3.1.3 of Chapter 3 that (C2) is violated starting in $s = k$. When the queue is full, an arrival followed by a departure leaves the queue length at $k$, while taking the events in the opposite order makes the queue length $k - 1$. Condition (CG) gets around this problem. State $k$ is a predecessor of both $k - 1$ and $k$ since

$$p(k-1; k, \beta) = p(k; k, \alpha) = 1;$$

but it is also a common successor since

$$p(k; k-1, \alpha) = p(k; k, \alpha) = 1.$$

Thus, (CG) is satisfied by taking $K(k, k-1) = k$. In short, the diagram

$$\begin{array}{ccc} k & \xrightarrow{\alpha} & k \\ \beta \downarrow & & \downarrow \alpha \\ k-1 & \xrightarrow{\alpha} & k \end{array}$$

satisfies (CG) though not (C2) because (CG) does not require that opposite sides correspond to the same event.

Let us investigate the implications of this structure for the construction of Section 6.3. We start from unit mean exponential random variables. These

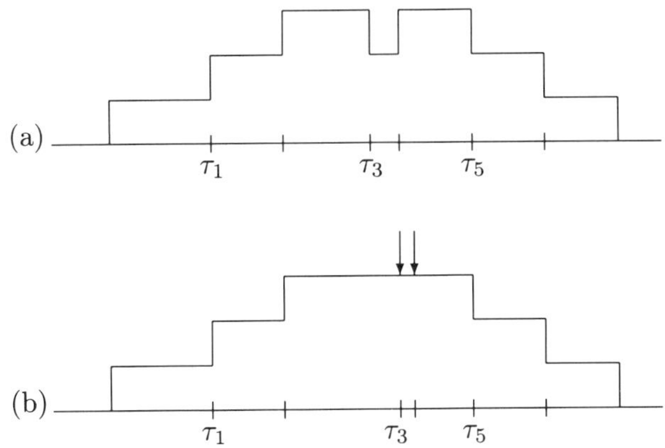


Figure 6.1: (a) Before the order change; (b) After.

are transformed to interarrival times and service times using Lemma 6.1 and the hazard rates $\mu_\alpha$ and $\mu_\beta$. Consider the $n$th transition, in the case where the current state $Y_{n-1}$ is less than $k$. If the $n$th event is an arrival, then $Y_n = Y_{n-1}+1$, and the residual clock time previously assigned to the transition $(Y_{n-1}, Y_{n-1}-1)$ is reassigned to $(Y_n, K(Y_n, Y_{n-1}-1)) = (Y_n, Y_{n-1}) = (Y_n, Y_n - 1)$. In words, the residual service time continues to be the residual service time. Similarly, if $a_n = \beta$, the residual interarrival time continues to be the residual interarrival time. This is also the case if $Y_{n-1} = k$ and $a_n = \beta$; but when $Y_{n-1} = k$ and $a_n = \alpha$, the construction calls for a different reassignment. The residual service time, previously assigned to $(Y_{n-1}, Y_{n-1} - 1) = (k, k-1)$, is reassigned to $(Y_n, K(Y_n, Y_{n-1} - 1)) = (k, K(k, k-1)) = (k, k)$. That is, *the residual service time is reassigned to the new interarrival time.* Of course, what we mean by this is that the residual time on the exponential sample $X_i$ assigned to the departure event, is reassigned to the arrival event, and transformed into an actual interarrival time via Lemma 6.1 and $\mu_\alpha$. Following this reassignment, we also draw a new sample and transform it to the residual service time.

Figure 6.1 illustrates the effect of an order change under this construction. In the figure, $k = 3$. The descending arrows indicate blocked arrivals; initially there are none. Suppose now that the service completion at $\tau_3$ is delayed to the point where the job that initially arrives at $\tau_4$ in fact arrives before the service completion. In this case, the job finds the queue full and is blocked, as indicated by the first descending arrow. At $\tau_3$, we now have a blocked arrival rather than a service completion. But under the construction described above, when an arrival is blocked, the residual service clock time is reassigned to the new interarrival time. More precisely,

$$c_3(k) = c_2(k-1) - M_\beta(\tau_3 - \tau_2, \tau_2, \theta),$$

and the next arrival is scheduled for

$$R_3(k) = \tau_3 + M_\alpha^{-1}(c_3(k), 0, \theta).$$

Just before the order change, however, the residual clock $c_2(k-1)$ was consumed by the server right at $\tau_3$; thus, just before the change

$$M_\beta(\tau_3 - \tau_2, \tau_2, \theta) = c_2(k-1).$$

This means that right at the point where the arrival occurs before the departure, $c_3(k) = 0$, and hence $R_3(k) = \tau_3$. In other words, when the departure and arrival change order, they become two blocked arrivals! (See the figure.) Changing the identity of an event in this way brings the process back to the right state, $k$, after both clocks run out:

$$k \xrightarrow{\beta} k-1 \xrightarrow{\alpha} k$$

$$k \xrightarrow{\alpha} k \xrightarrow{\alpha} k;$$

whereas merely reversing the order of events would not:

$$k \xrightarrow{\alpha} k \xrightarrow{\beta} k-1.$$

A small modification of the construction of Section 6.3 allows us to conclude even more:

**Proposition 6.3.** If $\mu_\alpha$ and $\mu_\beta$ exist, and are continuous and bounded on $(0, \infty)$, then (i), (ii) and (iii) of Lemma 6.3 hold almost surely. It follows that $L_{n_0}$ and $L_T$ are almost surely continuous in $\theta$.

We sketch the proof. The idea is to start with *two* streams $\{X_{0,i}, i = 1, 2, \ldots\}$ and $\{X_{1,i}, i = 1, 2, \ldots\}$ of unit mean exponential random variables. The streams are intermittently assigned to arrivals and service completions, but at a blocked arrival, the streams are exchanged. To verify Proposition 6.3, we need to show that when events change order, not only does the process return to the right state, but clocks are correctly assigned. In other words, we never need to correct the assignment of clocks (using Cases 1 and 2 of Section 5.3.2 of Chapter 5). Suppose that, initially, $Y_{n-1}, Y_n, Y_{n+1} = k, k-1, k$, and let $X_{i,j}$ and $X_{1-i,\ell}$ be the clocks assigned to $\alpha$ and $\beta$ respectively, just after the $n-1$st transition. Since $a_n = \beta$, just after the $n$th transition the clocks assigned are $X_{i,j}$ and $X_{1-i,\ell+1}$; and since $a_{n+1} = \alpha$, at the next transition they are $X_{i,j+1}$ and $X_{1-i,\ell+1}$.

Now suppose that under a small parameter change the arrival at the $n+$ 1st transition slips ahead of the departure at the $n$th transition. This makes $a_n = \alpha$ a blocked arrival. When we reassign clocks, we find that $X_{1-i,\ell}$ and $X_{i,j+1}$ are assigned to arrival and departure, respectively. As explained above, immediately after the order change there is another blocked arrival. Hence, at the $n+1$st transition we exchange streams again. This makes the clocks assigned to arrival and departure $X_{i,j+1}$ and $X_{1-i,\ell+1}$. This is the same assignment we had initially (i.e., before the order change), which is what we needed to check.

We end our investigation of the GI/G/1/K queue by looking in more detail at the epochs $\{T(\alpha,j), j=1,2,\ldots\}$ and $\{T(\beta,j), j=1,2,\ldots\}$ of arrivals and departures. We indicated at the end of Section 6.3 that these epochs may be discontinuous in $\theta$ even when all transition epochs $\tau_n$ are continuous. The present setting illustrates why. Consider, again, Figure 6.1. Our construction ensures that when events change order, the epochs of transitions change continuously. But to achieve this, we have to convert a service completion into an arrival. In the figure, when the arrival and service completion change order, $T(\beta,1)$ (the epoch of the first departure) jumps from $\tau_3$ to $\tau_5$. Thus, even almost sure continuity of the $\tau_n$'s does not guarantee continuity of the epochs of specific events.

## 6.5.2 Perturbing Routing Probabilities

Thus far, we have taken $\theta$ to be a parameter of the (actual) clock distributions $\{F_\alpha, \alpha \in \mathbf{A}\}$. But our construction of GSMPs via hazard rates depends on the distributions $\{F_\alpha\}$ only through the distributions $\{F_{s,s'}\}$ defined in Section 6.3 by

$$F_{s,s'}(x) = 1 - [1 - F_\alpha(x)]^{p(s';s,\alpha)}. \tag{6.17}$$

Hence, we could as well allow $p$ to depend on $\theta$ as $F_\alpha$. The construction and Algorithm 6.1 are essentially unchanged in this case; but the possibility that the routing depends on $\theta$ leads to interesting applications and some simplified expressions so we give it special attention. (Other approaches to estimating derivatives with respect to routing parameters are discussed in Gong [43] and in Vakili and Ho [80].)

For simplicity, we consider the case of a single class, closed Jackson-like network with routing matrix $(P_{ij})$. We assume that routing is state-independent, and that $P$—but not the $F_\alpha$'s—depends on $\theta$. We let $\beta_i$ denote service completion at node $i$, and represent states by vectors of queue lengths, which we denote by $s$, $s'$, etc. Thus, if $P_{ij} > 0$ and $s' = s - e_i + e_j$,

$$F_{s,s'}(x) = 1 - [1 - F_{\beta_i}(x)]^{P_{ij}}.$$

The construction is essentially the same as that described for Jackson networks in Section 5.4 of Chapter 5. In particular, we need a condition on $P$: For any nodes $j$ and $k$, if there is a node $i$ for which $P_{ij} > 0$ and $P_{ik} > 0$, then there must also be a node $i'$ for which $P_{ji'} > 0$ and $P_{ki'} > 0$. In this case, we let $K_P(j,k) = i'$, and call $i$ and $i'$ a common predecessor and a common successor of $j$ and $k$.

Given this condition on $P$, the construction proceeds from a state $s$, say, by assigning *potential* service times to every pair of nodes $i$ and $j$ for which $s_i > 0$ and $P_{ij} > 0$. If $s' = s - e_i + e_j$, then this potential service time is consumed at rate $\mu_{s,s'}$, the hazard rate of $F_{s,s'}$. If, of all the potential service times assigned in state $s$, this is the first to run out, then a service completion occurs at $i$, and a state transition is made to $s'$. Following the transition, residual potential service times for one pair of nodes may be reassigned to a new pair of nodes. The details of the reassignment (as determined by $K_P$) are exactly the same as in the Markov case; see Section 5.4 under the heading "Perturbing Routing Probabilities".

As is easily verified (see the remark at the end of Section 6.1), $\mu_{s,s'}(t) = P_{ij}\mu_{\beta_i}(t)$, where $\mu_{\beta_i}$ is the hazard rate of the actual service time distribution $F_{\beta_i}$ at node $i$. It is this relation, together with the fact that only $P$ may depend on $\theta$, that leads to special expressions. When we introduce the parameter $\theta$ in $P$, $\mu_{s,s'}$ becomes a *separable* function of $t$ and $\theta$, in that it is a product of a function that depends only on $\theta$ and one that depends only on $t$.

If $P$ is differentiable in $\theta$, then, of course, $\mu'_{s,s'} = P'_{ij}\mu_{\beta_i}$. Also, the shifted cumulative hazard for $\mu_{s,s'}$ is

$$\begin{aligned} M_{s,s'}(c,t,\theta) &= \int_0^c \mu_{s,s'}(t+u,\theta)du \\ &= P_{ij}\int_0^c \mu_{\beta_i}(t+u)du \\ &= P_{ij}M_{\beta_i}(c,t,\theta); \end{aligned} \tag{6.18}$$

so its partial derivative with respect to $\theta$ is $\partial_\theta M_{s,s'}(c,t,\theta) = P'_{ij}M_{\beta_i}(c,t)$. The inverse function is given by

$$\begin{aligned} M^{-1}_{s,s'}(x,t,\theta) &= \inf\{y \geq 0 : \int_0^y \mu_{s,s'}(t+u,\theta)du = x\} \\ &= \inf\{y \geq 0 : P_{ij}\int_0^y \mu_{\beta_i}(t+u)du = x\} \\ &= M^{-1}_{\beta_i}(x/P_{ij},t). \end{aligned}$$

If we let $y = M^{-1}_{s,s'}(x,t,\theta)$, then differentiation and Proposition 6.1 yield

$$\partial_\theta M^{-1}_{s,s'}(x,t,\theta) = \partial_x M^{-1}_{\beta_i}(x/P_{ij},t)[-x/P^2_{ij}]P'_{ij}$$

$$
\begin{aligned}
&= \frac{1}{\mu_{\beta_i}(t+y)}[-x/P_{ij}^2]P'_{ij} \\
&= \frac{-xP'_{ij}}{\mu_{s,s'}(t+y)P_{ij}}.
\end{aligned} \tag{6.19}
$$

The derivatives of $M_{s,s'}$ and $M_{s,s'}^{-1}$ can also be derived by introducing the parametric hazard rate. We have

$$
\begin{aligned}
\mu^*_{s,s'}(t,\theta) &= -\partial_\theta \ln(1 - F_{s,s'}(t,\theta)) \\
&= -\partial_\theta [P_{ij} \ln(1 - F_{\beta_i}(t))] \\
&= -P'_{ij} \ln(1 - F_{\beta_i}(t)) \\
&= P'_{ij} M_{\beta_i}(t, 0).
\end{aligned}
$$

Now using Proposition 6.1 we find that

$$
\begin{aligned}
\partial_\theta M_{s,s'}(c,t,\theta) &= \mu^*_{s,s'}(t+c,\theta) - \mu^*_{s,s'}(t,\theta) \\
&= P'_{ij}[M_{\beta_i}(t+c,0) - M_{\beta_i}(t,0)] \\
&= P'_{ij} M_{\beta_i}(c,t),
\end{aligned}
$$

which, of course, agrees with what we obtained by directly differentiating $M_{s,s'}$. For the inverse, if we let $y = M_{s,s'}^{-1}(x,t,\theta)$,

$$
\begin{aligned}
\partial_\theta M_{s,s'}^{-1}(x,t,\theta) &= -\frac{\mu^*_{s,s'}(t+y,\theta) - \mu^*_{s,s'}(t,\theta)}{\mu_{s,s'}(t+y)} \\
&= -P'_{ij} \frac{M_{\beta_i}(t+y,0) - M_{\beta_i}(t,0)}{\mu_{s,s'}(t+y)} \\
&= -\frac{P'_{ij} M_{s,s'}(y,t,\theta)}{P_{ij}\mu_{s,s'}(t+y)} \\
&= -\frac{P'_{ij} x}{P_{ij}\mu_{s,s'}(t+y)}.
\end{aligned}
$$

The third equality uses (6.18), the fourth follows from the general fact that $M^{-1}(\cdot,t,\theta)$ is, indeed, the inverse of $M(\cdot,t,\theta)$. This derivation naturally yields the same result as (6.19).

These expressions, for the case where only the routing depends on $\theta$, are much easier to implement that the general expressions of Algorithm 6.1, and form an important special case in applications.

* * *

## Notes and Comments

Hazard rates and cumulative hazard functions are the stock-in-trade of reliability theorists; Barlow and Proschan [2] is a standard reference. In a more general setting, they become intensities and compensators; see Brémaud [4]. Using intensities, the results of this chapter could be extended to cases in which the clock times are allowed to depend on the past of the process. For different applications of hazard rates and intensities in derivative estimation see Zazanis [87] and Chapter 7. The connection between Lemma 6.2 and proportional hazards was pointed out by David Yao.

As we have indicated (and as the reader will, no doubt, have noticed) the algorithm of Section 6.4.1 is far more involved than any we considered previously. Except in special cases (which include the examples of Section 6.5), implementation of the algorithm may be too computationally burdensome to be practical. An area for further investigation is finding an improved algorithm along the lines of Section 5.3.2 of Chapter 5; but the absence of the Markov property will certainly make any conditioning argument more difficult.

We have had to work hard here (in Section 6.5.2) and in Chapter 5 (Section 5.4) to obtain derivative estimates with respect to parameters of routing probabilities. In contrast, using the likelihood ratio method derivatives for routing probabilities fall out as easy special cases of the general theory; no special analysis is required. When one is interested in sensitivities with respect to a variety of parameters, it is reasonable to use different methods for different parameters. One might, for example, use likelihood ratios for routing probabilities and perturbation analysis for service rates.

# Chapter 7

# Smoothing

The verification of unbiasedness in Chapters 3, 5 and 6 followed the same pattern: we first established continuity of performance measures, then used continuity in justifying the interchange of differentiation and expectation. In each case, continuity hinged on showing that changes in the order of events (broadly construed) were in some sense negligible. In this chapter, we consider the problem of estimating derivatives when the performance measures of interest are discontinuous in the parameter of differentiation, $\theta$. More precisely, we consider performance measures whose *samples* are discontinuous, but whose expectations are differentiable. In this setting, the interchange of differentiation and expectation typically fails—the stochastic derivative of the performance measure is not an unbiased estimator of the derivative of the expected performance. The situation we consider is, in a sense, the extreme opposite of that considered in previous chapters. We focus on performance measures for which changes in the order of events are *not* negligible; indeed, these performance measures change values *only* when events change order.

## 7.1 Discontinuities and Conditioning

### 7.1.1 Two Types of Jumps

There are two ways in which a performance measure $R$, say, could fail to be continuous in $\theta$ for fixed random outcomes $\omega$. The discontinuities may arise because the underlying process $\{Z_t(\theta)\}$ (of which $R$ is a functional) is not "continuous", or they may be inherent to $R$. To put it another way, if a pair $(\theta, \omega)$ determines a sample path of $Z$, and if a sample path of $Z$ determines the value of the performance measure $R$, then $R$ is the composition of two maps,

$$(\theta, \omega) \longrightarrow \{Z_t(\theta, \omega)\} \longrightarrow R(\theta, \omega),$$

and either of these is potentially a source of discontinuities.[1]

Typically, the sample paths of $Z$ take values in an infinite dimensional space, so "continuity" of the first mapping is a delicate matter. At a minimum, we would like the transition epochs $\{\tau_n, n = 1, 2, \ldots\}$ to be continuous in $\theta$. Conditions (C2), (CM) and (CG) all imply that $Z$ is "sufficiently continuous" in $\theta$ for a certain class of performance; see Theorem 3.1, Lemma 5.2 and Lemma 6.3. But if the discontinuity is in the second mapping, no conditions on $Z$ can ensure that the composition is continuous.

Suppose, for example, that $R$ is defined by an indicator function $R(\theta) = \mathbf{1}\{Z_{t_0}(\theta) = s\}$, where $t_0$ is a fixed time and $s$ is a fixed state. Then, as a function of $\theta$, $R$ can only be constant or discontinuous. Wherever it exists, the stochastic derivative $dR(\theta)/d\theta$ *can only be zero.* On the other hand, the expectation $\mathbf{E}[R(\theta)]$ is just $P(Z_{t_0}(\theta) = s)$ which may well be differentiable and non-constant in $\theta$. Thus, we have a case where

$$\mathbf{E}[\frac{dR}{d\theta}] \neq \frac{d\mathbf{E}[R]}{d\theta}. \tag{7.1}$$

In fact, this will always be the case if $R$ takes values in a discrete set (in this case $\{0, 1\}$) and $\mathbf{E}[R(\theta)]$ is non-constant.

In this chapter, we only consider cases where the discontinuities are in the performance measure. The same ideas can, in principle, be applied to the harder problem of discontinuities in the process, but a complete theory for that case is not yet available. For our purposes, the underlying process (a GSMP) is "sufficiently continuous" in $\theta$ if the commuting condition (C2) holds. Analogous methods could be based on the common successor conditions (CM) and (CG).

We get around (7.1) by introducing an intermediate step: we take a conditional expectation before differentiating. The idea is that if $\mathbf{E}[R(\theta)]$ is continuous in $\theta$ then for a suitably chosen condition $\mathcal{H}$, $\mathbf{E}[R(\theta + h)|\mathcal{H}]$ may be continuous in a neighborhood of $h = 0$. If it is continuous, then it is reasonable to expect the derivative with respect to $h$, $\partial_h \mathbf{E}[R(\theta)|\mathcal{H}]$, evaluated at $h = 0$, to be an unbiased estimator of $\mathbf{E}[R(\theta)]'$. Taking a conditional expectation *smooths* the dependence on $\theta$. This is the method of *smoothed perturbation analysis* introduced in Gong [42] and Gong and Ho [45]. The formulation here is closer to that of Glasserman and Gong [33].

---

[1]Mathematics distinguishes two types of discontinuities: *simple,* or jump, discontinuities, and *oscillatory* discontinuities; see, e.g., Rudin [70]. This distinction has nothing to do with the one we make. All our discontinuities are jumps.

### 7.1.2 The Finite Dimensional Setting

Before we launch into smoothing for GSMPs, it may be helpful to introduce the basic ideas in a simpler setting. We replace the family of stochastic processes $\{Z_t(\theta), t \geq 0, \theta \in \Theta\}$ with a family of random vectors $\{\underline{X}(\theta), \theta \in \Theta\}$ taking values in $\mathbf{R}^d$, $d < \infty$. In this case, it is clear what we mean if we require that $\underline{X}$ be almost surely continuous in $\theta$ throughout $\theta$. We can think of a function $f : \mathbf{R}^d \to \mathbf{R}$ as a performance measure for $\underline{X}$. As an example, let $\underline{X}(\theta)$ be the state of stochastic network; the $i$th component of $\underline{X}$ is the (random) capacity of the $i$th arc. Then $f$ could be, e.g., the flow between a specified set of nodes. If both $f$ and $\underline{X}$ are differentiable, then the stochastic derivative of the composite map is $\nabla f \cdot \underline{X}'(\theta)$. Under additional regularity conditions (bounds on the derivatives) the expectation of this derivative is just $d\mathbf{E}[f(\underline{X}(\theta))]/d\theta$.

Suppose now that $f(\underline{X}(\theta))$ fails to be almost surely continuous in $\theta$. In this (finite dimensional) setting, the dichotomy proposed in Section 7.1.1 is clear; if the composite map $f(\underline{X}(\theta))$ is discontinuous, then either $\underline{X}$ or $f$ is discontinuous. Suppose that $\underline{X}$ is almost surely $C^1$ but that $f$ is piecewise constant and takes on only finitely many values. Let $\mathbf{R}^d$ be partitioned into sets $A_1, \ldots, A_n$ such that $f$ is constant and equal to $y_i$ throughout $A_i$, $i = 1, \ldots, n$. Then $df(\underline{X}(\theta))/d\theta$ is zero wherever it exists. Thus, in this case, the stochastic derivative is typically not an unbiased derivative estimator.

Let $\xi(\theta)$ be the index of the set that contains $\underline{X}(\theta)$; i.e., $\underline{X} \in A_\xi$. We obtain a smoothed derivative estimator by conditioning on $\xi$, as follows:

$$\begin{aligned}
\frac{d\mathbf{E}[f(\underline{X}(\theta))]}{d\theta} &= \lim_{h\downarrow 0} \frac{\mathbf{E}[f(\underline{X}(\theta+h)) - f(\underline{X}(\theta))]}{h} \\
&= \lim_{h\downarrow 0} \mathbf{E}\left[\frac{\mathbf{E}[f(\underline{X}(\theta+h)) - f(\underline{X}(\theta))|\xi(\theta)]}{h}\right] \\
&= \lim_{h\downarrow 0} \frac{1}{h}\mathbf{E}\Bigg[\sum_{i=1}^{n} P(\underline{X}(\theta+h) \in A_i|\xi(\theta)) \\
&\qquad \mathbf{E}[f(\underline{X}(\theta+h)) - f(\underline{X}(\theta))|\xi(\theta), \underline{X}(\theta+h) \in A_i]\Bigg]. \qquad (7.2)
\end{aligned}$$

Now let

$$Q(A_i|\xi) = \lim_{h\downarrow 0} \frac{1}{h} P(\underline{X}(\theta+h) \in A_i|\xi(\theta)), \qquad (7.3)$$

and

$$\begin{aligned}
\Delta(A_i|\xi) &= \lim_{h\downarrow 0} \mathbf{E}[f(\underline{X}(\theta+h)) - f(\underline{X}(\theta))|\xi(\theta), \underline{X}(\theta+h) \in A_i] \\
&= \mathbf{E}[f(\underline{X}(\theta+h)) - f(\underline{X}(\theta))|\xi(\theta), \underline{X}(\theta+h) \in A_i]. \qquad (7.4)
\end{aligned}$$

(The limit in $\Delta$ is superfluous because $f$ is constant on the $A_i$'s.) If the limit and the outer expectation in (7.2) can be interchanged, we get

$$\frac{d\mathbf{E}[f(\underline{X}(\theta))]}{d\theta} = \mathbf{E}\left[\sum_{i=1}^{n} Q(A_i|\xi)\Delta(A_i|\xi)\right]$$

and

$$\sum_{i=1}^{n} Q(A_i|\xi)\Delta(A_i|\xi) \tag{7.5}$$

yields an unbiased derivative estimator. We think of $Q(A_i|\xi)$ as the conditional "rate" at which $\underline{X}$ crosses from $A_\xi$ to $A_i$, and of $\Delta(A_i|\xi)$ as the size of the corresponding jump in $f(\underline{X})$. In the present example (and all others we consider) $\Delta$ is easy to evaluate: it is just $y_i - y_\xi$. More generally, it must be estimated. The ability to evaluate jump sizes is a key simplification that comes from having the discontinuities in the performance measure, rather than the underlying system.

This simple example charts the way for far more general cases. We first identify how changes in the underlying system introduce jumps in a performance measure, then evaluate the rate at which these jumps occur. Combining the jump rates with the jump sizes as in (7.5) yields our derivative estimator. Carrying this out involves two technical issues: justifying the interchange of limit and expectation in (7.2), and verifying the existence of the jump rate in (7.3).

### 7.1.3 Discontinuous Performance Measures

Although the finite dimensional setting of the previous section has interesting applications, our real interest is in derivative estimation for stochastic processes, especially GSMPs. We now introduce two classes of discontinuous performance measures for GSMPs which can be smoothed by conditioning. In calling these performance measures discontinuous, we mean that their *samples* are discontinous. Their expectations are assumed to be continuous and differentiable in $\theta$.

We use the GSMP notation of Chapters 2 and 3. Let $f : \mathbf{S} \to \mathbf{R}$ be a bounded function on the state space of $Z$. Fix an integer $n_0 > 0$ and a real $T > 0$, and define

$$R_{n_0}(\theta) = f(Y_{n_0}(\theta)) \text{ and } R_T(\theta) = f(Z_T(\theta)). \tag{7.6}$$

We call $R_{n_0}$ and $R_T$ *terminal reward* performance measures; they can be interpreted as the "reward" $f(\cdot)$ received at the end of some period of interest. The end of this period is either the epoch of the $n_0th$ transition or the fixed time $T$.

As we will see, there are some differences in just what needs to be smoothed, depending on whether we stop at a fixed transition or a fixed time.

Let $g : \mathbf{S} \times \mathbf{S} \to \mathbf{R}$ be bounded, and define

$$D_{n_0} = \sum_{i=1}^{n_0} g(Y_{i-1}, Y_i) \text{ and } D_T = \sum_{i=1}^{N_T} g(Y_{i-1}, Y_i), \tag{7.7}$$

where, as usual, $N_T$ is the number of transitions in $(0, T]$. We call $D_{n_0}$ and $D_T$ *discontinuous additive* performance measures. These, and terminal rewards, should be contrasted with the *cumulative* or *continuous additive* performance measures studied in previous chapters.

Some examples are in order to explain our interest in these performance measures. In the case of (7.6), $f$ is frequently an indicator function. For example, the system modeled by $Z$ may be subject to failure; i.e., there may be a subset $\mathbf{S}_F \subseteq \mathbf{S}$ which we call the failed states. Once $Z$ enters $\mathbf{S}_F$, it stays there. Then if $f(s) = \mathbf{1}\{s \notin \mathbf{S}_F\}$, $\mathbf{E}[R_T]$ is the probability that the system is still functioning at time $T$. In a queueing context, we might have $f(s) = \mathbf{1}\{s > k\}$, meaning that the queue length at a particular node exceeds $k$. Alternatively, if $s$ is a vector of queue lengths in a closed network and $f(s) = s_i$, then $\mathbf{E}[R_T]$ is the mean queue length at node $i$ at time $T$.

The performance measures $D_{n_0}$ and $D_T$ associate a "reward" $g(s, s')$ with a *transition* from $s$ to $s'$. They are useful in measuring quantities embedded at the transition epochs of $Z$. The examples are more familiar if we look first at steady-state performance. Consider a single queue. Let $g_k(s, s') = \mathbf{1}\{s' = s+1 = k+1\}$ and $g(s, s') = \mathbf{1}\{s' = s + 1\}$; then $g$ indicates that the transition from $s$ to $s'$ constitutes an arrival, and $g_k$ indicates that the arrival finds $k$ jobs in queue. If it exists, the limit

$$\lim_{n\to\infty} \frac{\sum_{i=1}^{n} g_k(Y_{i-1}, Y_i)}{\sum_{i=1}^{n} g(Y_{i-1}, Y_i)}$$

is the steay-state probability that an arrival finds $k$ jobs in queue. If we partition $\mathbf{S}$ into $\mathbf{S}_1, \mathbf{S}_2$ and let $g(s, s') = \mathbf{1}\{s \in \mathbf{S}_1, s' \in \mathbf{S}_2\}$, then

$$\lim_{t\to\infty} \frac{1}{t} \sum_{i=1}^{N_t} g(Y_{i-1}, Y_i)$$

is the "throughput" (asymptotic rate) of transitions from $\mathbf{S}_1$ to $\mathbf{S}_2$. The performance measures $D_{n_0}$ and $D_T$ yield finite-horizon counterparts of these steady-state quantities.

It is easy to see that even if the expectations of the performance measures in (7.6) and (7.7) are differentiable in $\theta$, the methods of earlier chapters do not provide a way of estimating their derivatives. In particular, the stochastic derivatives are useless as estimators. Since every $Y_i$ takes values in a discrete set, the same is true of the $R$'s and $D$'s. Hence, the stochastic derivatives of all these performance measures are identically zero wherever they exist. By smoothing first we will obtain unbiased derivative estimators.

## 7.2 The Rate of Event Order Changes

A change in a parameter $\theta$ can only introduce a change in the $R$'s and $D$'s by changing some $Y_n(\theta)$. For the $n$th state $Y_n$ to change, it is necessary that at least one of the first $n$ events $a_1(\theta), \ldots, a_n(\theta)$ change. Hence, to deal with discontinuities in the performance measures, we must examine changes in the sequence of events. In particular, to calculate the rate at which a performance measure jumps under a change in $\theta$, we must be able to calculate the rate at which events change order.

We begin with the simplest possible case. We take a fixed initial state, $Y_0$, with an event list of exactly two events, $\mathcal{E}(Y_0) = \{\alpha, \beta\}$, say. We suppose that there are states $s_\alpha$ and $s_\beta$ such that $p(s_\alpha; Y_0, \alpha) = p(s_\beta; Y_0, \beta) = 1$. The clock distributions $F_\alpha$ and $F_\beta$ depend on $\theta$. We want to evaluate

$$\lim_{h \downarrow 0} \frac{1}{h} P(a_1(\theta + h) = \alpha | a_1(\theta) = \beta; \tau_1(\theta)).$$

(and find conditions under which the limit exists). If we take $n_0 = 1$, then this is also the rate at which $R_{n_0}$ jumps from $f(s_\beta)$ to $f(s_\alpha)$.

Notice that $a_1(\theta) = \beta$ if and only if $X_\theta(\beta, 1) < X_\theta(\alpha, 1)$ (supposing ties are broken in favor of $\alpha$). Thus, the limit above is equivalent to

$$\lim_{h \downarrow 0} \frac{1}{h} P(X_{\theta+h}(\alpha) \leq X_{\theta+h}(\beta) | X_\theta(\alpha) > X_\theta(\beta); X_\theta(\beta)). \tag{7.8}$$

(We have dropped the index "1" from the clock samples to lighten the notation.) For this limit to exist, we need a condition on the clock distributions:

**(H1).** Every $F_\alpha, \alpha \in \mathbf{A}$, is continuously differentiable on $(0, \infty) \times \Theta$; and, for every $\theta \in \Theta$, $F_\alpha(0, \theta) = 0$.

We also need a condition on the dependence of the clock samples on $\theta$. At the end of Chapter 2, we remarked that it is sometimes useful to have a function $\psi_\alpha$ (we called it $G_\alpha$ in Chapter 2) such that

$$\frac{dX_\theta(\alpha, k)}{d\theta} = \psi_\alpha(X_\theta(\alpha, k), \theta) \tag{7.9}$$

for all $k$ and $\theta$. We can view (7.9) as differential equation in $X_\theta$ on $\Theta$. For concreteness, let us take $\Theta$ to be a compact interval $[\theta_1, \theta_2]$. We require

**(H2).** For every $\alpha \in \mathbf{A}$ there is a $\psi_\alpha$ satisfying (7.9). Moreover, every $\psi_\alpha$ satisfies the Lipschitz condition

$$|\psi_\alpha(x_2, \theta) - \psi_\alpha(x_1, \theta)| < K \cdot |x_2 - x_1|,$$

where $K$ is a constant, $0 < x_1, x_2 < \infty$, and $\theta_1 \le \theta \le \theta_2$.

Under condition (H2), (7.9) has a unique solution throughout $(\theta_1, \theta_2)$ for all initial conditions $X_{\theta_0}$, $\theta_1 \le \theta_0 \le \theta_2$. Let us denote the general solution to (7.9) by $\Psi_\alpha$; i.e., for all $\theta \in \Theta$ and all $h$ with $\theta + h \in \Theta$,

$$X_{\theta+h}(\alpha, k) = \Psi_\alpha(X_\theta(\alpha, k), \theta, h). \tag{7.10}$$

The solution $\Psi_\alpha$ extrapolates $X_{\theta+h}$ from $X_\theta$. As an example, let the dependence of $X_\theta$ on $\theta$ be of the form $X_{\theta'} = \theta' X_\theta / \theta$; i.e., $\theta$ is a *scale* parameter. Then $\psi(x, \theta) = x/\theta$ and $\Psi(x, \theta, h) = (\theta + h)x/\theta$.

The function $\Psi_\alpha$ is monotone: if $x_1 \le x_2$ then $\Psi_\alpha(x_1, \theta, h) \le \Psi_\alpha(x_2, \theta, h)$. (Trajectories from different initial conditions cannot cross.) It automatically has an inverse in the sense that

$$\Psi_\alpha(x, \theta, h) = y \text{ if and only if } \Psi_\alpha(y, \theta + h, -h) = x.$$

Denote by $\Psi_\alpha^{-1}(\cdot, \theta, h)$ the function $\Psi_\alpha(\cdot, \theta + h, -h)$. Additional properties of $\Psi_\alpha$ and $\Psi_\alpha^{-1}$ are summarized in

**Lemma 7.1.** Under condition (H2),
- **(i)** $\Psi_\alpha(x, \theta, 0) = x$, all $x$ and $\theta$.
- **(ii)** $\Psi_\alpha$ is continuously differentiable, and $\lim_{h\downarrow 0} h^{-1}[\Psi_\alpha(x, \theta, h) - \Psi_\alpha(x, \theta, 0)] = \psi_\alpha(x, \theta)$.
- **(iii)** $\lim_{h\downarrow 0} h^{-1}[\Psi_\alpha^{-1}(x, \theta, h) - \Psi_\alpha^{-1}(x, \theta, 0)] = -\psi_\alpha(x, \theta)$.

From now on we take a "nominal" value of $\theta$ as fixed and drop the second argument in $\Psi_\alpha$ and $\psi_\alpha$. With this in mind, the interpretation of part (ii) of the lemma is that

$$\partial_h \Psi_\alpha(x, 0) = \psi_\alpha(x) = \left.\frac{dX_\theta}{d\theta}\right|_{X_\theta = x}; \tag{7.11}$$

the last term is the derivative of $X_\theta$ evaluated at $X_\theta = x$.

We can now evaluate (7.8).

$$\begin{aligned}
&P(X_{\theta+h}(\alpha) \le X_{\theta+h}(\beta) | X_\theta(\beta); X_\theta(\alpha) > X_\theta(\beta)) \\
&\quad = P(\Psi_\alpha(X_\theta(\alpha), h) \le \Psi_\beta(X_\theta(\beta), h) | X_\theta(\beta); X_\theta(\alpha) > X_\theta(\beta)) \\
&\quad = P(X_\theta(\alpha) \le \Psi_\alpha^{-1}(\Psi_\beta(X_\theta(\beta), h), h) | X_\theta(\beta); X_\theta(\alpha) > X_\theta(\beta)) \\
&\quad = \frac{F_\alpha[\Psi_\alpha(\Psi_\beta(X_\theta(\beta), h), -h)] - F_\alpha[X_\theta(\beta)]}{1 - F_\alpha[X_\theta(\beta)]},
\end{aligned} \tag{7.12}$$

provided

$$X_\theta(\beta) \leq \Psi_\alpha^{-1}(\Psi_\beta(X_\theta(\beta), h), h), \tag{7.13}$$

and zero otherwise. Now divide (7.12) by $h$ and let $h \downarrow 0$. Letting $f_\alpha$ be the density of $F_\alpha$, the chain rule yields

$$\frac{f_\alpha(X_\theta(\beta))}{1 - F_\alpha(X_\theta(\beta))} \cdot \lim_{h \downarrow 0} \frac{\Psi_\alpha(\Psi_\beta(X_\theta(\beta), h), -h) - X_\theta(\beta)}{h},$$

if (7.13) holds for all small $h$, and zero otherwise. We recognize the first factor as the hazard rate, $\mu_\alpha$, of $F_\alpha$, evaluated at $X_\theta(\beta)$. To evaluate the second factor, we add and subtract $\Psi_\beta(X_\theta(\beta), h)$ to rewrite the numerator as

$$[\Psi_\alpha(\Psi_\beta(X_\theta(\beta), h), -h) - \Psi_\beta(X_\theta(\beta), h)] + [\Psi_\beta(X_\theta(\beta), h) - X_\theta(\beta)].$$

We consider the derivatives of the two bracketed terms separately. The second term yields

$$\lim_{h \downarrow 0} \frac{\Psi_\beta(X_\theta(\beta), h) - X_\theta(\beta)}{h} = \psi_\beta(X_\theta(\beta)) = \frac{dX_\theta(\beta)}{d\theta}.$$

For the first term, we use the fact that $\Psi_\alpha$ is $C^1$ to get

$$\begin{aligned}
&\lim_{h \downarrow 0} \frac{\Psi_\alpha(\Psi_\beta(X_\theta(\beta), h), -h) - \Psi_\beta(X_\theta(\beta), h)}{h} \\
&\quad = \lim_{h \downarrow 0} \frac{\Psi_\alpha(\Psi_\beta(X_\theta(\beta), 0), -h) - \Psi_\beta(X_\theta(\beta), 0)}{h} \\
&\quad = \lim_{h \downarrow 0} \frac{\Psi_\alpha(X_\theta(\beta), -h) - X_\theta(\beta)}{h} \\
&\quad = -\psi_\alpha(X_\theta(\beta)).
\end{aligned}$$

Combining these terms, we get

$$\mu_\alpha(X_\theta(\beta)) \left( \frac{dX_\theta(\beta)}{d\theta} - \psi_\alpha(X_\theta(\beta)) \right). \tag{7.14}$$

But this is only correct if (7.13) holds for all sufficiently small $h$; otherwise, the limit is zero. And (7.13) holds for all sufficiently small $h$ if and only if the term in parentheses in (7.14) is non-negative. Thus, putting all the pieces together, we have evaluated (7.8) to be

$$\mu_\alpha(X_\theta(\beta)) \left[ \frac{dX_\theta(\beta)}{d\theta} - \psi_\alpha(X_\theta(\beta)) \right]^+, \tag{7.15}$$

where $[x]^+ = \max(x, 0)$. We state this as

**Lemma 7.2.** Under conditions (H1)-(H2), the rate (7.8) at which $\alpha$ and $\beta$ change order is given by (7.15).

This result is worth contemplating because it is the fundamental building block for the smoothed derivative estimators of Sections 7.3 and 7.4. Its meaning may be clearer if we rewrite (7.15) as

$$\mu_\alpha(X_\theta(\beta)) \left[ \frac{dX_\theta(\beta)}{d\theta} - \left. \frac{dX_\theta(\alpha)}{d\theta} \right|_{X_\theta(\alpha)=X_\theta(\beta)} \right]^+ . \tag{7.16}$$

If $X_\theta(\alpha)$ is much larger than $X_\theta(\beta)$, then a small change in $\theta$ will not cause $\alpha$ and $\beta$ to change order. The hazard rate $\mu_\alpha$ evaluated at $X_\theta(\beta)$ yields the "probability" that $X_\theta(\alpha)$ is just greater than $X_\theta(\beta)$, given that it is no smaller. In other words, this is the conditional "probability" that $\alpha$ occurs just after $\beta$. When this happens, the derivative of $X_\theta(\alpha)$ is $\psi_\alpha(X_\theta(\beta))$; i.e., the derivative is evaluated at $X_\theta(\alpha) = X_\theta(\beta)$. If $X_\theta(\beta)$ is increasing faster than $X_\theta(\alpha)$, then the two events will change order; otherwise, they will not. Thus, the "$[\cdot]^+$" term is the rate at which the two events change order, given that $\alpha$ occurs right after $\beta$.

When $\alpha$ and $\beta$ change order, $R_{n_0}$ jumps from $f(s_\beta)$ to $f(s_\alpha)$, so (7.16) is the jump rate for $R_{n_0}$. The size of this jump is $[f(s_\beta) - f(s_\alpha)]$. By considering also the case where intially $a_1 = \alpha$ and $a_2 = \beta$, we are led to the following derivative estimate for $R_{n_0}$:

$$\mu_{a_2}(X_\theta(a_1)) \left[ \frac{dX_\theta(a_1)}{d\theta} - \psi_{a_2}(X_\theta(a_1)) \right]^+ \cdot [f(s_{a_2}) - f(s_{a_1})].$$

The next section generalizes this estimate.

# 7.3 Stopping at a Fixed Transition

We are now ready to develop smoothed derivative estimates for $R_{n_0}$ and $D_{n_0}$. We proceed in three steps. First, we identify the critical event order changes and the conditional expectation that smooths them. We then calculate the rate of order changes, generalizing Lemma 7.2. Finally, we justify the interchange of limit and expectation to show that our estimators are unbiased. Our derivation focuses first on $R_{n_0}$, then carries out the extension to $D_{n_0}$.

## 7.3.1 Setting up the Estimates

The first order of business is deciding what to condition on. In the previous section, we simply conditioned on $X_\theta(a_1, 1)$, the length of the first clock to run out. In fact, this constitutes the entire past up to the first transition. For $R_{n_0}$ we essentially condition on the entire past up to the $n_0th$ transition. For all $n$, define

$$\mathcal{H}_n = (Y_1, \ldots, Y_n; a_1, \ldots, a_n; \tau_1, \ldots, \tau_n); \tag{7.17}$$

then $\mathcal{H}_n$ records states, events and transition epochs up to the $n$th transition. The inclusion of $Y_0$ is unnecessary because we always take the initial state to be fixed (and known). Also define

$$\underline{a}_n = (a_1, \ldots, a_n),$$

so that $\underline{a}_n$ records the first $n$ events.

As we noted earlier, $R_{n_0}$ only changes value when there is a change among the first $n_0$ events—i.e., when there is a change in $\underline{a}_{n_0}$. Hence, our estimator will take the general form

$$\sum_{\underline{a}'} Q_{n_0}(\underline{a}'|\mathcal{H}_{n_0})\Delta_{n_0}(\underline{a}'|\mathcal{H}_{n_0}), \tag{7.18}$$

where the sum is over all sequences of $n_0$ events, and

$$Q_{n_0}(\underline{a}'|\mathcal{H}_{n_0}) = \lim_{h\downarrow 0} \frac{1}{h} P(\underline{a}_{n_0}(\theta + h) = \underline{a}'|\mathcal{H}_{n_0}) \tag{7.19}$$

is the rate at which $\underline{a}_{n_0}(\theta)$ jumps to $\underline{a}'$, and

$$\begin{aligned}\Delta_{n_0}(\underline{a}'|\mathcal{H}_{n_0}) &= \lim_{h\downarrow 0} \mathbf{E}[f(Y_{n_0}(\theta + h)) - f(Y_{n_0}(\theta))|\mathcal{H}_{n_0}, \underline{a}_{n_0}(\theta + h) = \underline{a}'] \\ &= \mathbf{E}[f(Y_{n_0}(\theta + h)) - f(Y_{n_0}(\theta))|\mathcal{H}_{n_0}, \underline{a}_{n_0}(\theta + h) = \underline{a}']\end{aligned}$$

is the expected jump in $R_{n_0}$ when $\underline{a}_{n_0}$ jumps to $\underline{a}'$. (Since conditioning on $(\mathcal{H}_{n_0}, \underline{a}_{n_0}(\theta + h) = \underline{a}')$ determines the distribution of $Y_{n_0}(\theta + h)$ independently of $h$, it allows us to drop the limit in the definition of $\Delta_{n_0}$.)

As it stands, (7.18) is a sum over all sequences of length $n_0$ that can be formed from $\mathbf{A}$. Of course, most of these are not physically possible and will consequently have a zero $Q_{n_0}(\underline{a}'|\mathcal{H}_{n_0})$. In fact, under reasonable conditions on the clock times, only a small subset of the possible $\underline{a}'$'s will have a non-zero jump rate. We will be able to restrict attention to those $\underline{a}'$'s which are obtained from $\underline{a}_{n_0}(\theta)$ through a *single* order change. For this we need the following conditions:

**(H3).** There is a constant $B > 0$ such that for all $\theta$, $\alpha$ and $k$, $|dX_\theta(\alpha, k)/d\theta| \leq B \cdot X_\theta(\alpha, k)$.

**(H4).** For each $\theta \in \Theta$ there is a $q^*(\theta)$ such that for all $t \geq 0$ and all $\alpha$, $\mu_\alpha(t) \leq q^*(\theta)$.

Condition (H3) is a strengthening of (A3) of Chapter 3. Condition (H4) is just the requirement that the clock hazard rates be bounded.

Define

$$\begin{aligned}\mathcal{A}_n(\theta) &= \{\underline{a}_n(\theta)\} \\ &\cup \{\underline{a}' : \underline{a}' = (a_1(\theta), \ldots, a_i(\theta), a_{i-1}(\theta), \ldots, a_n(\theta)),\ \text{some } i = 2, \ldots, n\} \\ &\cup \{\underline{a}' : \underline{a}' = (a_1(\theta), \ldots, a_{n-1}(\theta), \alpha),\ \text{some } \alpha \in \mathcal{E}(Y_{n-1})\}.\end{aligned}$$

Thus, $\mathcal{A}_n(\theta)$ is the set of $\underline{a}'$'s that can be obtained from $\underline{a}_n(\theta)$ through (at most) one order change. The order change could simply permute $(a_1, \ldots, a_n)$ or, more significantly, it might involve replacing $a_n$ with some $a_i$, $i > n$ in the event list of $Y_{n-1}$.

The following result allows us to consider only $\underline{a}'$'s in $\mathcal{A}_n$:

**Lemma 7.3.** For a GSMP satisfying (C2), under conditions (H1), (H3) and (H4),

$$P(\underline{a}_n(\theta + h) \notin \mathcal{A}_n(\theta) | \mathcal{H}_n) = o(h), \tag{7.20}$$

with probability one. Moreover, if $\mathbf{E}[\tau_n^2(\theta)] < \infty$ then

$$P(\underline{a}_n(\theta + h) \notin \mathcal{A}_n(\theta)) = o(h). \tag{7.21}$$

**Proof.** The proof is similar to that of Lemma 5.1, which bounds the rate of multiple order changes in the Markov case. The uniformization argument used there can be adapted here because we have assumed that all hazard rates are bounded.

For any $\underline{a}_n(\theta + h)$ which is a permutation of $\underline{a}_n(\theta)$, (7.20) holds trivially: under (C2), given $\mathcal{H}_n$, the values of the clock samples for $a_1(\theta), \ldots, a_n(\theta)$ are determined at $\theta + h$ by the $\Psi_{a_i}$'s. Hence, the left side of (7.20) is, in fact, zero for all $|h| < h_1$ for some $h_1$ determined by $\mathcal{H}_n$. The interesting case is if $\underline{a}_n(\theta + h) = (a_1(\theta), \ldots, a_{n-1}(\theta), \alpha)$ for $\alpha \neq a_n(\theta)$. For this case to arise, $a_n$ and some $a_{n+k}$, $k \geq 1$, must change order. It is enough to show that the case $k > 1$ is negligible.

From the proof of Theorem 3.2 in Chapter 3, we know that (H3) implies $|d\tau_i/d\theta| \leq B \cdot \tau_i$. By integrating this inequality, we get

$$|\tau_i(\theta + h) - \tau_i(\theta)| \leq (e^{Bh} - 1)\tau_i(\theta).$$

It follows (after some algebra) that, in order that $a_n$ and some $a_{n+k}$, $k > 1$ change order when $\theta$ is changed to $\theta + h$, we must have

$$\tau_{n+k}(\theta) - \tau_n(\theta) \leq (e^{2Bh} - 1)\tau_n(\theta).$$

Thus, we want to bound

$$P(\tau_{n+k}(\theta) - \tau_n(\theta) \le (e^{2Bh} - 1)\tau_n(\theta)|\mathcal{H}_n).$$

Let $\Lambda(\theta) = |\mathbf{A}| \cdot q^*(\theta)$ with $q^*$ provided by (H4). Then $\Lambda(\theta)$ bounds the maximum rate of transitions of $\{Z_t(\theta), t \ge 0\}$. For $\epsilon \ge 0$,

$$P(\tau_{n+1}(\theta) - \tau_n(\theta) \le \epsilon|\mathcal{H}_n) \le \Lambda(\theta)\epsilon,$$

and, for $k > 1$,

$$P(\tau_{n+k}(\theta) - \tau_n(\theta) \le \epsilon|\mathcal{H}_n) \le \Lambda(\theta)\epsilon^2.$$

Evaluating these at $\epsilon = (e^{2Bh} - 1)\tau_n$, we get

$$P(\tau_{n+1} - \tau_n \le [e^{2Bh} - 1]\tau_n|\mathcal{H}_n) \le \Lambda \cdot ([e^{2Bh} - 1]\tau_n), \tag{7.22}$$

and

$$\begin{aligned} P(\tau_{n+k} - \tau_n \le [e^{2Bh} - 1]\tau_n|\mathcal{H}_n) &\le \Lambda \cdot ([e^{2Bh} - 1]\tau_n)^2 \\ &\le \Lambda \cdot (2Be^{2Bh}\tau_n)^2 \cdot h^2. \end{aligned} \tag{7.23}$$

Since $\tau_n$ is almost surely finite, this bound is $o(h)$. Equation (7.21) follows by taking expectations of both sides. □

**Remark.** In the proof of this lemma (and throughout this chapter) we are, strictly speaking, relying on the slightly stronger condition (C2'), rather than (C2). In words, the property we need is that changing the order of events does not change the state reached. Proposition 3.1 of Chapter 3 connects (C2) and (C2') for abstract GSMPs. For physically meaningful GSMPs, (C2) alone is typically sufficient for there to be a $\phi$ satisfying (C2'); see Section 4.1.1 of Chapter 4. For this reason, we do not distinguish between the conditions here.

Lemma 7.3 greatly reduces the number of order changes we need to consider in (7.18); for any $\underline{a}' \notin \mathcal{A}_{n_0}$ we have $Q_{n_0}(\underline{a}'|\mathcal{A}_{n_0}) = 0$. We can further reduce the number of cases by looking at $\Delta_{n_0}$. To avoid trivialities, we exclude those $\underline{a}'$ which are infeasible, in that they do not represent a physically possible sequence of events. A sequence $\alpha_0, \ldots, \alpha_n$ is *feasible* (starting in $Y_0$) if there are states $s_1, \ldots, s_n$ and $s_0 = Y_0$ such that $\alpha_i \in \mathcal{E}(s_i)$, $i = 0, \ldots, n$, and

$$p(s_1; s_0, \alpha_0)p(s_2; s_1, \alpha_1) \cdots p(s_n; s_{n-1}, \alpha_{n-1}) > 0.$$

For any $\underline{a}'$ which is not feasible, we simply define $Q_n(\underline{a}'|\mathcal{H}_n) = \Delta_n(\underline{a}'|\mathcal{H}_n) = 0$. Of the physically possible sequences, only $|\mathcal{E}(Y_{n-1})| - 1$ of them contribute a non-zero $\Delta_n$:

**Lemma 7.4.** For a GSMP satisfying (C2), if $\underline{a}'$ is a permutation of $\underline{a}_n(\theta)$ then $\Delta_n(\underline{a}'|\mathcal{H}_n) = 0$. If $\underline{a}' \in \mathcal{A}_n$ and $\underline{a}'$ is not a permutation of $\underline{a}_n$, then $\underline{a}'$ has the form $(a_1, \ldots, a_{n-1}, \alpha)$ and

$$\Delta_n(\underline{a}'|\mathcal{H}_n) = \sum_{s'} p(s'; Y_{n-1}, \alpha)[f(s') - f(Y_n)]. \tag{7.24}$$

**Proof.** The first part is an immediate consequence of (C2). Permuting the elements of $\underline{a}_n$ leaves $Y_n$ unchanged and makes the change in $f(Y_n)$ zero. The second part is obtained by considering the possible values of $Y_n(\theta + h)$ when $\underline{a}_n(\theta + h) = \underline{a}'$. □

If we write simply $Q_{n_0}(\alpha|\mathcal{H}_{n_0})$ for $Q_{n_0}((a_1, \ldots, a_{n_0-1}, \alpha)|\mathcal{H}_{n_0})$ when $\alpha \in \mathcal{E}(Y_{n_0-1})$, then (7.18) becomes

$$\sum_{\alpha} Q_{n_0}(\alpha|\mathcal{H}_{n_0}) \sum_{s'} p(s'; Y_{n_0-1}, \alpha)[f(s') - f(Y_{n_0})]. \tag{7.25}$$

It remains to calculate the jump rate $Q_{n_0}(\alpha|\mathcal{H}_{n_0})$.

### 7.3.2 Calculating Jump Rates

We need two more pieces of notation. Let

$$S(\alpha, k) = \text{epoch of } k\text{th setting of a clock for } \alpha;$$

for any non-interruptive GSMP, this is just $T(\alpha, k) - X(\alpha, k)$. For $\alpha \in \mathcal{E}(Y_n)$, let

$$t_n(\alpha) = \text{age of } \alpha\text{-clock at } n\text{th transition}.$$

This was used in Chapter 6. If the clock for $\alpha$ which is active at $\tau_n$ is the $k$th clock, then $t_n(\alpha) = \tau_n - S(\alpha, k)$.

In Section 7.2, the function $\psi_\alpha$ allowed us to interpret "the derivative of $X_\theta(\alpha, k)$ evaluated at $x$" via

$$\left.\frac{dX_\theta(\alpha, k)}{d\theta}\right|_{X_\theta = x} = \psi_\alpha(x).$$

We can generalize this to the derivatives of transition epochs. Suppose that $\alpha \in \mathcal{E}(Y_n)$ and that the number of instances of $\alpha$ among $a_1, \ldots, a_n$ is $k - 1$; in the notation of Chapter 2, $k$ is given by $N(\alpha, n) + 1$. Then, given $\mathcal{H}_n$,

$$\{a_{n+1} = \alpha, \tau_{n+1} = t\} \text{ if and only if } \{a_{n+1} = \alpha, X_\theta(\alpha, k) = t - S_\theta(\alpha, k)\}.$$

Moreover, in this case,

$$\begin{aligned}\frac{d\tau_{n+1}}{d\theta} &= \frac{dS_\theta(\alpha,k)}{d\theta} + \left.\frac{dX_\theta(\alpha,k)}{d\theta}\right|_{X=t-S} \\ &= \frac{dS_\theta(\alpha,k)}{d\theta} + \psi_\alpha(t - S_\theta(\alpha,k)).\end{aligned}$$

We could reasonably denote this quantity by

$$\left.\frac{d\tau_{n+1}}{d\theta}\right|_{a_{n+1}=\alpha,\tau_{n+1}=t}.$$

To lighten the notation, we denote it simply by

$$\left.\frac{d\tau_{n+1}}{d\theta}\right|_{\alpha,t}.$$

Using the triggering indicators, $\eta$, and the event order pairs, $r_i$, (see Chapter 2, Section 2.2, especially Definition 2.1 and Lemma 2.2) we can write this as

$$\left.\frac{d\tau_{n+1}}{d\theta}\right|_{\alpha,t} = \sum_{i=1}^{n} X_\theta(r_i)\eta(\alpha,k;r_i) + \psi_\alpha(t - S_\theta(\alpha,k)). \tag{7.26}$$

This implicitly provides an algorithm for evaluating the derivative on the left.

**Proposition 7.1.** For a GSMP satisfying (C2), under conditions (H1)-(H4),

$$Q_n(\alpha|\mathcal{H}_n) = \mu_\alpha(t_n(\alpha))\left[\frac{d\tau_n}{d\theta} - \left.\frac{d\tau_{n+1}}{d\theta}\right|_{\alpha,\tau_n}\right]^+,$$

if $\alpha \in \mathcal{E}(Y_{n-1})$, and zero otherwise.

**Proof.** $Q_n(\alpha|\mathcal{H}_n)$ is the derivative with respect to $h$, at $h = 0$, of

$$P(\underline{a}_n(\theta+h) = (a_1(\theta),\ldots,a_{n-1}(\theta),\alpha)|\mathcal{H}_n).$$

(Clearly, this is zero if $\alpha$ is not in $\mathcal{E}(Y_{n-1})$; henceforth assume it is.) By Lemma 7.3, this derivative is the same as that of the conditional probability that $\alpha$ and $a_n$ change order; i.e., the derivative of

$$P(T_{\theta+h}(\alpha,k_\alpha) < T_{\theta+h}(a_n,k_{a_n})|\mathcal{H}_n)$$

where $k_\alpha = N(\alpha,n)+1$ and $k_{a_n} = N(a_n,n)$. This probability can be rewritten as

$$\begin{aligned}&P(X_{\theta+h}(\alpha,k_\alpha) < T_{\theta+h}(a_n,k_{a_n}) - S_{\theta+h}(\alpha,k_\alpha)|\mathcal{H}_n) \\ &= P(\Psi_\alpha(X_\theta(\alpha,k_\alpha),h) < T_{\theta+h}(a_n,k_{a_n}) - S_{\theta+h}(\alpha,k_\alpha)|\mathcal{H}_n) \\ &= P(X_\theta(\alpha,k_\alpha) < \Psi_\alpha(T_{\theta+h}(a_n,k_{a_n}) - S_{\theta+h}(\alpha,k_\alpha),-h)|\mathcal{H}_n).\end{aligned}$$

Recall that $T_\theta(a_n, k_{a_n}) = \tau_n(\theta)$. Given $\mathcal{H}_n$, $X_\theta(\alpha, k_\alpha)$ is distributed as a sample from $F_\alpha$ conditioned on being greater than or equal to $\tau_n(\theta) - S_\theta(\alpha, k_\alpha)$. Thus, this probability is equal to

$$\frac{F_\alpha(\Psi_\alpha(T_{\theta+h}(a_n, k_{a_n}) - S_{\theta+h}(\alpha, k_\alpha), -h)) - F_\alpha(\tau_n - S_\theta(\alpha, k_\alpha))}{1 - F_\alpha(\tau_n - S_\theta(\alpha, k_\alpha))},$$

provided

$$\Psi_\alpha(T_{\theta+h}(a_n, k_{a_n}) - S_{\theta+h}(\alpha, k_\alpha), -h) \geq \tau_n - S_\theta(\alpha, k_\alpha), \tag{7.27}$$

and zero otherwise. If (7.27) holds for all small $h$, then passing to the limit as in Lemma 7.2, we get

$$\mu_\alpha(\tau_n - S_\theta(\alpha, k_\alpha)) \cdot \partial_h \Psi_\alpha(T_{\theta+h}(a_n, k_{a_n}) - S_{\theta+h}(\alpha, k_\alpha), -h)|_{h=0}.$$

In the first factor, we can substitute $t_n(\alpha)$ for $\tau_n - S_\theta(\alpha, k_\alpha)$; for the second factor, we proceed as in Lemma 7.2 to get

$$\begin{aligned}
&\mu_\alpha(t_n(\alpha))\left(\frac{dT_\theta(a_n, k_{a_n})}{d\theta} - \frac{dS_\theta(\alpha, k)}{d\theta} - \psi_\alpha(T_\theta(a_n, k_{a_n}) - S_\theta(\alpha, k))\right) \\
= \;&\mu_\alpha(t_n(\alpha))\left(\frac{d\tau_n}{d\theta} - \frac{dS_\theta(\alpha, k)}{d\theta} - \psi_\alpha(\tau_n - S_\theta(\alpha, k))\right) \\
= \;&\mu_\alpha(t_n(\alpha))\left(\frac{d\tau_n}{d\theta} - \left.\frac{d\tau_{n+1}}{d\theta}\right|_{\alpha, \tau_n}\right).
\end{aligned}$$

If (7.27) does not hold for all sufficiently small $h$, the limit is zero. Combining the two cases we get the result. □

The interpretation of this result is similar to that of Lemma 7.2. The rate at which $\underline{a}_n$ becomes $(a_1, \ldots, a_{n-1}, \alpha)$ is the rate at which $a_n$ and $\alpha$ change order. This rate is the "probability" $\mu_\alpha(t_n(\alpha))$ that $\alpha$ occurs just after $a_n$, times the amount by which $a_n$ is advancing faster than $\alpha$.

### 7.3.3 Unbiasedness

Proposition 7.1 provides an explicit expression for the rate in the estimator (7.25). We now verify that this estimator is unbiased—that is, we verify that the limit as $h \downarrow 0$ and the expectation (over $\mathcal{H}_{n_0}$) can be interchanged.

**Theorem 7.1.** Suppose that (H1)-(H4) and (C2) hold, that every $X_\theta(\alpha, k)$ has finite second moment, that $f$ is bounded, and that $\mathbf{E}[R_{n_0}(\theta)] = \mathbf{E}[f(Y_{n_0})]$ is differentiable in $\theta$. Then

$$\mathbf{E}\left[\sum_\alpha Q_{n_0}(\alpha|\mathcal{H}_{n_0}) \sum_{s'} p(s'; Y_{n_0-1}, \alpha)[f(s') - f(Y_{n_0})]\right] = \frac{d\mathbf{E}[R_{n_0}]}{d\theta}.$$

**Proof.** Let $\delta_h f = f(Y_{n_0}(\theta+h)) - f(Y_{n_0}(\theta))$. Then $d\mathbf{E}[R_{n_0}]/d\theta$ is equal to $\lim_{h\downarrow 0} h^{-1}\mathbf{E}[\delta_h f]$. Abbreviating $\underline{a}_{n_0}$ to $\underline{a}$, we may rewrite this last expectation as

$$\mathbf{E}[\delta_h f \cdot \mathbf{1}\{\underline{a}(\theta+h) \in \mathcal{A}_{n_0}(\theta)\}] + \mathbf{E}[\delta_h f \cdot \mathbf{1}\{\underline{a}(\theta+h) \notin \mathcal{A}_{n_0}(\theta)\}].$$

Since $f$ is bounded, the second term is $o(h)$, by Lemma 7.3, and we may restrict attention to the first term. (If every $X_\theta(\alpha, k)$ has finite second moment then so does every $\tau_n(\theta)$.) By considering only those $\underline{a}' \in \mathcal{A}_{n_0}$ for which $\Delta_{n_0}(\underline{a}'|\mathcal{H}_{n_0}) \neq 0$, we may express it as

$$\mathbf{E}[\delta_h f \cdot \mathbf{1}\{\underline{a}(\theta+h) \in \mathcal{A}_{n_0}(\theta)\}] = \mathbf{E}[\mathbf{E}[\delta_h f \cdot \mathbf{1}\{\underline{a}(\theta+h) \in \mathcal{A}_{n_0}(\theta)\}|\mathcal{H}_{n_0}]]$$

$$= \mathbf{E}\left[\sum_{\alpha,s'} p(s'; Y_{n_0-1}, \alpha)[f(s') - f(Y_{n_0-1})] \cdot P_h(\alpha|\mathcal{H}_{n_0})\right],$$

where $P_h(\alpha|\mathcal{H}_{n_0})$ is short for $P(\underline{a}(\theta+h) = (a_1(\theta), \ldots, a_{n_0-1}(\theta), \alpha)|\mathcal{H}_{n_0})$, and where the sum runs over $\alpha \in \mathcal{E}(Y_{n_0-1})$ and all $s$. Since $Q_{n_0}(\alpha|\mathcal{H}_{n_0})$ is equal to $\lim_{h\downarrow 0} h^{-1}P_h(\alpha|\mathcal{H}_{n_0})$, we need only show that the limit and expectation can be interchanged.

If $\|f\|$ is the supremum of $|f|$, then

$$h^{-1}\left|\sum_{\alpha}\sum_{s'} p(s'; Y_{n_0-1}, \alpha)[f(s') - f(Y_{n_0-1})] \cdot P_h(\alpha|\mathcal{H}_{n_0})\right|$$

$$\leq 2\|f\| \sum_{\alpha} \frac{P(\underline{a}(\theta+h) \in \{(a_1, \ldots, a_{n_0-1}, \alpha)\}|\mathcal{H}_{n_0})}{h}. \quad (7.28)$$

Equation (7.22) provides a bound on the conditional probability that $a_n$ and some $\alpha$ change order:

$$\begin{aligned} P(\underline{a}(\theta+h) \in \{(a_1, \ldots, a_{n_0-1}, \alpha)\}|\mathcal{H}_{n_0}) &\leq (e^{2Bh} - 1)\Lambda(\theta)\tau_{n_0}(\theta), \\ &\leq 2Bhe^{2Bh}\Lambda(\theta)\tau_{n_0}(\theta), \end{aligned}$$

where $B$ is the bound in (H3) and $\Lambda(\theta) = |\mathbf{A}| \cdot q^*(\theta)$ bounds the total rate of transitions. Substituting this into the right side of (7.28) shows that the left side of (7.28) is bounded by

$$(2\|f\|) \cdot |\mathbf{A}| \cdot 2Be^{2Bh}\Lambda(\theta)\tau_{n_0}(\theta).$$

If every $X_\theta(\alpha, k)$ has finite expectation, so does $\tau_{n_0}$; hence, this bound is integrable and we may apply the dominated convergence theorem to conclude that the limit and expectation can be interchanged. □

### 7.3.4 Example: The GI/G/1 Queue

To illustrate the results of the previous section, we consider the sensitivity of $\mathbf{E}[R_{n_0}] = \mathbf{E}[f(Y_{n_0})]$ in a single-server queue. As usual, $\alpha$ denotes arrival, $\beta$ denotes departure, and $\{Y_n\}$ is the sequence of queue lengths. To evaluate $Q_{n_0}$, we distinguish two cases, $a_{n_0} = \beta$ or $\alpha$. If $a_{n_0} = \beta$, then $\tau_{n_0}$ is the sum of the serivce times in the current busy period plus the sum of the interarrival times that end before or at the initiation of the current busy period. Let us write this as

$$\tau_{n_0} = \sum_{j \geq b_1} X_\theta(\beta, j) + \sum_{i \leq b_2} X_\theta(\alpha, i),$$

where $j \geq b_1$ is shorthand for the indices of service times in the current busy period (up to the $n_0th$ transition), and $i \leq b_2$ denotes the indices of arrivals before the current busy period. Differentiating,

$$\frac{d\tau_{n_0}}{d\theta} = \sum_{j \geq b_1} \frac{dX_\theta(\beta, j)}{d\theta} + \sum_{i \leq b_2} \frac{dX_\theta(\alpha, i)}{d\theta}.$$

For $a_{n_0}$ and $a_{n_0+1}$ to change order, it must happen that $a_{n_0+1} \neq a_{n_0} = \beta$; i.e., $a_{n_0+1} = \alpha$. If $a_{n_0+1}$ is the $k_\alpha + 1$st arrival, then its epoch is given by

$$\tau_{n_0+1} = \sum_{i \leq k_\alpha + 1} X_\theta(\alpha, i),$$

and

$$\left.\frac{d\tau_{n_0+1}}{d\theta}\right|_{\alpha, \tau_{n_0}} = \sum_{i \leq k_\alpha} \frac{dX_\theta(\alpha, i)}{d\theta} + \psi_\alpha(t_{n_0}(\alpha)).$$

Subtracting, we get

$$\frac{d\tau_{n_0}}{d\theta} - \left.\frac{d\tau_{n_0+1}}{d\theta}\right|_{\alpha, \tau_{n_0}} = \sum_{j \geq b_1} \frac{dX_\theta(\beta, j)}{d\theta} - \sum_{b_2 < i \leq k_\alpha} \frac{dX_\theta(\alpha, i)}{d\theta} - \psi_\alpha(t_{n_0}(\alpha)).$$

The sum over $j$ is over completed service times in the current busy period; the sum over $i$ is over interarrival times in the current busy period.

If instead $a_{n_0} = \alpha$, then for $a_{n_0}$ and $a_{n_0+1}$ to change order we must have $a_{n_0+1} = \beta$ and $\beta \in \mathcal{E}(Y_{n_0-1})$, which requires $Y_{n_0-1} > 0$, hence $Y_{n_0} > 1$. In this case, we get

$$\frac{d\tau_{n_0}}{d\theta} - \left.\frac{d\tau_{n_0+1}}{d\theta}\right|_{\beta, \tau_{n_0}} = \sum_{b_2 < i \leq k_\alpha} \frac{dX_\theta(\alpha, i)}{d\theta} - \sum_{j \geq b_1} \frac{dX_\theta(\beta, j)}{d\theta} - \psi_\beta(t_{n_0}(\beta)).$$

Thus, for the two cases we get

$$Q_{n_0}(\alpha | \mathcal{H}_{n_0}) = \mathbf{1}\{a_{n_0} = \beta\} \mu_\alpha(t_{n_0}(\alpha)) \left[ \frac{d\tau_{n_0}}{d\theta} - \left.\frac{d\tau_{n_0+1}}{d\theta}\right|_{\alpha, \tau_{n_0}} \right]^+,$$

and

$$Q_{n_0}(\beta|\mathcal{H}_{n_0}) = \mathbf{1}\{a_{n_0} = \alpha, Y_{n_0} > 1\}\mu_\beta(t_{n_0}(\alpha)) \left[ \frac{d\tau_{n_0}}{d\theta} - \left.\frac{d\tau_{n_0+1}}{d\theta}\right|_{\beta,\tau_{n_0}} \right]^+ .$$

An interesting special case is when

$$\frac{dX_\theta(\alpha,i)}{d\theta} = m_\alpha(\theta)X_\theta(\alpha,i) \text{ and } \frac{dX_\theta(\beta,j)}{d\theta} = m_\beta(\theta)X_\theta(\beta,j)$$

for all $i$ and $j$, for some functions $m_\alpha$, $m_\beta$. If we let $\tau_b$ be the epoch of initiation of the current busy period, then the expressions above simplify to

$$\begin{aligned} \left[ \frac{d\tau_{n_0}}{d\theta} - \left.\frac{d\tau_{n_0+1}}{d\theta}\right|_{\alpha,\tau_{n_0}} \right]^+ &= [(m_\beta(\theta) - m_\alpha(\theta))(\tau_{n_0} - \tau_b)]^+ \\ &= [m_\beta(\theta) - m_\alpha(\theta)]^+(\tau_{n_0} - \tau_b). \end{aligned}$$

and

$$\left[ \frac{d\tau_{n_0}}{d\theta} - \left.\frac{d\tau_{n_0+1}}{d\theta}\right|_{\beta,\tau_{n_0}} \right]^+ = [m_\alpha(\theta) - m_\beta(\theta)]^+(\tau_{n_0} - \tau_b).$$

This case is particularly convenient for implementation.

The jump size $\Delta_{n_0}$ is easy to evaluate. Reversing an arrival and a departure always changes $Y_{n_0}$ by exactly two. Thus,

$$\Delta_{n_0}(\alpha|\mathcal{H}_{n_0}) = f(Y_{n_0} + 2) - f(Y_{n_0}),$$

and

$$\Delta_{n_0}(\beta|\mathcal{H}_{n_0}) = f(Y_{n_0} - 2) - f(Y_{n_0});$$

recall that in the second case $Y_{n_0} > 1$. In the special case where $f(x) = x$ (so that $\mathbf{E}[R_{n_0}]$ is the mean queue length at the $n_0th$ transition), these differences further simplify to $+2$ and $-2$. This case is not covered by Theorem 7.1, which requires that $f$ be bounded. In practice, one could truncate $f$ at a large value.

### 7.3.5 The Discontinuous Additive Case

We now turn to a smoothed derivative estimate for

$$D_{n_0}(\theta) = \sum_{i=1}^{n_0} g(Y_{i-1}(\theta), Y_i(\theta)).$$

The derivation is similar to that for $R_{n_0}$; the main difference is that now a jump in $\underline{a}_{i+1}$ from $(a_1, \ldots, a_i, a_{i+1})$ to $(a_1, \ldots, a_{i+1}, a_i)$ introduces *two* kinds of changes in the performance measure, $D_{n_0}$. A switch in the order of $a_i$ and $a_{i+1}$ changes $Y_i$ to, say, $\hat{Y}_i$, but under (C2) leaves $Y_{i+1}$ unchanged. Hence, the switch changes

$$g(Y_i, Y_{i+1}) \text{ to } g(\hat{Y}_i, Y_{i+1})$$

and

$$g(Y_{i-1}, Y_i) \text{ to } g(Y_{i-1}, \hat{Y}_i).$$

Lemma 7.3 allows us to consider only individual order changes of consecutive events. We may ignore the possibility that both $Y_i$ and $Y_{i+1}$ (or $Y_i$ and $Y_{i-1}$) change as $\theta$ is perturbed to $\theta + h$.

We begin by focusing on a single term, $g(Y_{i-1}, Y_i)$. Our basic building block is still the rate, $Q_i(\alpha|\mathcal{H}_i)$, at which $\underline{a}_i$ jumps to $(a_1, \ldots, a_{i-1}, \alpha)$. As with $R_{n_0}$, we may ignore the case $\underline{a}_i(\theta + h) \notin \mathcal{A}_i(\theta)$. But now, when $a_i$ and $a_{i+1}$ change order, we have two jump sizes to consider, depending on whether $Y_i$ appears as the first or second argument of $g$. We denote these by $\Delta_i^{(1)}$ and $\Delta_i^{(2)}$, respectively. We begin with $\Delta_i^{(2)}$. To simplify the notation, write $(\mathcal{H}_i, \alpha)$ to indicate that $\mathcal{H}_i$ holds and $\underline{a}_i(\theta + h) = (a_1(\theta), \ldots, a_{i-1}(\theta), \alpha)$.

$$\begin{aligned} \Delta_i^{(2)}(\alpha|\mathcal{H}_i) &= \lim_{h \downarrow 0} \mathbf{E}[g(Y_{i-1}(\theta + h), Y_i(\theta + h)) - g(Y_{i-1}(\theta), Y_i(\theta))|\mathcal{H}_i, \alpha] \\ &= \mathbf{E}[g(Y_{i-1}(\theta), Y_i(\theta + h)) - g(Y_{i-1}(\theta), Y_i(\theta))|\mathcal{H}_i, \alpha] \\ &= \sum_{s'} p(s'; Y_{i-1}, \alpha)[g(Y_{i-1}, s') - g(Y_{i-1}, Y_i)], \end{aligned}$$

with our earlier convention that $\Delta_i^{(2)}(\alpha|\mathcal{H}_i) = 0$ if $\alpha \notin \mathcal{E}(Y_{i-1})$; i.e., if the sequence $(a_1, \ldots, a_{i-1}, \alpha)$ is not feasible.

The case of $\Delta_i^{(1)}(\alpha|\mathcal{H}_i)$ is a bit more involved. We must examine the change in $g(Y_i, Y_{i+1})$ under a change in $a_i$, given $\mathcal{H}_i$. The (minor) complication is that $Y_{i+1}$ is not completely determined by $\mathcal{H}_i$ the way $Y_{i-1}$ is; thus,

$$\begin{aligned} &\lim_{h \downarrow 0} \mathbf{E}[g(Y_i(\theta + h), Y_{i+1}(\theta + h)) - g(Y_i(\theta), Y_{i+1}(\theta))|\mathcal{H}_i, \alpha] \\ &\quad \neq \sum_{s'} p(s'; Y_{i-1}, \alpha)[g(s', Y_{i+1}) - g(Y_i, Y_{i+1})]. \end{aligned}$$

Instead, we need the conditional expectation of the right-hand side over the conditional distribution of $Y_{i+1}$. For this, we need to look more closely at the possible values of $Y_{i+1}$.

Recall the condition that for all $s, s', s''$ and all $\alpha \in \mathcal{E}(s)$

$$p(s'; s, \alpha) > 0, p(s''; s, \alpha) > 0 \text{ and } p(s'; s, \alpha) = p(s''; s, \alpha) \Rightarrow s' = s'' \tag{7.29}$$

used in Chapter 3, Section 3.1.1 to derive (C2') from (C2). This condition states simply that no two transitions out of a state due to the same event may

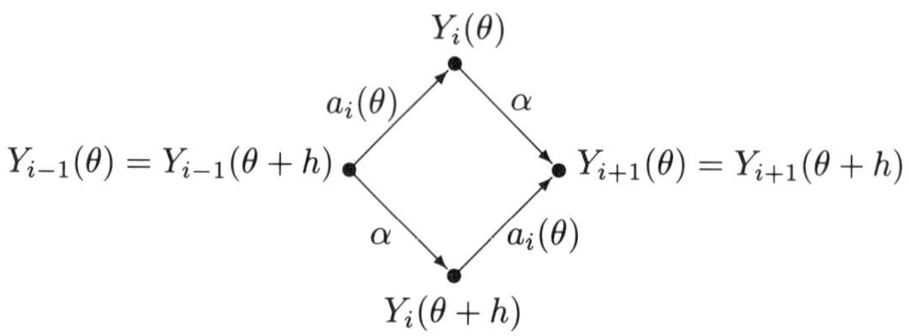


Figure 7.1: Given $Y_{i-1}(\theta), Y_i(\theta)$ and $Y_i(\theta + h)$, $Y_{i+1}(\theta)$ is determined

have exactly the same probability. Let us assume it is in effect. (Alternatively, just assume that our choice of $\phi$ satisfies (C2').)

Given $(\mathcal{H}_i, \alpha)$, $Y_{i-1}(\theta + h) = Y_{i-1}(\theta)$ so the conditional distribution of $Y_i(\theta + h)$ is $p(\cdot; Y_{i-1}(\theta), \alpha)$. If we further condition on $Y_i(\theta + h) = s$, what is the distribution of $Y_{i+1}(\theta)$? Under (C2) and (7.29), $Y_{i+1}(\theta)$ is completely determined by $(\mathcal{H}_i, \alpha, Y_i(\theta + h) = s)$. For under (C2), when $a_i(\theta)$ and $\alpha$ change order, $Y_i(\theta + h)$ is $s$ only if

$$p(s; Y_{i-1}(\theta), \alpha) = p(Y_{i+1}(\theta); Y_i(\theta), \alpha); \tag{7.30}$$

and under (7.29), this uniquely determines $Y_{i+1}(\theta)$. (See Figure 7.1.) Let us denote the $Y_{i+1}(\theta)$ so determined by $\phi_{s,\alpha}(Y_i(\theta))$.

(The GSMP framework here obscures what is really a simple fact. Consider the case of a closed Jackson-like network. Let $Y_{i-1}(\theta) = (n_1, \ldots, n_M)$, a vector of queue lengths. Let

$$\begin{aligned} Y_i(\theta) &= Y_{i-1}(\theta) - e_j + e_k, \\ Y_i(\theta + h) &= Y_{i-1}(\theta) - e_\ell + e_m. \end{aligned}$$

At $\theta$, the $i$th event corresponds to the movement of a job from $j$ to $k$, at $\theta + h$, when $a_i$ and $a_{i+1}$ change order, it corresponds to the movement of a job from $\ell$ to $m$. We deduce that if only the $i$th and $i+1$st events change order, then at $\theta$ the $i+1$st event is the movement of a job from $\ell$ to $m$, so $Y_{i+1}(\theta) = Y_i(\theta) - e_\ell + e_m$.)

We now have

$$\begin{aligned} &\Delta_i^{(1)}(\alpha|\mathcal{H}_i) \\ &= \mathbf{E}[g(Y_i(\theta + h), Y_{i+1}(\theta + h)) - g(Y_i(\theta), Y_{i+1}(\theta))|\mathcal{H}_i, \alpha] \\ &= \sum_s p(s; Y_{i-1}, \alpha)\mathbf{E}[g(s, Y_{i+1}) - g(Y_i, Y_{i+1})|\mathcal{H}_i, \alpha, Y_i(\theta + h) = s] \\ &= \sum_s p(s; Y_{i-1}, \alpha)[g(s, \phi_{s,\alpha}(Y_i)) - g(Y_i, \phi_{s,\alpha}(Y_i))]. \end{aligned}$$

This analysis treats a single term of the form $g(Y_i, Y_{i+1})$. To treat $D_{n_0}$, we simply sum over the transitions $1, \ldots, n_0$. With the convention that $Q_0 \equiv \Delta_0^{(\ell)} \equiv 0$, $\ell = 1, 2$, we obtain the derivative estimate

$$\sum_{i=1}^{n_0} \sum_{\alpha} Q_{i-1}(\alpha|\mathcal{H}_{i-1})\Delta_{i-1}^{(1)}(\alpha|\mathcal{H}_{i-1}) + Q_i(\alpha|\mathcal{H}_i)\Delta_i^{(2)}(\alpha|\mathcal{H}_i). \tag{7.31}$$

**Theorem 7.2.** Suppose that (H1)-(H4) and (C2) hold, that every $X_\theta(\alpha, k)$ has finite second moment, that $g$ is bounded, and that $\mathbf{E}[D_{n_0}(\theta)]$ is differentiable in $\theta$. Then the expectation of (7.31) is $d\mathbf{E}[D_{n_0}(\theta)]/d\theta$.

**Proof.** Let $\delta_h g_i = g(Y_{i-1}(\theta+h), Y_i(\theta+h)) - g(Y_{i-1}(\theta), Y_i(\theta))$, and consider the interchange of limit and expectation for terms of the form $\lim_{h\downarrow 0} h^{-1}\mathbf{E}[\delta_h g_i]$. Write this expectation as

$$\begin{aligned}&\mathbf{E}[\delta_h g_i \cdot \mathbf{1}\{\underline{a}_{i-1}(\theta+h) \in \mathcal{A}_{i-1}(\theta), \underline{a}_i(\theta+h) \in \mathcal{A}_i(\theta)\}]+\\ &\quad +\mathbf{E}[\delta_h g_i \cdot \mathbf{1}\{\underline{a}_{i-1}(\theta+h) \notin \mathcal{A}_{i-1}(\theta) \text{ or } \underline{a}_i(\theta+h) \notin \mathcal{A}_i(\theta)\}].\end{aligned}$$

The second term is negligible by Lemma 7.3. The first term simplifies to

$$\begin{aligned}&\mathbf{E}[\delta_h g \cdot \mathbf{1}\{\text{only } a_{i-1}, a_i \text{ change order}\}]\\ +\;&\mathbf{E}[\delta_h g \cdot \mathbf{1}\{\text{only } a_i, a_{i+1} \text{ change order}\}] + o(h)\end{aligned}$$

because multiple order changes are negligible and $\delta_h g$ is zero if no events change order. If we rewrite these two terms as expectations of conditional expectations given $\mathcal{H}_{i-1}$ and $\mathcal{H}_i$ respectively, then divide by $h$ and let $h \downarrow 0$, we get the $i$th summand in (7.31), provided the limit can be interchanged with the expectations over $\mathcal{H}_{i-1}$ and $\mathcal{H}_i$. The justification for this interchange is the same as that for $R_{n_0}$. □

## 7.4 Stopping at a Fixed Time

This section applies smoothing to $\mathbf{E}[R_T(\theta)]$ and $\mathbf{E}[D_T(\theta)]$ for a fixed time $T > 0$. The overall approach is similar to that of Section 7.3, but there is one important difference. Whereas previously we only needed to consider changes in the order of (consecutive) events, we must now also consider order changes between event epochs and the fixed time $T$. Take the case of $\mathbf{E}[R_T(\theta)] = \mathbf{E}[f(Z_T(\theta))]$. If a small change in $\theta$ allows one more or one less event to occur in $(0, T]$, then $Z_T$ (hence, $R_T$) jumps.

### 7.4.1 The Rate of Crossing a Fixed Time

We begin by carrying out a derivation analogous to that of Section 7.2 (for event order changes), but for the rate at which the epoch of an event crosses a fixed time. As in Section 7.2, we first consider individual clock samples. In the next section we generalize to event epochs.

Let $X_\theta$ be a generic clock sample, and consider the the conditional rate at which it crosses time $T$. If $X_\theta \leq T$, then

$$\begin{aligned}&\lim_{h\downarrow 0}\frac{1}{h}P(X_{\theta+h} > T|X_\theta \leq T)\\ &= \lim_{h\downarrow 0}\frac{1}{h}P(\Psi(X_\theta, h) > T|X_\theta \leq T)\\ &= \lim_{h\downarrow 0}\frac{1}{h}P(X_\theta > \Psi(T,-h)|X_\theta \leq T).\end{aligned}$$

If $X_\theta$ has distribution $F$, then this is

$$\lim_{h\downarrow 0}\frac{1}{h}\frac{F(T) - F(\Psi(T,-h))}{F(T)},$$

if $\Psi(T,-h) \leq T$, and zero otherwise. Passing to the limit, we get

$$\frac{F'(T)}{F(T)}[\psi(T)]^+ = \frac{F'(T)}{F(T)}\left[\left.\frac{dX_\theta}{d\theta}\right|_{X_\theta=T}\right]^+, \tag{7.32}$$

where $F'$ is the density of $F$.

If, instead, $X_\theta > T$, we have

$$\lim_{h\downarrow 0}\frac{1}{h}P(X_{\theta+h} \leq T|X_\theta > T) \tag{7.33}$$

$$\begin{aligned}&= \lim_{h\downarrow 0}\frac{1}{h}P(\Psi(X_\theta, h) \leq T|X_\theta > T)\\ &= \lim_{h\downarrow 0}\frac{1}{h}P(X_\theta \leq \Psi(T,-h)|X_\theta > T)\\ &= \lim_{h\downarrow 0}\frac{1}{h}\frac{F(\Psi(T,-h)) - F(T)}{1-F(T)},\end{aligned}$$

if $\Psi(T,-h) \geq T$, zero otherwise. This is

$$\frac{F'(T)}{1-F(T)}[-\psi(T)]^+ = \frac{F'(T)}{1-F(T)}\left[-\left.\frac{dX_\theta}{d\theta}\right|_{X_\theta=T}\right]^+.$$

Interestingly, equation (7.33) also defines the parametric hazard rate of $X_\theta$; see the derivation following Defintion 6.2, Chapter 6.

### 7.4.2 The Smoothed Derivative Estimators

Recall that $N_T(\theta)$ is the number of transitions made by the process $\{Z_t(\theta), t \geq 0\}$ in $(0,T]$. To simplify notation, for the rest of this section we take $T > 0$ and $\theta$ to be fixed, and write $N$ for $N_T(\theta)$. Thus, $R_T = f(Z_T) = f(Y_N)$. We will argue that, as far as the derivative of $\mathbf{E}[R_T(\theta)]$ is concerned, we only need to consider jumps in $N_T(\theta)$. In particular, it is enough to consider the rate at which $\tau_N$ (the last event in $(0,T]$) and $\tau_{N+1}$ (the first event after $T$) cross $T$.

Let

$$\mathcal{H}_{N-1} = (N; a_1, \ldots, a_N; Y_1, \ldots, Y_{N-1}; \tau_1, \ldots, \tau_{N-1}),$$

and

$$\mathcal{H}_N = (N; a_1, \ldots, a_N; Y_1, \ldots, Y_N; \tau_1, \ldots, \tau_N).$$

Note that $\mathcal{H}_{N-1}$ is not obtained from $\mathcal{H}_N$ by replacing the index $N$ with $N-1$. In particular, $\mathcal{H}_{N-1}$ includes the event $a_N$ (though not its epoch).

Define $\underline{a}_T$ to be the sequence of events in $(0,T]$; $\underline{a}_T$ differs from $\underline{a}_N = (a_1, \ldots, a_N)$ in that

$$\underline{a}_T(\theta + h) = (a_1(\theta + h), \ldots, a_{N_T(\theta+h)}(\theta + h))$$

whereas

$$\underline{a}_N(\theta + h) = (a_1(\theta + h), \ldots, a_{N_T(\theta)}(\theta + h)),$$

because without arguments $N \equiv N_T(\theta)$.

We focus on changes in $\underline{a}_T$ that correspond to the occurrence of one more or one less event in $(0,T]$. However, order changes among $a_1, \ldots, a_N$ are still possible and must be considered. Call $\underline{a}'$ a *transposition* of $\underline{a}_T$ if it is obtained from $\underline{a}_T$ by transposing at most one pair of events $a_i, a_{i+1}$. Let

$$\begin{aligned} \mathcal{A}_T(\theta) &= \{\text{transpositions of } \underline{a}_T(\theta)\} \\ &\cup \{\underline{a}_{N-1}(\theta)\} \\ &\cup \{(a_1(\theta), \ldots, a_N(\theta), \alpha), \alpha \in \mathcal{E}(Y_N)\}. \end{aligned}$$

An argument similar to that of Lemma 7.3 shows that, under the conditions in that lemma,

$$P(\underline{a}_T(\theta + h) \notin \mathcal{A}_T(\theta) | \mathcal{H}_{N-1}) = o(h),$$
$$P(\underline{a}_T(\theta + h) \notin \mathcal{A}_T(\theta) | \mathcal{H}_N) = o(h),$$

with probability one, and

$$P(\underline{a}_T(\theta + h) \notin \mathcal{A}_T(\theta)) = o(h).$$

Moreover, for a GSMP satisfying (C2), permuting $(a_1, \ldots, a_N)$ leaves $Y_N$ unchanged; thus, only event sequences in the second and third rows of the definition of $\mathcal{A}_T$ are canditates for $\underline{a}_T(\theta + h)$ that introduce a jump in $R_T$.

For $\alpha \in \mathcal{E}(Y_N)$, denote by $Q_T(\alpha|\mathcal{H}_N)$ the conditional rate at which $\underline{a}_T$ jumps to $(a_1(\theta), \ldots, a_N(\theta), \alpha)$. Previously, we let $t_n(\alpha)$ be the age of the $\alpha$ clock active (if any) at $\tau_n$. Now let $t_T(\alpha)$ be the age at $T$, provided $\alpha \in \mathcal{E}(Z_T)$; since $t_T(\alpha) = t_N(\alpha) + (T - \tau_N)$, it is determined by $\mathcal{H}_N$. Adapting the argument of Section 7.3.2, we find that

$$Q_T(\alpha|\mathcal{H}_N) = \mu_\alpha(t_T(\alpha)) \left[ - \left. \frac{d\tau_{N+1}}{d\theta} \right|_{\alpha,T} \right]^+,$$

if $\alpha \in \mathcal{E}(Y_N)$ and zero otherwise. To evaluate the term in square brackets, let $k_\alpha$ be one more than the number of instances of $\alpha$ among $a_1, \ldots, a_N$. Then

$$\left. \frac{d\tau_{N+1}}{d\theta} \right|_{\alpha,T} = \sum_{i=1}^{N} \frac{dX_\theta(r_i)}{d\theta} \eta(\alpha, k_\alpha; r_i) + \psi_\alpha(t_T(\alpha)). \tag{7.34}$$

We also need the rate at which $\underline{a}_T$ jumps to $(a_1, \ldots, a_{N-1})$. Clearly, this cannot be represented as the addition of any event $\alpha$ to $\underline{a}_T$. This makes it awkward to represent the corresponding jump rate using notation analogous to $Q_T(\alpha|\mathcal{H}_N)$. Purely as a notational device, let us refer to this case using the argument $-a_N$, indicating that the last event is dropped. We condition on $\mathcal{H}_{N-1}$; thus $Q_T(-a_N|\mathcal{H}_{N-1})$ is the rate at which $\underline{a}_T$ jumps from $(a_1, \ldots, a_{N-1}, a_N)$ to $(a_1, \ldots, a_{N-1})$. Recall that $a_N$ is determined by $\mathcal{H}_{N-1}$.

The jump rate $Q_T(-a_N|\mathcal{H}_{N-1})$ looks a little different from the others. As in (7.33), the probability element is given not by a hazard rate but by the ratio of a density to a distribution function. In the case of (7.33), we have $F'(T)/F(T)$. The density element for $Q_T(-a_N|\mathcal{H}_{N-1})$ generalizes this.

Let $\alpha$ be any event in $\mathcal{E}(Y_{N-1})$. Given that its clock has attained an age of $t_{N-1}(\alpha)$, the density of the time until the next occurrence of $\alpha$ is

$$\frac{f_\alpha(t_{N-1}(\alpha) + x)}{1 - F_\alpha(t_{N-1}(\alpha))}.$$

The corresponding distribution is

$$\frac{F_\alpha(t_{N-1}(\alpha) + x) - F_\alpha(t_{N-1}(\alpha))}{1 - F_\alpha(t_{N-1}(\alpha))}.$$

For $Q_T(-a_N|\mathcal{H}_{N-1})$ we need the conditional probability density that $a_N$ occurs right at $T$. Take $\alpha = a_N$. Then the conditional probability density that $\alpha$ occurs at $T$, given that it occurs somewhere in $(\tau_{N-1}, T]$, is the ratio of the density and distribution above, evaluated at $x = T - \tau_{N-1}$. Since $t_{N-1}(\alpha) + (T - \tau_{N-1}) = t_T(\alpha)$, this is

$$\frac{f_\alpha(t_T(\alpha))}{F_\alpha(t_T(\alpha)) - F_\alpha(t_{N-1}(\alpha))}.$$

We now have

$$Q_T(-a_N|\mathcal{H}_{N-1}) = \frac{f_{a_N}(t_T(a_N))}{F_{a_N}(t_T(a_N)) - F_{a_N}(t_{N-1}(a_N))}\left[\left.\frac{d\tau_N}{d\theta}\right|_{a_N,T}\right]^+. \quad (7.35)$$

The bracketed term is evaluated as in (7.26) and (7.34).

To complete our derivative estimators we need the corresponding jump sizes. For $R_T$, the $\Delta_T$'s are easy to evaluate:

$$\begin{aligned}
&\Delta_T(\alpha|\mathcal{H}_N) \\
&= \lim_{h\downarrow 0}\mathbf{E}[f(Z_T(\theta+h)) - f(Z_T(\theta))|\mathcal{H}_N, \underline{a}_T(\theta+h) = (a_1,\ldots,a_N,\alpha)] \\
&= \sum_{s'} p(s';Y_N,\alpha)[f(s') - f(Y_N)],
\end{aligned}$$

and similarly

$$\Delta_T(-a_N|\mathcal{H}_N) = \sum_{s'} p(s';Y_{N-1},a_N)[f(Y_{N-1}) - f(s')].$$

Since

$$D_T(\theta) = \sum_{i=1}^{N_T(\theta)} g(Y_{i-1}(\theta), Y_i(\theta)),$$

if $\underline{a}_T$ jumps to $(a_1,\ldots,a_N,\alpha)$, we gain an additional term, $g(Y_N, Y_{N+1})$. Similarly, if $\underline{a}_T$ jumps to $(a_1,\ldots,a_{N-1})$ we lose the last term $g(Y_{N-1}, Y_N)$. Thus, let

$$\Delta_T^{(1)}(\alpha|\mathcal{H}_N) = \sum_{s'} p(s';Y_N,\alpha)g(Y_N,s'),$$

and

$$\Delta_T^{(2)}(-a_N|\mathcal{H}_N) = -\sum_{s'} p(s';Y_{N-1},a_N)g(Y_{N-1},s').$$

Putting the pieces together, we get the derivative estimate

$$Q_T(-a_N|\mathcal{H}_{N-1})\Delta_T(-a_N|\mathcal{H}_{N-1}) + \sum_{\alpha} Q_T(\alpha|\mathcal{H}_N)\Delta_T(\alpha|\mathcal{H}_N) \quad (7.36)$$

for $R_T$. For $D_T$, we must also consider order changes between $a_i$ and $a_{i+1}$, $i = 1,\ldots,N_T - 1$, so we get

$$\begin{aligned}
&\sum_{i=1}^{N_T-1}\sum_{\alpha} Q_{i-1}(\alpha|\mathcal{H}_{i-1})\Delta_{i-1}^{(1)}(\alpha|\mathcal{H}_{i-1}) + Q_i(\alpha|\mathcal{H}_i)\Delta_i^{(2)}(\alpha|\mathcal{H}_i) \\
&+ \sum_{\alpha} Q_T(\alpha|\mathcal{H}_N)\Delta_T^{(1)}(\alpha|\mathcal{H}_N) \\
&+ Q_T(-a_N|\mathcal{H}_{N-1})\Delta_T^{(2)}(-a_N|\mathcal{H}_{N-1}). \quad (7.37)
\end{aligned}$$

As in the previous cases, we have

**Theorem 7.3.** Suppose that (H1)-(H4) and (C2) hold, that every $X_\theta(\alpha, k)$ has finite second moment, that $f$ and $g$ are bounded, and that $\mathbf{E}[R_T(\theta)]$ and $\mathbf{E}[D_T(\theta)]$ are differentiable in $\theta$. Then the expectation of (7.36) is $d\mathbf{E}[R_T(\theta)]/d\theta$ and that of (7.37) is $d\mathbf{E}[D_T(\theta)]/d\theta$.

## 7.5 The Dynkin and Lévy Formulas

For *Markovian* GSMPs—meaning GSMPs in which all clock distributions are exponential—an alternative approach to smoothing is available, based on two Markov chain identities called the Dynkin and Lévy formulas. This approach also smooths by introducing a conditional expectation; but now it is the dependence of the process on the time parameter, $t$, rather than the differentiation parameter, $\theta$, that is smoothed. Conditioning enters through the definition of the infinitesimal generator of the process, which plays a central role.

To avoid confusion with the jump rates of Sections 7.3 and 7.4, let us use $\mathbf{Q}_\theta$ rather than $Q_\theta$ to denote the generator of $\{Z_t(\theta), t \geq 0\}$. If each clock distribution $F_\alpha(x, \theta)$ is exponential with mean $\mu_\alpha^{-1}(\theta)$, the entries of $\mathbf{Q}_\theta$ are given by

$$\mathbf{Q}_\theta(s, s') = \sum_\alpha \mathbf{1}\{\alpha \in \mathcal{E}(s)\}\mu_\alpha(\theta)p(s'; s, \alpha) \tag{7.38}$$

for $s' \neq s$, and

$$\mathbf{Q}_\theta(s, s) = -\sum_{s' \neq s} \mathbf{Q}_\theta(s, s'). \tag{7.39}$$

Also let $\mathbf{Q}_\theta(s) = -\mathbf{Q}_\theta(s, s)$. While identity in (7.38) is clear on physical grounds, the *definition* of $\mathbf{Q}_\theta(s, s')$ is, more precisely,

$$\mathbf{Q}_\theta(s, s') = \lim_{h \downarrow 0} \frac{P(Z_{t+h}(\theta) = s' | Z_t(\theta) = s)}{h}. \tag{7.40}$$

The limit on the right is independent of $t$ because the $\mu_\alpha$'s are independent of $t$. From (7.40), an analogy between the jump rates $Q_n$ and $\mathbf{Q}_\theta$ becomes clear; compare with (7.3) and (7.19). The key difference is that $Q_n$ describes conditional jump rates under changes in $\theta$, while the jumps reflected in $\mathbf{Q}_\theta$ are due to changes in $t$. If we simply differentiate $Z$ with respect to $t$ without conditioning (as in (7.40)) we get $dZ_t(\theta)/dt = 0$ wherever the derivative exists; so here, too, conditioning is necessary for a non-trivial limit.

Though we usually think of $\mathbf{Q}_\theta(s, s')$ as a jump rate, $\mathbf{Q}_\theta$ itself is also an operator acting on functions. If $f : \mathbf{S} \to \mathbf{R}$ let

$$\widehat{f}(\theta, s) = \sum_{s'} \mathbf{Q}_\theta(s, s')f(s'), \tag{7.41}$$

and if $g : \mathbf{S} \times \mathbf{S} \to \mathbf{R}$ let

$$\widehat{g}(\theta, s) = \sum_{s' \neq s} \mathbf{Q}_\theta(s, s')g(s, s'). \tag{7.42}$$

If we think of $f$ as a column vector indexed by $s$, then $\widehat{f}$ is just the vector obtained from multiplication of $f$ by the matrix $\mathbf{Q}_\theta$. For $\widehat{g}$, think of each $g(s, \cdot)$ as a column vector. In light of (7.40) we also have

$$\widehat{f}(\theta, s) = \lim_{h \downarrow 0} \frac{\mathbf{E}[f(Z_{t+h}(\theta)) - f(Z_t(\theta))|Z_t(\theta) = s]}{h} \tag{7.43}$$

and an analogous expression for $\widehat{g}$.

With these definitions, we can state the fundamental identities of this section. Drop, for the moment, the parameter $\theta$. Let $f$ and $g$ be bounded, and let $\sigma$ be any (almost surely finite) *stopping time* for $Z$. Then *Dynkin's formula* is

$$\mathbf{E}[f(Z_\sigma)] = \mathbf{E}[f(Z_0)] + \mathbf{E}\left[\int_0^\sigma \widehat{f}(Z_t)dt\right], \tag{7.44}$$

and *Lévy's formula* is

$$\mathbf{E}[\sum_{i:\tau_i \leq \sigma} g(Y_{i-1}, Y_i)] = \mathbf{E}\left[\int_0^\sigma \widehat{g}(Z_t)dt\right]. \tag{7.45}$$

These expressions make sense intuitively because of the interpretation in (7.43) of $\widehat{f}$ and $\widehat{g}$ as derivatives. By specializing to $\sigma = \tau_{n_0}$ and $\sigma = T$, the left-hand sides of (7.44) and (7.45) provide alternative expressions for $\mathbf{E}[R_{n_0}], \mathbf{E}[R_T]$, $\mathbf{E}[D_{n_0}]$ and $\mathbf{E}[D_T]$. Furthermore, if we reintroduce the parameter $\theta$, these identities hold for all $\theta$.

Let

$$\widehat{R}_{n_0}(\theta) = \int_0^{\tau_{n_0}} \widehat{f}(\theta, Z_t(\theta))dt,$$

$$\widehat{D}_{n_0}(\theta) = \int_0^{\tau_{n_0}} \widehat{g}(\theta, Z_t(\theta))dt,$$

and define $\widehat{R}_T$ and $\widehat{D}_T$ similarly by integrating up to $T$. It follows from Lévy's formula that if $\mathbf{E}[D_{n_0}(\theta)]$ and $\mathbf{E}[D_T(\theta)]$ are differentiable, then

$$\frac{d\mathbf{E}[D_{n_0}]}{d\theta} = \frac{d\mathbf{E}[\widehat{D}_{n_0}]}{d\theta} \text{ and } \frac{d\mathbf{E}[D_T]}{d\theta} = \frac{d\mathbf{E}[\widehat{D}_T]}{d\theta}.$$

If, as we always assume, the initial state $Z_0$ is independent of $\theta$, then Dynkin's formula implies

$$\frac{d\mathbf{E}[R_{n_0}]}{d\theta} = \frac{d\mathbf{E}[\widehat{R}_{n_0}]}{d\theta} \text{ and } \frac{d\mathbf{E}[R_T]}{d\theta} = \frac{d\mathbf{E}[\widehat{R}_T]}{d\theta}.$$

These identities provide an alternative route to derivative estimation. Rather than try to estimate, say, $d\mathbf{E}[R_{n_0}]/d\theta$ directly, we can work with $\widehat{R}_{n_0}$ to estimate $d\mathbf{E}[\widehat{R}_{n_0}]/d\theta$. The performance measures $\widehat{R}_{n_0}$ and $\widehat{D}_{n_0}$ have (almost) the form of $L_{n_0}$, $\widehat{R}_T$ and $\widehat{D}_T$ (almost) the form of $L_T$, so the results of Chapter 3 are applicable. We say "almost" because of one small difference: even if $f$ and $g$ do not depend on $\theta$, $\widehat{f}$ and $\widehat{g}$ *do* depend on $\theta$ because $\mathbf{Q}_\theta$ does. Hence, the integrands of the $\widehat{R}$'s and $\widehat{D}$'s depend explicitly on $\theta$ as well as implicitly through $Z$. This puts us in the setting of Corollary 3.5 in Section 3.1.2 of Chapter 3.

To facilitate interchanges of derivatives with sums over $\mathbf{S}$, let us take $\mathbf{S}$ to be finite. Assume, also, that every $\mu_\alpha(\theta)$ is continuously differentiable in $\theta$ throughout $\Theta$. This makes every $\mathbf{Q}_\theta(s,s')$ and $\widehat{f}$ and $\widehat{g}$ continuously differentiable in $\theta$ as well. As usual, take the exponential clock samples for $\theta_1, \theta_2 \in \Theta$ to be related by

$$X_{\theta_1}(\alpha,k)\mu_\alpha(\theta_1) = X_{\theta_2}(\alpha,k)\mu_\alpha(\theta_2);$$

then $X_\theta(\alpha,k)$ inherits differentiability from $\mu_\alpha(\theta)$. As in Chapter 3, by differentiating we get

$$\frac{d\widehat{R}_{n_0}}{d\theta} = \int_0^{\tau_{n_0}} \partial_\theta \widehat{f}(\theta, Z_t)dt + \sum_{i=0}^{n_0-1} \widehat{f}(\theta, Y_i)[\frac{d\tau_{i+1}}{d\theta} - \frac{d\tau_i}{d\theta}] \tag{7.46}$$

and

$$\frac{d\widehat{R}_T}{d\theta} = \int_0^T \partial_\theta \widehat{f}(\theta, Z_t)dt + \sum_{i=1}^{N_T} \frac{d\tau_i}{d\theta}[\widehat{f}(\theta, Y_{i-1}) - \widehat{f}(\theta, Y_i)], \tag{7.47}$$

where

$$\partial_\theta \widehat{f}(\theta, s) = \sum_{s'} \partial_\theta \mathbf{Q}_\theta(s,s')f(s').$$

Similarly

$$\frac{d\widehat{D}_{n_0}}{d\theta} = \int_0^{\tau_{n_0}} \partial_\theta \widehat{g}(\theta, Z_t)dt + \sum_{i=0}^{n_0-1} \widehat{g}(\theta, Y_i)[\frac{d\tau_{i+1}}{d\theta} - \frac{d\tau_i}{d\theta}] \tag{7.48}$$

and

$$\frac{d\widehat{D}_T}{d\theta} = \int_0^T \partial_\theta \widehat{g}(\theta, Z_t)dt + \sum_{i=1}^{N_T} \frac{d\tau_i}{d\theta}[\widehat{g}(\theta, Y_{i-1}) - \widehat{g}(\theta, Y_i)], \tag{7.49}$$

where

$$\partial_\theta \widehat{g}(\theta, s) = \sum_{s' \neq s} \partial_\theta \mathbf{Q}_\theta(s,s')g(s,s').$$

If we can show that the expectations of these derivatives are the derivatives of the corresponding $\mathbf{E}[\widehat{R}]$'s and $\mathbf{E}[\widehat{D}]$'s, we will have found unbiased derivative estimators for the $\mathbf{E}[R]$'s and $\mathbf{E}[D]$'s.

**Theorem 7.4.** Suppose (C2) holds. Suppose $\mathbf{S}$ is finite and suppose every $\mu_\alpha(\theta)$, $\alpha \in \mathbf{A}$, is strictly positive and continuously differentiable in $\theta$ on an interval $[\theta_1, \theta_2]$. Then the expectations of (7.46)-(7.49) are, respectively, $d\mathbf{E}[R_{n_0}]/d\theta$, $d\mathbf{E}[R_T]/d\theta$, $d\mathbf{E}[D_{n_0}]/d\theta$ and $d\mathbf{E}[D_T]/d\theta$, at every $\theta_1 \le \theta \le \theta_2$.

**Proof.** We only need to validate the interchange of derivative and expectation for the $\widehat{R}$'s and $\widehat{D}$'s. Since $\widehat{f}$ and $\widehat{g}$ are $C^1$ in $\theta$ and $\mathbf{S}$ is finite, $\partial_\theta \widehat{f}$ and $\partial_\theta \widehat{g}$ are bounded on $[\theta_1, \theta_2] \times \mathbf{S}$. Thus, the conditions of Corollary 3.5 are satisfied. For the interchange, we check the conditions of Corollary 3.3. Conditions (A1) and (A2) of Chapter 2 are automatically satisfied because all clocks are exponential and $\mu_\alpha(\theta)$ is continuously differentiable. Since $[\theta_1, \theta_2]$ is closed, $\sup_\theta \mu_\alpha(\theta)$ is finite, and that ensures that $P(\inf_\theta X_\theta(\alpha, k) = 0) = 0$. Similarly, $\inf_\theta \mu_\alpha(\theta) > 0$ so $\mathbf{E}[\sup_\theta X_\theta(\alpha, k)] = (\inf_\theta \mu_\alpha(\theta))^{-1}$ is finite. Thus, the conditions of Corollary 3.3 hold. □

This result raises the following question: When the methods of Sections 7.3-7.4 and this section are both available (i.e., when the GSMP is Markovian and (C2) holds), which method is better? More precisely, what is the relative computational effort and relative variance for the two methods? We do not have answers to these questions; it seems unlikely that simple answers exist. It is certainly not the case that either method is universally preferable.

There is, however, a special class of processes for which the two methods can be compared—for which, in fact, they *coincide*. Recall from Chapter 5, Section 5.4 that $\theta$ is a *scale parameter* for $\mathbf{Q}_\theta$ if $\mathbf{Q}_\theta(s, s') > 0$ implies that $[\partial_\theta \mathbf{Q}_\theta(s, s')/\mathbf{Q}_\theta(s, s')] = [\partial_\theta \mathbf{Q}_\theta(s)/\mathbf{Q}_\theta(s)]$. Recall that this implies that $\partial_\theta[\mathbf{Q}_\theta(s, s')/\mathbf{Q}_\theta(s)] = 0$, so a scale parameter affects the holding times of $\{Z_t(\theta), t \ge 0\}$ but not the sequence of states $Y_1, Y_2, \ldots$.

In the GSMP setting, scale parameters arise if we have *speeds* that depend on the state only; i.e., if in state $s$, all events in $\mathcal{E}(s)$ have the same speed $\lambda(s, \theta)$. Suppose that the rates $\mu_\alpha$ do not depend on $\theta$; only the speeds do. This automatically makes $\theta$ a scale parameter. Clearly, in this case, changes in $\theta$ cannot change the order of events. The introduction of speeds makes

$$\mathbf{Q}_\theta(s, s') = \sum_\alpha \mathbf{1}\{\alpha \in \mathcal{E}(s)\}\lambda(s, \theta)\mu_\alpha p(s'; s, \alpha).$$

As shown in Section 5.4, when $\theta$ is a scale parameter (and the integrand, $f$, of $L_{n_0}$ and $L_T$ does not depend on $\theta$),

$$\frac{dL_{n_0}}{d\theta} = -\int_0^{\tau_{n_0}} f(Z_t)\frac{\mathbf{Q}'_\theta(Z_t)}{\mathbf{Q}_\theta(Z_t)}dt \tag{7.50}$$

and

$$\frac{dL_T}{d\theta} = -\int_0^{\tau_N} f(Z_t)\frac{\mathbf{Q}'_\theta(Z_t)}{\mathbf{Q}_\theta(Z_t)}dt + f(Z_T)\int_0^{\tau_N} \frac{\mathbf{Q}'_\theta(Z_t)}{\mathbf{Q}_\theta(Z_t)}dt, \tag{7.51}$$

where $\mathbf{Q}'_\theta = \partial_\theta \mathbf{Q}_\theta$ and $N = N_T(\theta)$. As noted in Section 5.4, these identities do not depend on (CM) or (C2) or any other such restriction on the structure of the state space. Using (7.50) and (7.51) we can now show that for scale parameters, the estimators based on the Dynkin and Lévy formulas coincide with those of Sections 7.3-7.4:

**Theorem 7.5.** Suppose $\mathbf{S}$ is finite, every $\lambda(s,\theta)$ is continuously differentiable, and $f$ and $g$ are bounded. Then $d\widehat{R}_{n_0}/d\theta$ and $d\widehat{D}_{n_0}/d\theta$ are equal to the estimators of Theorems 7.1 and 7.2; all are zero. If, in addition, every $\mathbf{Q}'_\theta(s) \geq 0$, then $d\widehat{R}_T/d\theta$ and $d\widehat{D}_T/d\theta$ are equal to the estimators of Theorem 7.3, and are not necessarily zero.

**Proof.** First note that since a scale parameter does not change the sequence $\{Y_n\}$, $\mathbf{E}[R_{n_0}]$ and $\mathbf{E}[D_{n_0}]$ do not depend on $\theta$; hence, the derivative estimates *should* yield zero. In more detail, consider first the smoothed estimates of Section 7.3. Since events cannot change order, every $Q_i(\alpha|\mathcal{H}_i)$ is identically zero. It follows that the estimates for $R_{n_0}$ and $D_{n_0}$ are zero. Now consider $d\widehat{R}_{n_0}/d\theta$. Combining (7.46), (7.50) and the definition of $\widehat{f}$ we find that $d\widehat{R}_{n_0}/d\theta$ is

$$\int_0^{\tau_{n_0}} \sum_s \mathbf{Q}'_\theta(Z_t,s)f(s)dt - \int_0^{\tau_{n_0}} \sum_s \mathbf{Q}_\theta(Z_t,s)f(s)\frac{\mathbf{Q}'_\theta(Z_t)}{\mathbf{Q}_\theta(Z_t)}dt.$$

Since, for a scale parameter

$$\mathbf{Q}'_\theta(Z_t,s) - \mathbf{Q}_\theta(Z_t,s)\frac{\mathbf{Q}'_\theta(Z_t)}{\mathbf{Q}_\theta(Z_t)} = 0, \tag{7.52}$$

the difference of the integrals is zero. The case of $\widehat{D}_{n_0}$ is similar.

Now consider $R_T$. Under the assumption that every $\mathbf{Q}'_\theta(s) \geq 0$, every $\tau_i$ is decreasing in $\theta$ because, for a scale parameter

$$\frac{d\tau_i}{d\theta} = -\int_0^{\tau_i} \frac{\mathbf{Q}'_\theta(Z_t)}{\mathbf{Q}_\theta(Z_t)}dt; \tag{7.53}$$

this is a special case of (7.50). Consequently, we need to consider the rate at which $\tau_{N+1}$ crosses $T$, but not the rate at which $\tau_N$ crosses $T$; the bracketed term in (7.35) is zero. The smoothed estimator for $R_T$ is

$$\sum_s \mathbf{Q}_\theta(Z_T,s)[f(s) - f(Z_T)]\left(-\left.\frac{d\tau_{N+1}}{d\theta}\right|_T\right)$$

$$\begin{aligned}
&= \sum_s \mathbf{Q}_\theta(Z_T, s) f(s) \left( - \left. \frac{d\tau_{N+1}}{d\theta} \right|_T \right) \\
&= \sum_s \mathbf{Q}_\theta(Z_T, s) f(s) \left( \int_0^T \frac{\mathbf{Q}'_\theta(Z_t)}{\mathbf{Q}_\theta(Z_t)} dt \right). \qquad (7.54)
\end{aligned}$$

The first equality follows from $\mathbf{Q}(s,s) = -\sum_{s' \neq s} \mathbf{Q}(s, s')$, the second from (7.53).

Consider, next, the estimator based on Dynkin's formula. From (7.47) and (7.51) we have $d\widehat{R}_T/d\theta$ equal to

$$\begin{aligned}
\int_0^T \partial_\theta \widehat{f}(\theta, Z_t) dt \quad &- \quad \int_0^{\tau_N} \widehat{f}(\theta, Z_t) \frac{\mathbf{Q}'_\theta(Z_t)}{\mathbf{Q}_\theta(Z_t)} dt \\
&+ \quad \widehat{f}(\theta, Z_T) \int_0^{\tau_N} \frac{\mathbf{Q}'_\theta(Z_t)}{\mathbf{Q}_\theta(Z_t)} dt.
\end{aligned}$$

Substituting for $\partial_\theta \widehat{f}$ and cancelling according to (7.52) this simplifies to

$$\widehat{f}(\theta, Z_T) \int_0^T \frac{\mathbf{Q}'_\theta(Z_t)}{\mathbf{Q}_\theta(Z_t)} dt,$$

which is the same as (7.54). For $D_T$ and $\widehat{D}_T$, both derivations yield

$$\widehat{g}(\theta, Z_T) \int_0^T \frac{\mathbf{Q}'_\theta(Z_t)}{\mathbf{Q}_\theta(Z_t)} dt.$$

□

*　　　*　　　*

## Notes and Comments

Smoothing-by-conditioning as a general way of estimating derivatives from discontinuous samples was invented by W.B. Gong [42]. The first examples are in Gong and Ho [45]; an application to a routing problem is considered in Gong [43]. Glasserman and Gong [33] developed the connection between smoothing and the commuting condition. This condition *localizes* the effect of order changes, making it possible to give fairly explicit expressions for the jump rates and jump sizes. These can be implemented with reasonable effort.

A type of smoothing was used earlier in Zazanis [85] to solve a specific problem: estimating the *second* derivative of the mean waiting time in a queue. This suggests another possible application of the approach developed here. Suppose (C2) holds and that $L(\theta)$ is almost surely continuous in $\theta$. Suppose that $L$ is piecewise linear; then its derivative is piecewise constant. To estimate the second derivative of $L$, we might try to evaluate the rate at which the first derivative jumps and the size of its jumps. Finding a general class of systems and performance indices for which this can be carried out is a worthwhile topic for investigation.

The central question in smoothing through conditional expectations is finding what to condition on—or, equivalently, what to integrate out. Generally speaking, the more you integrate the more you smooth. But evaluating conditional expectations (and conditional jump rates) can be difficult, so one might choose to integrate out as little as necessary to achieve continuity. On the other hand, from the method of *conditional Monte Carlo* we know that integrating more always reduces variance, and it is reasonable to expect this to hold in the limit of smoothed derivative estimates. Thus, there is a trade-off between the extra work to evaluate conditional expectations and the reduced variance from integration. The estimators discussed here condition on less (integrate out more) than those in Glasserman and Gong [33]. New estimators that apparently integrate out even more are developed in Gong, Cassandras, Kallmes, and Wardi [44] and in Wardi, Gong, Cassandras, and Kallmes [83].

The Dynkin and Lévy formulas are discussed in, for example, Brémaud [4], where they are related to point process compensators. The use of Dynkin's formula for smoothing was proposed in Glasserman and Gong [32]. A smoothing technique similar to that of Section 7.5, based on compensators, was introduced earlier in Zazanis [87].

In estimating derivatives via likelihood ratios, the discontinuities studied here are entirely irrelevant. The applicability and implementation of the likelihood ratio method do not depend on properties of the particular performance measure under consideration. However, even in this setting, the Dynkin and Lévy formulas lead to alternative estimates. It would be interesting to find conditions under which either an ordinary likelihood ratio estimator or one based on Dynkin's or Lévy's formula is guaranteed to perform better than the other.

# Chapter 8

# Steady-State Derivative Estimation

Up to this point, we have focused exclusively on estimating derivatives of finite-horizon performance measures. But in many applications—some would say most—performance in *steady state* is of primary interest. In this chapter, we take up the question of steady-state derivative estimation. Results from earlier chapters, especially Chapters 2 and 3, will serve us well; indeed, without a thorough analysis of the finite-horizon setting it would be premature to look at steady state. That is one reason for making this chapter the last; another is that the theory for steady state—still an active area of research—is less fully developed. We consider, here, a class of GSMPs for which a fairly complete theory exists.

We begin with a general theorem (Theorem 8.1, due to Hu and Strickland) on convergence to the derivative of steady-state performance. This result is independent of the specific form of the estimator. In order to apply Theorem 8.1 to specific cases, in Section 8.2 we develop some background on convergence to steady state, without reference to derivatives. We consider GSMPs with an explicit *regenerative* structure, and state some known results on regenerative processes in continuous and in discrete time. Using this framework, in Section 8.3 we verify the hypotheses of Theorem 8.1 to establish (for a class of GSMPs) the *strong consistency* of finite-horizon derivative estimates as the time horizon grows. Finally, in Section 8.4, we consider steady-state derivative estimation making explicit use of regenerative cycles, and discuss an associated procedure for producing confidence intervals.

## 8.1 General Conditions for Strong Consistency

Let $\{Z_t(\theta), t \geq 0, \theta \in \Theta\}$ be a family of GSMPs with state space $\mathbf{S}$. Let $f : \mathbf{S} \to \mathbf{R}$ be a real-valued function on this state space. By the steady-state performance of $\{Z_t(\theta), t \geq 0\}$ with cost function $f$ we mean

$$\ell_f(\theta) \triangleq \lim_{t\to\infty} \frac{1}{t} \int_0^t f(Z_u(\theta))du. \tag{8.1}$$

Of course, to talk about steady-state performance we must verify that this limit exists (almost surely). In addition, we will want to ensure that $\ell_f(\theta)$ is a constant (for each $\theta$) which does not depend on the initial state $Z_0$. Conditions that ensure these properties form a traditional, but still active, area of research in applied probability. A large part of the theory of simulation is devoted to finding point and interval estimators of $\ell_f(\theta)$ when such conditions are in force. We therefore have at our disposal many results regarding (8.1); some are reviewed in Section 8.2.

Our main concern is with conditions under which (the deterministic function) $\ell_f(\cdot)$ is differentiable in $\theta$, and with estimation of the *derivative* of steady-state performance, $d\ell_f(\theta)/d\theta$. As usual, let

$$L_T(\theta) = \int_0^T f(Z_t(\theta))dt \quad \text{and} \quad L_n(\theta) = \int_0^{\tau_n} f(Z_t(\theta))dt,$$

where $T > 0$ is real, $n > 0$ is integer, and $\tau_n$ is the epoch of the $n$th transition of $\{Z_t\}$. Then if the limit in (8.1) exists, a.s., $L_T(\theta)/T$ and $L_n(\theta)/\tau_n(\theta)$ converge with probability one to $\ell_f(\theta)$ as $T \to \infty$ and $n \to \infty$. Thus, natural candidate estimators of $\ell'_f(\theta)$ are $\{L'_T(\theta)/T, T \geq 0\}$ and $\{[L_n(\theta)/\tau_n(\theta)]', n = 1, 2, \ldots\}$. From Chapters 2-6 we know how to evaluate these types of estimators and have conditions under which they are unbiased over finite horizons. In this chapter, we focus on their convergence as $T \to \infty$ and $n \to \infty$; i.e., on the validity of

$$\lim_{T\to\infty} \frac{d}{d\theta}\frac{1}{T}\int_0^T f(Z_u(\theta))du = \frac{d}{d\theta}\lim_{T\to\infty}\frac{1}{T}\int_0^T f(Z_u(\theta))du,$$

and

$$\lim_{n\to\infty} \frac{d}{d\theta}\frac{1}{\tau_n}\int_0^{\tau_n} f(Z_u(\theta))du = \frac{d}{d\theta}\lim_{n\to\infty}\frac{1}{\tau_n}\int_0^{\tau_n} f(Z_u(\theta))du.$$

It is to be expected that stronger conditions are needed for these limits than for (8.1). What may be surprising is the difference between the conditions for convergence of these two types of derivative estimators. We will argue that the one based on $L_n$ is typically better.

In this section, we give general conditions for the convergence of abstract derivative estimates to the derivative of steady-state performance, using a result of Hu and Strickland [58]. Questions of the existence of a steady-state mean are postponed to the next section. Most of the rest of this chapter is devoted to finding verifiable conditions under which the hypotheses of Theorem 8.1, below, hold.

Let $\{G_n(\theta), n \geq 0, \theta \in \Theta\}$ be a parametric family of sequences of random variables. We think of each $\{G_n(\theta), n \geq 0\}$ as a sequence of performance values along a sample path of $\{Z_t(\theta), t \geq 0\}$, but this interpretation is not needed for the result. Let $\{g_n(\theta), n \geq 0, \theta \in \Theta\}$ be another family of sequences, interpreted as derivative estimates for $\{G_n(\theta), n \geq 0, \theta \in \Theta\}$. (It is possible, though not necessary, that $g_n(\theta) = G_n'(\theta)$.) Suppose that, for each $\theta$, $\lim_{n\to\infty} G_n(\theta) = \ell(\theta)$, almost surely, with $\ell(\theta)$ deterministic, or, alternatively, suppose that $\lim_{n\to\infty} \mathbf{E}[G_n(\theta)] = \ell(\theta)$. Call $\{g_n(\theta), n \geq 0\}$ a *strongly consistent* estimator of $\ell'(\theta)$ if, with probability one,

$$\lim_{n\to\infty} g_n(\theta) = \ell'(\theta). \tag{8.2}$$

Call the *family* of estimators $\{g_n(\theta), n \geq 0, \theta \in \Theta\}$ strongly consistent almost everywhere (a.e.) if (8.2) holds for almost every $\theta \in \Theta$. The following result gives sufficient conditions for strong consistency a.e.

**Theorem 8.1.** (Hu and Strickland [58].) Let $\{G_n(\theta), n \geq 0, \theta \in \Theta\}$ and $\{g_n(\theta), n \geq 0, \theta \in \Theta\}$ be families of sequences as above. Suppose $\Theta$ is a compact interval, and suppose the following hold for all $\theta \in \Theta$:

**(i)** $\mathbf{E}[g_n(\theta)] = d\mathbf{E}[G_n(\theta)]/d\theta + b_n(\theta)$, where $b_n(\theta) \to 0$ as $n \to \infty$;

**(ii)** $\lim_{n\to\infty} \mathbf{E}[G_n(\theta)] = \lim_{n\to\infty} G_n(\theta) = \ell(\theta)$, a.s., for some $\ell(\theta)$;

**(iii)** $\lim_{n\to\infty} \mathbf{E}[g_n(\theta)] = \lim_{n\to\infty} g_n(\theta) = \gamma(\theta)$, a.s., for some $\gamma(\theta)$;

**(iv)** the functions $\{|\mathbf{E}[g_n(\cdot)] - b_n(\cdot)|, n \geq 0\}$ are bounded by a (deterministic) function $\bar{g}(\cdot)$ which is integrable over $\Theta$.

Then $\ell$ is differentiable and $\{g_n(\theta), n \geq 0, \theta \in \Theta\}$ is a strongly consistent estimator of $\ell'(\theta)$, a.e.; that is, (8.2) holds for almost every $\theta \in \Theta$.

We prove this result using Lemma 8.1, below, but first comment on the hypotheses. Condition (i) says that the *bias* $\mathbf{E}[g_n(\theta)] - d\mathbf{E}[G_n(\theta)]/d\theta$ vanishes as $n \to \infty$. Results from previous chapters are relevant here because they provide conditions under which the bias is zero for each $n$; in our applications of the theorem (in Section 8.3), $b_n(\theta) \equiv 0$. Conditions (ii) and (iii) ensure, in part, that the question of steady-state derivative estimation is well-posed; something like these conditions is necessary. This particular pair of assumptions ties in well with Lemma 8.1 and the regenerative framework of Section 8.2. Condition (iv) allows us to interchange limits over $n$ with integrals over $\theta$, and is reasonable in practice.

The following is an intermediate result regarding deterministic functions:

**Lemma 8.1.** ([58].) Let $\{h_n, n \geq 0\}$ and $h$ be functions on a compact interval $\Theta = [\theta_1, \theta_2]$, $h_n(\theta) \to h(\theta)$ for all $\theta \in \Theta$. Suppose that each $h_n$ is differentiable and that, for all $n$, $|h'_n| \leq \bar{h}$, where $\bar{h}$ is integrable over $\Theta$. Suppose that $\lim_{n\to\infty} h'_n$ exists almost everywhere on $\Theta$. Then, at almost every $\theta \in \Theta$, $h$ is differentiable and $\lim_{n\to\infty} h'_n(\theta) = h'(\theta)$.

**Proof.** A result in real analysis states that, for every $n$ and almost every $\theta_0 \in [\theta_1, \theta_2]$,

$$h_n(\theta_0) = h_n(\theta_1) + \int_{\theta_1}^{\theta_0} h'_n(\theta)d\theta;$$

see, e.g., Theorem 8.21 of Rudin [71]. Take the limit as $n \to \infty$ of both sides. Since the $\{|h'_n|\}$ are dominated by an integrable function, we may interchange the limit with the integral over $\theta$ to get, for almost every $\theta_0$,

$$h(\theta_0) = h(\theta_1) + \int_{\theta_1}^{\theta_0} \lim_{n\to\infty} h'_n(\theta)d\theta.$$

This reveals $h$ to be differentiable at almost every $\theta_0$; see, e.g., Theorem 8.17 of Rudin [71]. Differentiating, we get

$$h'(\theta_0) = \lim_{n\to\infty} h'_n(\theta_0),$$

which is what we needed to show. □

**Proof of Theorem 8.1.** By hypotheses (i)-(iii),

$$\lim_{n\to\infty} g_n(\theta) = \lim_{n\to\infty} \mathbf{E}[g_n(\theta)] = \lim_{n\to\infty} d\mathbf{E}[G_n(\theta)]/d\theta,$$

for every $\theta \in \Theta$, almost surely. The functions $|d\mathbf{E}[G_n(\cdot)]/d\theta|$, $n \geq 0$, are dominated by a function integrable over $\Theta$, by conditions (i) and (iv). From Lemma 8.1 we conclude that, a.e. on $\Theta$,

$$\lim_{n\to\infty} \frac{d}{d\theta}\mathbf{E}[G_n(\theta)] = \frac{d}{d\theta} \lim_{n\to\infty} \mathbf{E}[G_n(\theta)],$$

including the fact that the right side exists. Combining the equations above and substituting according to (iii), we get

$$\lim_{n\to\infty} g_n(\theta) = \ell'(\theta),$$

almost surely, for almost every $\theta$. □

## 8.2 Some Limit Theory for GSMPs

We now detour from derivative estimation to develop some necessary background on the existence of the steady-state mean

$$\ell_f(\theta) = \lim_{t\to\infty} \frac{1}{t} \int_0^t f(Z_u(\theta))du, \tag{8.3}$$

and on the convergence of GSMPs to steady state, in general. The *regenerative* framework developed in this section is well-suited to consideration of (8.3), and extends nicely to the convergence of stochastic derivatives for GSMPs.

### 8.2.1 Regeneration

The intuitive notion of a regenerative process is of one that occasionally "starts afresh", in the sense that at certain random time instants the future evolution of the process becomes independent of its past. The most familiar example of this mechanism is the sequence of returns of a discrete-state Markov chain to a fixed state, but regeneration applies far beyond this setting. In great generality, the almost sure existence of an infinite sequence of regeneration points guarantees the existence of a "steady state". An added feature of the regenerative approach is that it simplifies the attendant analysis by dividing an infinite sample path into a sequence of i.i.d. *cycles*, consisting of finite path segments between regeneration points.

Recent work on the stability of stochastic processes has led to variations and generalizations of the basic sense of regeneration, showing that many systems with no "obvious" regenerative structure are regenerative in a broader sense. But our goal here is modest: to review the simplest notion of a regenerative GSMP and to discuss its implications for (8.3). We devote more effort to extending the basic theory for $\ell_f(\theta)$ to $\ell'_f(\theta)$ than to considering the most general conditions for the existence of a steady-state mean.

Drop, for the time being, the parameter $\theta$. Recall that a random time $\sigma$, $\sigma \geq 0$, is a *stopping time* for $\{Z_t, t \geq 0\}$ if the occurrence or non-occurrence of $\sigma$ by time $T \geq 0$ is known, given knowledge of $Z_t$ up to time $T$; i.e., if the event $\{\sigma \leq T\}$ is determined by $\{Z_t, 0 \leq t \leq T\}$. (As always, we take $Z$ to have almost surely right-continuous, piecewise constant sample paths.) The following is not restricted to GSMPs:

**Definition 8.1.** The process $\{Z_t, t \geq 0\}$ is *regenerative* if there exists a sequence $\{\sigma_k, k \geq 0\}$ of stopping times such that

**(i)** $\{\sigma_{k+1} - \sigma_k, k \geq 0\}$ are independent and identically distributed;

**(ii)** for every sequence of times $0 < t_1 < \cdots < t_n$ and every $k \geq 0$, the random vectors $(Z_{t_1}, \ldots, Z_{t_n})$ and $(Z_{\sigma_k+t_1}, \ldots, Z_{\sigma_k+t_m})$ have the same distribution, and the processes $\{Z_t, 0 \leq t < \sigma_k\}$ and $\{Z_{\sigma_k+t}, t \geq 0\}$ are independent.

Thus, in a regenerative process, the regeneration points $\{\sigma_k, k \geq 0\}$ cut $\{Z_t, t \geq 0\}$ into independent and identically distributed *cycles* of the form $\{Z_t, \sigma_{k-1} \leq t < \sigma_k\}$. The occurrence or non-occurrence prior to $T$ of the $k$th regeneration, $\sigma_k$, is determined by $Z_t$ up to $T$.

We now summarize, without proofs, some consequences of regeneration. References to more complete treatments are provided in the notes at the end of this chapter. To state the results, we need some notation and definitions. For $t \geq 0$ and $s \in \mathbf{S}$, let $\pi_t(s) = P(Z_t = s)$. The distribution $\pi_t$ depends on the initial state $Z_0$, but we do not indicate this dependence explicitly. In general, say that a family $\{\nu_t, t \geq 0\}$ of probabilities on $\mathbf{S}$ converges *weakly* to a probability $\nu$ if for all bounded $f : \mathbf{S} \to \mathbf{R}$

$$\lim_{t\to\infty} \sum_{s\in\mathbf{S}} f(s)\nu_t(s) = \sum_{s\in\mathbf{S}} f(s)\nu(s).$$

Denote this mode of convergence by $\nu_t \Rightarrow \nu$. Say that $Z_t$ converges to $Z_\infty$ *in distribution* (denoted $Z_t \Rightarrow Z_\infty$) if there is a probability $\pi$ such that $\pi_t \Rightarrow \pi$ and if $Z_\infty$ is a random variable on $\mathbf{S}$ with distribution $\pi$. Recall that a distribution function on $[0, \infty)$ is called *lattice* if it assigns probability one to a set of the form $\{0, \delta, 2\delta, \ldots, \}$, for some $\delta > 0$, (and *non-lattice* otherwise). The largest such $\delta$ is called the *span* of the distribution. Finally, let $F_c$ be the common distribution and $m_c$ the common expectation of all cycle lengths $\sigma_k - \sigma_{k-1}$, $k = 1, 2, \ldots$.

**Proposition 8.1.** Suppose that $\{Z_t, t \geq 0\}$ is regenerative and that, in the notation above, $m_c < \infty$. Then the following hold:

**(i)** If $F_c$ is non-lattice, there is a $\pi$ such that $\pi_t \Rightarrow \pi$ as $t \to \infty$; hence, there is a $Z_\infty$ such that $Z_t \Rightarrow Z_\infty$.

**(ii)** If $F_c$ is lattice with span $\delta$, there is a $\pi$ such that $\pi_{n\delta} \Rightarrow \pi$ as $n \to \infty$; hence, there is a $Z_\infty$ such that $Z_{n\delta} \Rightarrow Z_\infty$.

**(iii)** In either case, $\pi$ is given, almost surely, by

$$\pi(s) = \lim_{t\to\infty} \frac{1}{t} \int_0^t \mathbf{1}\{Z_u = s\} du.$$

**Proposition 8.2.** If $m_c < \infty$, then in the non-lattice case $\pi$ is also given by

$$\pi(s) = \frac{1}{m_c} \mathbf{E}\left[\int_{\sigma_{k-1}}^{\sigma_k} \mathbf{1}\{Z_t = s\} dt\right],$$

for all $k = 1, 2, \ldots$. Moreover, if $\sum_{s\in\mathbf{S}} |f(s)|\pi(s) < \infty$ then, almost surely,

$$\begin{aligned} \lim_{t\to\infty} \frac{1}{t} \int_0^t f(Z_u) du &= \lim_{t\to\infty} \frac{1}{t} \mathbf{E}\left[\int_0^t f(Z_u) du\right] \\ &= \frac{1}{m_c} \mathbf{E}\left[\int_{\sigma_{k-1}}^{\sigma_k} f(Z_t) dt\right], \end{aligned} \tag{8.4}$$

for all $k = 1, 2, \ldots$.

Thus, regeneration plus a finite mean cycle length between regeneration points ensures the existence of a limiting distribution $\pi$. It also implies that $\ell_f$ (as defined in (8.3), but without parameter) exists and is equal to the mean in steady state $\ell_f = \mathbf{E}[f(Z_\infty)]$, provided that

$$\mathbf{E}[|f(Z_\infty)|] = \sum_{s \in \mathbf{S}} |f(s)|\pi(s)$$

is finite. A modification of Proposition 8.2 for the lattice case, along the lines of Proposition 8.1(ii), also holds.

Equation (8.4) expresses the steady-state mean as a ratio of expectations over random but *finite* horizons, and serves as a starting point for the role of regenerative processes in simulation. Let

$$\tilde{L}_k = \int_{\sigma_{k-1}}^{\sigma_k} f(Z_t)dt \quad \text{and} \quad \tilde{T}_k = \sigma_k - \sigma_{k-1};$$

then $\{(\tilde{L}_k, \tilde{T}_k), k = 1, 2, \ldots\}$ is an i.i.d. sequence and

$$\lim_{n \to \infty} \frac{\sum_{k=1}^n \tilde{L}_k}{\sum_{k=1}^n \tilde{T}_k} = \mathbf{E}[f(Z_\infty)] = \ell_f \tag{8.5}$$

if $\mathbf{E}[|f(Z_\infty)|] < \infty$. Thus, (8.5) provides a natural point estimator for $\ell_f$. The fact that the $(\tilde{L}_k, \tilde{T}_k)$'s are i.i.d. makes (8.5) convenient for forming confidence interval estimates as well. We will have more to say about this in Section 8.4.

Re-introduce the parameter $\theta$ and assume that $\ell_f(\theta)$ is differentiable. Equations (8.4) and (8.5) immediately suggest an approach to estimating $\ell_f'(\theta)$. Since $\ell_f(\theta) = \mathbf{E}[\tilde{L}_k(\theta)]/\mathbf{E}[\tilde{T}_k(\theta)]$, $k = 1, 2, \ldots$, differentiating we obtain

$$\ell_f'(\theta) = \frac{\mathbf{E}[\tilde{L}_k(\theta)]'}{\mathbf{E}[\tilde{T}_k(\theta)]} - \frac{\mathbf{E}[\tilde{L}_k(\theta)]}{\mathbf{E}[\tilde{T}_k(\theta)]} \frac{\mathbf{E}[\tilde{T}_k(\theta)]'}{\mathbf{E}[\tilde{T}_k(\theta)]}. \tag{8.6}$$

Suppose that over the $k$th cycle we can evaluate $\tilde{L}_k'$ and $\tilde{T}_k'$, and that the vectors $\{(\tilde{L}_k', \tilde{T}_k', \tilde{L}_k, \tilde{T}_k)\}$ are i.i.d. By summing over cycles we obtain, in the limit,

$$\frac{\sum_k \tilde{L}_k'}{\sum_k \tilde{T}_k} - \frac{\sum_k \tilde{L}_k}{\sum_k \tilde{T}_k} \frac{\sum_k \tilde{T}_k'}{\sum_k \tilde{T}_k} \to \frac{\mathbf{E}[\tilde{L}_k']}{\mathbf{E}[\tilde{T}_k]} - \frac{\mathbf{E}[\tilde{L}_k]}{\mathbf{E}[\tilde{T}_k]} \frac{\mathbf{E}[\tilde{T}_k']}{\mathbf{E}[\tilde{T}_k]}. \tag{8.7}$$

A *sufficient* condition for the right side of (8.7) to equal the right side of (8.6) is that

$$\mathbf{E}[\tilde{L}_k'] = \mathbf{E}[\tilde{L}_k]' \quad \text{and} \quad \mathbf{E}[\tilde{T}_k'] = \mathbf{E}[\tilde{T}_k]'; \tag{8.8}$$

and in this case the left side of (8.7) is a strongly consistent estimator of $\ell'_f$. On the surface, (8.8) appears to be the sort of problem dealt with in earlier chapters—the interchange of a derivative and an expectation. Unfortunately, except in very special cases, our earlier results cannot help us here. Even with the structural conditions of, for example, Chapter 3, we will see that the regeneration points $\{\sigma_k\}$ are typically discontinuous in $\theta$; hence, $\tilde{L}_k$ and $\tilde{T}_k$ are typically discontinuous, too. (Notice that $\tilde{L}_k$ does not have the form of $L_n$ or $L_T$; the limits of integration are different.) Fortunately, unbiasedness of $\tilde{L}'_k$ and $\tilde{T}'_k$ (in the sense of (8.8)) is not *necessary* for the left side of (8.7) to converge to $\ell'_f(\theta)$. We will, in fact, provide conditions under which the left side of (8.7) is a valid estimator of $\ell'_f(\theta)$, even if $\mathbf{E}[\tilde{L}'_k] \neq \mathbf{E}[\tilde{L}_k]'$ and $\mathbf{E}[\tilde{T}'_k] \neq \mathbf{E}[\tilde{T}_k]'$.

### 8.2.2 GSMPs with One-Clock States

When is a GSMP a regenerative process? This question is a topic of current research. Recent results center on conditions on clock distributions and transition probabilities under which an often subtle regenerative structure exists. There is, however, a class of GSMPs in which regeneration is clear. We focus on this restricted class partly because of its simplicity, but also because of how well it meshes with derivative estimation, as explained in Section 8.3.1.

Compare the evolution of a GSMP $\{Z_t, t \geq 0\}$ with that of a Markov process on the same state space, $\mathbf{S}$. Upon entry to a state $s$, the Markov process "forgets" its past and pursues its course just as it would upon any other entry to $s$. But the GSMP, upon entering $s$, is still influenced by its past through the clock mechanism that drives it. Future transitions of $Z_t$ depend on the *residual* clock times upon entry to $s$, and these residual clock times depend on the course taken by $Z_t$ before entry to $s$.

These considerations lead us to look for transitions of $Z_t$ at which *all clocks are reset*; i.e., at which all events are "new". At such transitions, the clock mechanism probabilistically restarts. In more detail, we want a pair of states $s_1, s_2$ and an event $\alpha$ such that $p(s_2; s_1, \alpha) > 0$ and $\mathcal{E}(s_2) \cap (\mathcal{E}(s_1) - \{\alpha\}) = \emptyset$. However, since we only consider *non-interruptive* GSMPs (condition (C1) of Chapter 2), we also need $(\mathcal{E}(s_1) - \{\alpha\}) \subseteq \mathcal{E}(s_2)$. Thus, what we want is $\mathcal{E}(s_1) - \{\alpha\} = \emptyset$, or simply $\mathcal{E}(s_1) = \{\alpha\}$. Call such an $s_1$ a *one-clock state.* Upon *exit* from a one-clock state, $\{Z_t, t \geq 0\}$ regenerates; the value of $Z_t$ is $s$ with probability $p(s; s_1, \alpha)$, independent of the past, and new clocks are set, independent of the past. Of course, in general, there may be no one-clock states or there may be several. When there is at least one, let us pick one arbitrarily and call it $s^*$. Let $\alpha^*$ be the unique element of $\mathcal{E}(s^*)$.

From the remarks above and Definition 8.1, we have

**Proposition 8.3.** If there is a one-clock state $s^*$ to which $\{Z_t, t \geq 0\}$ returns infinitely often with probability one, then $\{Z_t, t \geq 0\}$ is regenerative. The epochs of exits from $s^*$ form a sequence of regeneration points $\{\sigma_k, k \geq 0\}$.

Thus, GSMPs with a one-clock state form an easily recognized class of potentially regenerative processes. (Conditions are still needed to ensure infinitely many returns to $s^*$.) Typical examples of such systems arise in Jackson-like open and closed queueing networks. In the open case, if there is just one external arrival event (which is always active) then the state in which all queues are empty is a one-clock state. In a closed network, the state in which all jobs are at one queue is a one-clock state.

Various authors have investigated conditions on clock distributions and transition probabilities under which a GSMP (or similar process) returns infinitely often to a particular state. Results in Glynn [40] and Haas and Shedler [47] are particularly relevant to our setting. Typical conditions ensure, roughly, that the process is sufficiently "random" that all transitions that can occur, do occur. Rather than review all existing results, we state just one which, though not especially general, is easily understood.

Recall (from Chapter 2) that a GSMP is *irreducible* if for every pair of states $s$ and $s'$ there exist events $\alpha_0, \ldots, \alpha_m$ and states $s_1, \ldots, s_m$ such that $\alpha_0 \in \mathcal{E}(s)$, $\alpha_i \in \mathcal{E}(s_i)$, $i = 1, \ldots, m$, and

$$p(s_1; s, \alpha_0)p(s_2; s_1, \alpha_1) \cdots p(s_m; s_{m-1}, \alpha_{m-1})p(s'; s_m, \alpha_m) > 0.$$

Irreducibility gives us

**Proposition 8.4.** In a finite-state, irreducible GSMP in which every clock distribution has a density which is strictly positive on $(0, \infty)$, every state is visited infinitely often, a.s.

In our application of regeneration to derivative estimates, we will not use this or any other specific result. Instead, given a one-clock state $s^*$, we will *assume* that the GSMP makes infinitely many visits to that state, almost surely.

## 8.2.3 The View in Discrete Time

The regeneration points provided by Proposition 8.3 have a special property: they coincide with transition epochs. Associated with the sequence $\{\sigma_k, k \geq 0\}$ of epochs of exits from $s^*$ is a (random) sequence $\{K_k, k \geq 0\}$ of indices for which $\sigma_k = \tau_{K_k}$, a.s.; the $k$th regeneration occurs at the $K_k$th transition. This makes it convenient to examine the regenerative structure of $\{Z_t, t \geq 0\}$ on the discrete time scale of state transitions. The discrete-time view is useful when we consider derivatives in Section 8.3.1.

In the notation of Chapters 2 and 3, $Y_n$ is the $n$th state visited by $\{Z_t, t \geq 0\}$, and $c_n$ is the vector of clock readings just after the $n$th transition. As discussed in Chapter 2, $\{(Y_n, c_n), n = 0, 1, 2, \ldots\}$ is a general state space Markov chain. Its transition law is displayed near the end of Section 2.1 of Chapter 2. Regenerative properties of $\{Z_t, t \geq 0\}$ are closely tied to those of this underlying Markov chain. In general, we cannot expect $\{(Y_n, c_n)\}$ to make repeated returns to the same state; however, given a one-clock state we may have something just as useful, as we now explain.

The state space $\Sigma$ of $\{(Y_n, c_n)\}$ is the set of pairs $(s, c)$ with $s \in \mathbf{S}$ and $c = (c(\alpha), \alpha \in \mathbf{A})$, where $c(\alpha) \geq 0$ and $c(\alpha) > 0$ only if $\alpha \in \mathcal{E}(s)$. A subset of $\Sigma$ is called *recurrent* if it is reached from any state in $\Sigma$ in finitely many transitions, with probability one. Suppose there is a one-clock state $s^*$, $\mathcal{E}(s^*) = \{\alpha^*\}$. Define $\Sigma^* = \{(s, c) \in \Sigma : s = s^*\}$; then if $(Y_n, c_n)$ is in $\Sigma^*$, $Y_{n+1}$ has distribution $p(\cdot; s^*, \alpha^*)$, every $c_{n+1}(\alpha)$, $\alpha \notin \mathcal{E}(Y_{n+1})$ is zero, and every $c_{n+1}(\alpha)$, $\alpha \in \mathcal{E}(Y_{n+1})$ has the distribution $F_\alpha$ of a new $\alpha$-clock. In other words, for any $n \geq 0$, for any $(s^*, c) \in \Sigma^*$, for any $s' \in \mathbf{S}$, and for any vector $x = (x(\alpha), \alpha \in \mathbf{A})$, with every $x(\alpha) \geq 0$,

$$P\Big(Y_{n+1} = s', c_{n+1} \leq x | Y_n = s^*, c_n = c\Big) = p(s'; s^*, \alpha^*) \prod_{\alpha \in \mathcal{E}(s')} F_\alpha(x(\alpha)). \quad (8.9)$$

This transition probability has an important property: it does not depend on $(Y_n, c_n)$ except through the fact that $(Y_n, c_n) \in \Sigma^*$. To put it another way, the function $P(Y_{n+1} = s', c_{n+1} \leq x | (Y_n, c_n) = (\cdot, \cdot))$ is constant on $\Sigma^*$, for all $s'$ and $x$. If $\Sigma^*$ is, in addition, recurrent, then this property makes it a *regeneration set*. A Markov chain $\{(Y_n, c_n), n \geq 0\}$ with a regeneration set is called *Harris recurrent*. Intuitively, $\{(Y_n, c_n)\}$ restarts, independent of its past, at each transition out of $\Sigma^*$. (The form of (8.9) is actually stronger than what is needed for regeneration; see the notes.)

Harris recurrence implies the existence of a "steady state". In the setting above, let $K_0$ be the smallest index for which $(Y_{K_0 - 1}, c_{K_0 - 1}) \in \Sigma^*$ and let $K_1$ be the next smallest such index. (Thus, these are indices of consecutive exits from $s^*$.) Let $m_d = \mathbf{E}[K_1 - K_0]$.

**Proposition 8.5.** If $m_d < \infty$ then, for all $(Y_0, c_0)$, $(Y_n, c_n)$ converges in distribution. The limit distribution, $\nu$, is defined on (measurable) subsets $A$ of $\Sigma$ by

$$\nu(A) = \frac{1}{m_d} \mathbf{E}\left[\sum_{i=K_0}^{K_1 - 1} \mathbf{1}\{(Y_i, c_i) \in A\}\right].$$

Moreover, for any $g : \Sigma \to \mathbf{R}$ with

$$\mathbf{E}\left[\sum_{i=K_0}^{K_1 - 1} |g(Y_i, c_i)|\right] < \infty, \quad (8.10)$$

and any $(Y_0, c_0)$, we have

$$\begin{aligned}\lim_{n\to\infty} \frac{1}{n}\sum_{i=0}^{n-1} g(Y_i, c_i) &= \lim_{n\to\infty} \frac{1}{n}\sum_{i=0}^{n-1} \mathbf{E}[g(Y_i, c_i)] \\ &= \frac{1}{m_d}\mathbf{E}\left[\sum_{i=K_0}^{K_1-1} g(Y_i, c_i)\right]. \end{aligned} \tag{8.11}$$

Thus, if $\{(Y_n, c_n), n \geq 0\}$ returns infinitely often to $s^*$, time averages of functions of the state converge to the steady-state mean. Moreover, the steady-state mean is a ratio of averages over cycles between exits from $s^*$.

It is worth noting that there is no loss of information in passing from $\{Z_t, t \geq 0\}$ to $\{(Y_n, c_n), n \geq 0\}$; indeed, in Chapter 2 we defined the former from the latter. Recall that $t^*(s, c) = \min\{c(\alpha) : \alpha \in \mathcal{E}(s)\}$ is the holding time in $s$ when the clock reading upon entry to $s$ is $c$. Using this definition we have

$$\frac{1}{\tau_n}\int_0^{\tau_n} f(Z_t)dt = \sum_{i=0}^{n-1} f(Y_i)t^*(Y_i, c_i) \Big/ \sum_{i=0}^{n-1} t^*(Y_i, c_i),$$

and an analogous expression for $\int_0^T f(Z_t)dt/T$.

## 8.3 Strongly Consistent Derivative Estimates

### 8.3.1 Regenerative Formulation

Much of the limit theory for stochastic processes is devoted to the convergence of time averages. In Section 8.2, we focused on limits of continuous-time averages of functions of $\{Z_t, t \geq 0\}$ and discrete-time averages of functions of $\{(Y_n, c_n), n \geq 0\}$. Applying these results to derivative estimates requires some reformulation to get the estimates in a convenient form—stochastic derivatives of time averages are not, in general, time averages themselves. In particular, recall that under conditions on the clock distributions we have (Lemma 2.4)

$$\frac{dL_{n_0}}{d\theta} = \sum_{i=0}^{n_0-1} f(Y_i)[\frac{d\tau_{i+1}}{d\theta} - \frac{d\tau_i}{d\theta}] \tag{8.12}$$

and

$$\frac{dL_T}{d\theta} = \sum_{i=0}^{N_T-1} f(Y_i)[\frac{d\tau_{i+1}}{d\theta} - \frac{d\tau_i}{d\theta}] - \frac{d\tau_{N_T}}{d\theta} f(Z_T); \tag{8.13}$$

neither of these is immediately amenable to the results of the previous section. However, by augmenting the process $\{(Y_n, c_n)\}$ with information about event epoch derivatives, we can put (8.12) and (8.13) into a more suitable form.

For the rest of this chapter, we work exclusively with the construction of Chapter 2—the standard construction of a GSMP—and the resulting stochastic derivatives. Unless otherwise stated, $\Theta$ is a compact interval. To ensure differentiability in $\theta$, we impose

**(D1).(i)** Every $F_\alpha(x,\theta)$, $\theta \in \Theta$, is continuous in $x$ and is zero at $x = 0$.

**(ii)** Every clock sample $X_\theta(\alpha, k)$ is continuously differentiable in $\theta$ throughout $\Theta$, a.s., and for each $\alpha$ there is a function $\psi_\alpha$ such that $dX_\theta(\alpha,k)/d\theta = \psi_\alpha(X_\theta(\alpha,k),\theta)$.

The last part of these conditions is not needed for differentiability of $L_{n_0}$ and $L_T$, but is convenient for our reformulation. When there is no chance of confusion, we drop the second argument of $\psi_\alpha$.

Algorithm 2.1 of Chapter 2 tracks "perturbations" in scheduled event epochs and calculates, in particular, the sequence $\{d\tau_i/d\theta, i = 1, 2, \ldots\}$. We can model the progress of this algorithm by appending a new sequence $\{\delta_n, n = 0, 1, 2, \ldots\}$ to $\{(Y_n, c_n), n = 0, 1, 2, \ldots\}$. Each $\delta_n$ is a vector indexed by $\mathbf{A}$, $\delta_n = (\delta_n(\alpha), \alpha \in \mathbf{A})$, where $\delta_n(\alpha)$ represents the derivative in the next scheduled occurrence of $\alpha$, as of $\tau_n$. More precisely, define $\{\delta_n\}$ by $\delta_0(\alpha) = 0$ for $\alpha \notin \mathcal{E}(Y_0)$, $\delta_0(\alpha) = \psi_\alpha(c_0(\alpha))$ for $\alpha \in \mathcal{E}(Y_0)$, and by the recursion

$$\delta_{n+1}(\alpha) = \begin{cases} \delta_n(a_{n+1}) + \psi_\alpha(c_{n+1}(\alpha)), & \alpha \in \mathcal{N}(Y_{n+1}; Y_n, a_{n+1}), \\ \delta_n(\alpha), & \alpha \in \mathcal{O}(Y_{n+1}; Y_n, a_{n+1}), \\ 0, & \alpha \notin \mathcal{E}(Y_{n+1}). \end{cases}$$

(Recall that $\mathcal{N}(s'; s, \alpha)$ and $\mathcal{O}(s'; s, \alpha)$ are the sets of new and old events following a transition from $s$ to $s'$ triggered by $\alpha$.)

Straightforward comparison of this recursion with Algorithm 2.1 shows that, for all $n$, $d\tau_n/d\theta = \delta_{n-1}(a_n)$; hence, the $i$th summand in (8.12) can be written as

$$f(Y_i)[\delta_i(a_{i+1}) - \delta_{i-1}(a_i)]. \tag{8.14}$$

(For the case $i = 0$, take $\delta_{-1} \equiv 0$.) Furthermore, it is not hard to see that $\{(Y_n, c_n, \delta_n), n \geq 0\}$ is a Markov chain; this follows from the fact that $\delta_{n+1}$ is a deterministic function of $(Y_n, c_n, \delta_n)$ and $(Y_{n+1}, c_{n+1})$. The drawback in working with this chain is that the $i$th summand (8.14) depends on $\delta_{i-1}$ as well as $\delta_i$. To get around this, define, for n=0,1,2,. . . ,

$$\Delta_n(\alpha) = \begin{cases} \delta_n(\alpha) - \delta_{n-1}(a_n), & \alpha \in \mathcal{E}(Y_n), \\ 0 & \alpha \notin \mathcal{E}(Y_n); \end{cases} \tag{8.15}$$

thus, $\Delta_n(a_{n+1}) = \tau'_{n+1} - \tau'_n$ is the derivative of the holding time in $Y_n$.

Let $\mathbf{X}_n = (Y_n, c_n, \Delta_n)$. The process $\{\mathbf{X}_n\}$ tracks both the underlying GSMP data $\{(Y_n, c_n)\}$ and the auxilliary information $\{\Delta_n\}$ needed to evaluate derivatives. In general, augmenting the state of a Markov chain in this way can destroy the Markov property; and even if the Markov property is preserved, Harris recurrence may be destroyed. It *happens* that, with a one-clock state, the augmented process $\{\mathbf{X}_n\}$ is both Markov and Harris recurrent. This is a remarkable fact: the same structure (the presence of a one-clock state) that makes $\{Z_t, t \geq 0\}$ and $\{(Y_n, c_n), n \geq 0\}$ plainly regenerative, ensures that $\{\mathbf{X}_n\}$ remains regenerative:

**Theorem 8.2.** With the notation above,

**(i)** $\{\mathbf{X}_n, n = 0, 1, \ldots\}$ is a Markov chain.

**(ii)** If the GSMP has a one-clock state with respect to which $\{(Y_n, c_n)\}$ is Harris recurrent, then $\{\mathbf{X}_n\}$ is Harris recurrent.

**(iii)** For each function $f$ on $\mathbf{S}$, there is a function $\hat{f}$ such that

$$\frac{dL_{n_0}}{d\theta} = \sum_{i=0}^{n_0-1} \hat{f}(\mathbf{X}_i).$$

**Proof. (i).** To establish that $\{\mathbf{X}_n\}$ is Markov, we give an explicit recursion for $\{\Delta_n\}$, without reference to $\{\delta_n\}$. Initialize $\Delta_0(\alpha)$ the way we initialized $\delta_0(\alpha)$. The sequence $\{\Delta_n\}$ obeys

$$\Delta_{n+1}(\alpha) = \begin{cases} \psi_\alpha(c_{n+1}(\alpha)), & \alpha \in \mathcal{N}(Y_{n+1}; Y_n, a_{n+1}), \\ \Delta_n(\alpha) - \Delta_n(a_{n+1}), & \alpha \in \mathcal{O}(Y_{n+1}; Y_n, a_{n+1}), \\ 0, & \alpha \notin \mathcal{E}(Y_{n+1}). \end{cases}$$

Recall (from Chapter 2) that $a_{n+1}$ is a deterministic function of $(Y_n, c_n)$; it is the event in $\mathcal{E}(Y_n)$ with the smallest $c_n(\alpha)$. Thus, $\Delta_{n+1}(\alpha)$ is a deterministic function of $(Y_n, c_n, \Delta_n)$ and $(Y_{n+1}, c_{n+1})$. Since $\{(Y_n, c_n)\}$ is Markov, it follows that $\mathbf{X}_{n+1}$ depends on $\mathbf{X}_0, \ldots, \mathbf{X}_n$ only through $\mathbf{X}_n$; thus, $\{\mathbf{X}_n\}$ is Markov.

**(ii).** Suppose the GSMP has a one-clock state $s^*$. At each exit from $s^*$, $\Delta_{n+1}$ is a deterministic function of $(Y_{n+1}, c_{n+1})$ *only*; for $\alpha \notin \mathcal{E}(Y_{n+1})$, $\Delta_{n+1}(\alpha) = 0$, and for $\alpha \in \mathcal{E}(Y_{n+1})$, $\Delta_{n+1}(\alpha) = \psi_\alpha(c_{n+1}(\alpha))$. Hence, at each exit from $s^*$, $\mathbf{X}_{n+1}$ depends on $\mathbf{X}_n$ only through the fact that $(Y_n, c_n) \in \Sigma^*$. By assumption, $\Sigma^*$ is recurrent for $\{(Y_n, c_n)\}$, so $\{(y, c, \Delta) : y = s^*\}$ is a regeneration set for $\{\mathbf{X}_n\}$.

**(iii).** Given $f$, define $\hat{f}$ by $\hat{f}(y, c, \Delta) = f(y)\Delta(a^*(y, c))$, where $a^*(y, c)$ is the event that triggers the transition out of $y$ when the clock reading is $c$. Then $\hat{f}(Y_i) = f(Y_i)\Delta(a_{i+1}) = f(Y_i)[\tau'_{i+1} - \tau'_i]$. □

With this reformulation, we are able to express $dL_{n_0}/d\theta$ as an unnormalized time average of a regenerative process, and $dL_T/d\theta$ as such a time average, but for the term $f(Z_T) \cdot d\tau_{N_T}/d\theta$. (This discrepancy turns out to be significant.) We are now poised to apply the results of Section 8.2 to derivatives.

### 8.3.2 Convergence Through Transition Epochs

Using the results of Section 8.2 and the formulation of Section 8.3.1, we can now give conditions for the differentiability of $\ell_f(\theta)$, and for the convergence of $\{(L_n/\tau_n)', n > 0\}$ to $\ell'_f$. Estimation of the steady-state derivative via $\{L'_T/T, T > 0\}$ is taken up in Section 8.3.3.

In the ratio $L_n/\tau_n$, both numerator and denominator depend on $\theta$; hence, differentiation yields a rather cumbersome expression. Our analysis is streamlined if we consider separately terms of the form $L_n/n$ and use the following result. (As usual, $f$ is the integrand of $L_n$; $\tau_n$ is a special case of $L_n$ with $f \equiv 1$.)

**Lemma 8.2.** If, for every bounded function $f : \mathbf{S} \to \mathbf{R}$, $\lim_{n\to\infty}(L'_n/n) = (\lim_{n\to\infty} L_n/n)'$ (including the fact that the limits and derivatives exist), and if $\lim_{n\to\infty} \tau_n/n > 0$, then, for every such $f$, $\lim_{n\to\infty}(L_n/\tau_n)' = (\lim_{n\to\infty} L_n/\tau_n)'$ (including the fact that the limits and derivatives exist).

**Proof.** Abbreviating "$\lim_{n\to\infty}$" to "lim", we have

$$\begin{aligned}
[\lim L_n/\tau_n]' &= [(\lim L_n/n)(\lim n/\tau_n)]' \\
&= (\lim L_n/n)'(\lim n/\tau_n) + (\lim L_n/n)(\lim n/\tau_n)' \\
&= (\lim L'_n/n)(\lim n/\tau_n) + (\lim L_n/n)\lim(n/\tau_n)' \\
&= \lim L'_n/\tau_n - (\lim L_n/n)\lim(n\tau'_n/\tau_n^2) \\
&= \lim[(L'_n/\tau_n) - (L_n/\tau_n)(\tau'_n/\tau_n)] \\
&= \lim[L_n/\tau_n]'.
\end{aligned}$$

□

The estimator $\{L'_n/n, n > 0\}$ is better suited to Theorem 8.1 because, under appropriate conditions, $\mathbf{E}[L'_n/n] = \mathbf{E}[L_n/n]'$, whereas, under the same conditions, $(L_n/\tau_n)'$ is biased because it is a ratio of random variables. To ensure unbiasedness of $L'_n/n$, we use results from Chapter 3. One of the conditions we use (in addition to (D1)) is

**(D2).** There is a constant $B > 0$ such that, for all $\alpha$ and $\theta$, $|\psi_\alpha(x,\theta)| \le B(x+1)$.

Under this condition, we get a simple bound on the holding time perturbations $\{\Delta_n(a_{n+1})\}$:

**Lemma 8.3.** Suppose that, for all $\theta$, $\{Z_t(\theta), t \ge 0\}$ is regenerative with respect to the sequence $\{\sigma_k(\theta), k \ge 0\}$ of exits from a one-clock state $s^*$. Suppose also that (D1) and (D2) hold. Let $\{K_i, i \ge 0\}$ be the random indices defined (uniquely, a.s.) by $\tau_{K_i} = \sigma_i$. Then for any $n = 0, 1, \ldots$, if $K_{i-1} \le n < K_i$ (i.e., the $n$th transition falls in the $i$th cycle),

$$|\Delta_n(a_{n+1})| \le 2B[(K_i - K_{i-1}) + (\sigma_i - \sigma_{i-1})], \tag{8.16}$$

where $B$ is the constant in (D2).

**Proof.** The sequence $\{\Delta_n\}$ regenerates at each exit from $s^*$ so, without loss of generality, we may take $K_0 \le n < K_1$. We may further assume that the process $\{Z_t, t \ge 0\}$ starts just after such an exit; i.e., $\sigma_0 = \tau_0 = 0$ and $K_0 = 0$. We showed in the proof of Theorem 3.2 (Chapter 3, equation (3.7)) that (D2) implies that $|d\tau_i/d\theta| \le B(\tau_i + i)$, for all $i$. Since $\Delta_n(a_{n+1}) = \tau'_{n+1} - \tau'_n$, we have

$$\begin{aligned} |\Delta_n| &\le |\tau'_{n+1}| + |\tau'_n| \\ &\le 2B(\tau_{n+1} + n + 1) \\ &\le 2B(\tau_{K_1} + K_1) \\ &\le 2B[(\sigma_1 - \sigma_0) + (K_1 - K_0)]. \end{aligned}$$

□

The significance of this result is that it bounds the growth of the perturbations $\{\Delta_n(a_{n+1})\}$ purely in terms of quantities associated with the nominal process $\{Z_t, t \ge 0\}$. These quantities are the length of the regenerative cycles in discrete and continuous time. This bound is the main reason for imposing (D2). Under the conditions of the lemma, the individual terms $\tau'_n$ potentially go to infinity as $n$ increases, but the *differences* $[\tau'_{n+1} - \tau'_n]$ remain stable.

For the main result of this section, let us impose

**(D3).** For every event $\alpha$, and every $k = 1, 2, \ldots$, $\mathbf{E}[\sup_\theta X_\theta(\alpha, k)] < \infty$.

**Theorem 8.3.** Suppose that, for all $\theta \in \Theta$, $\{Z_t(\theta), t \ge 0\}$ is regenerative with respect to the sequence $\{\sigma_k(\theta), k \ge 0\}$ of exits from a one-clock state $s^*$. Suppose that $f : \mathbf{S} \to \mathbf{R}$ is bounded and that

**(a)** the clocks satisfy (D1)-(D3);

**(b)** the GSMP satisfies condition (C2') of Chapter 3;

**(c)** the cycles lengths satisfy $\sup_\theta \mathbf{E}[(\sigma_{i+1} - \sigma_i)^2] < \infty$ and $\sup_\theta \mathbf{E}[(K_{i+1} - K_i)^2] < \infty$.

Then, for all $\theta \in \Theta$, $\ell_f(\theta)$ exists, and for almost every $\theta \in \Theta$

$$\ell'_f(\theta) = \lim_{n\to\infty} \frac{d}{d\theta}\left[\frac{L_n(\theta)}{\tau_n(\theta)}\right].$$

**Proof.** It is enough to show that $\lim_{n\to\infty}(L'_n/n) = (\lim_{n\to\infty} L_n/n)'$; the result then follows from Lemma 8.2. We use Theorem 8.1, mapping $L_n/n$ to $G_n$ and $L'_n/n$ to $g_n$. Condition (i) of Theorem 8.1 follows from Corollary 3.3 of Chapter 3; $L'_n$ is unbiased because the hypotheses of that corollary are implied by (a) and (b). Condition (ii) follows from Proposition 8.2 and the fact that $f$ is bounded, via (8.4); continuity of the clock distributions implies the cycle lengths are non-lattice. For (iii) we use Proposition 8.5, so we first need to show that

$$\mathbf{E}\left[\sum_{n=K_{i-1}}^{K_i-1} |\hat{f}(\mathbf{X}_n)|\right] < \infty, \tag{8.17}$$

where $\hat{f}$ is the function provided by Theorem 8.2(iii). Let $\|f\| = \sup_s |f(s)|$; then, using Lemma 8.3 for the second inequality,

$$\begin{aligned}
\sum_{n=K_{i-1}}^{K_i-1} |\hat{f}(\mathbf{X}_n)| &= \sum_{n=K_{i-1}}^{K_i-1} |f(Y_n)\Delta_n(a_{n+1})| \\
&\leq \|f\| \sum_{n=K_{i-1}}^{K_i-1} |\Delta_n(a_{n+1})| \\
&\leq 2B\|f\| \sum_{n=K_{i-1}}^{K_i-1} [(K_i - K_{i-1}) + (\sigma_i - \sigma_{i-1})] \\
&= 2B\|f\|(K_i - K_{i-1})[(K_i - K_{i-1}) + (\sigma_i - \sigma_{i-1})]. \quad (8.18)
\end{aligned}$$

This bound has finite expectation because the discrete and continuous time cycle lengths have finite second moments under condition (c) above. Proposition 8.5, though stated only for $\{(Y_n, c_n)\}$, applies equally well to the chain $\{\mathbf{X}_n\}$ (and function $\hat{f}$). We conclude that, almost surely, the following exist and are equal:

$$\lim_{n\to\infty} \frac{L'_n}{n} = \lim_{n\to\infty} \frac{1}{n}\sum_{i=0}^{n-1} \hat{f}(\mathbf{X}_i) = \lim_{n\to\infty} \frac{1}{n}\sum_{i=0}^{n-1} \mathbf{E}[\hat{f}(\mathbf{X}_i)] = \lim_{n\to\infty} \mathbf{E}\left[\frac{L'_n}{n}\right];$$

so condition (iii) of Theorem 8.1 holds. Finally, let $K^*_n$ be the number of cycles started up to the $n$th transition. Arguing as in (8.18), we get

$$\begin{aligned}
\mathbf{E}\left[\left|\frac{L'_n}{n}\right|\right] &= \mathbf{E}\left[\frac{1}{n}\sum_{i=0}^{n-1} |\hat{f}(\mathbf{X}_i)|\right] \\
&\leq 2B\|f\|\mathbf{E}\left[\frac{1}{n}\sum_{i=0}^{K^*_{n-1}} [(K_i - K_{i-1}) + (\sigma_i - \sigma_{i-1})]\right] \\
&\leq 2B\|f\|\mathbf{E}\left[\frac{1}{n}\sum_{i=0}^{n-1} [(K_i - K_{i-1}) + (\sigma_i - \sigma_{i-1})]\right] \\
&= 2B\|f\|\mathbf{E}[(K_1 - K_0) + (\sigma_1 - \sigma_0)],
\end{aligned}$$

because the cycles are i.i.d. Condition (iv) of Theorem 8.1 is now satisfied because this expectation is a bounded function of $\theta$ on $\Theta$ when (c) holds. □

### 8.3.3 Convergence Through Continuous Time

We now consider the convergence of $L'_T/T$ as $T \to \infty$. While the convergence of this estimator to $\ell'_f(\theta)$ is closely related to that of $(L_n/\tau_n)'$, new issues arise because of the extra term $\tau'_{N_T} f(Z_T)/T$ in (8.13). The limit as $T \to \infty$ of $L'_T/T$ is not an immediate extension of Theorem 8.3. In this regard, it is worth noting that whereas $\{L_n/\tau_n, n > 0\}$ is a subsequence of $\{L_T/T, T > 0\}$, $\{(L_n/\tau_n)', n > 0\}$ is not a subsequence of $\{L'_T/T, T > 0\}$. Compare (8.12) and (8.13); notice that they do not coincide when evaluated at $\tau_n = T$.

**Theorem 8.4.** Suppose that the conditions of Theorem 8.3 are in effect. Then, for almost every $\theta \in \Theta$,

$$\lambda_{\tau'}(\theta) \triangleq \lim_{n\to\infty} \tau'_n(\theta)/n$$

exists and is finite, a.s. For almost every $\theta \in \Theta$,

**(i)** if $\lambda_{\tau'}(\theta) = 0$, then, a.s.,

$$L'_T(\theta)/T \to \ell'_f(\theta), \quad \text{as } T \to \infty; \tag{8.19}$$

**(ii)** if $\lambda_{\tau'}(\theta) \neq 0$, there exists a random variable $L'_\infty(\theta)$ such that

$$L'_T(\theta)/T \Rightarrow L'_\infty(\theta), \quad \text{as } T \to \infty \tag{8.20}$$

and $\mathbf{E}[L'_\infty(\theta)] = \ell'_f(\theta)$.

**Proof.** The conditions of Theorem 8.3 imply the existence of $\lambda_{\tau'}$ because $\tau'_n$ is a special case of $L'_n$. Regeneration (with a finite mean cycle length) implies the existence, (a.s. for every $\theta \in \Theta$) of $\lambda_\tau = \lim_{n\to\infty} \tau_n/n > 0$. Rewriting the conclusion of Theorem 8.3, we have (a.e. on $\Theta$)

$$\begin{aligned} \ell'_f = \lim_{n\to\infty}\left(\frac{L_n}{\tau_n}\right)' &= \lim_{n\to\infty} \frac{n}{\tau_n}\frac{L'_n}{n} - \frac{\tau'_n/n}{\tau_n/n}\frac{L_n}{\tau_n} \\ &= \frac{1}{\lambda_\tau}\lim_{n\to\infty}\left(\frac{L'_n}{n}\right) - \frac{\lambda_{\tau'}}{\lambda_\tau}\ell_f. \end{aligned} \tag{8.21}$$

By Proposition 8.1, $Z_t \Rightarrow Z_\infty$, which implies $f(Z_t) \Rightarrow f(Z_\infty)$. From (8.13) and Theorem 8.3, taking the limit (in distribution), we get

$$\begin{aligned} \lim_{T\to\infty}\frac{L'_T}{T} &= \lim_{T\to\infty}\frac{1}{T}\sum_{i=0}^{N_T-1} f(Y_i)[\frac{d\tau_{i+1}}{d\theta} - \frac{d\tau_i}{d\theta}] - \frac{\tau'_{N_T}}{T} f(Z_T) \\ &= \lim_{T\to\infty}\frac{N_T}{T}\cdot\lim_{T\to\infty}\frac{1}{N_T}\sum_{i=0}^{N_T-1} f(Y_i)[\frac{d\tau_{i+1}}{d\theta} - \frac{d\tau_i}{d\theta}] \\ &\quad - \lim_{T\to\infty}\frac{N_T}{T}\cdot\lim_{T\to\infty}\frac{\tau'_{N_T}}{N_T} f(Z_T) \end{aligned}$$

$$
\begin{aligned}
&= \frac{1}{\lambda_\tau} \cdot \lim_{n\to\infty} \frac{1}{n} \sum_{i=0}^{n-1} f(Y_i)[\frac{d\tau_{i+1}}{d\theta} - \frac{d\tau_i}{d\theta}] \\
&\quad - \frac{1}{\lambda_\tau} \cdot \lim_{n\to\infty} \frac{\tau_n'}{n} \cdot \lim_{T\to\infty} f(Z_T) \\
&= \frac{1}{\lambda_\tau} \lim_{n\to\infty} \left(\frac{L_n'}{n}\right) - \frac{\lambda_{\tau'}}{\lambda_\tau} f(Z_\infty). \qquad (8.22)
\end{aligned}
$$

Each of these terms is a deterministic constant, except $f(Z_\infty)$. Two cases emerge: If $\lambda_{\tau'} = 0$, the limit in (8.22) holds almost surely; the second term vanishes (recall that $f$ is bounded), leaving

$$\frac{1}{\lambda_\tau} \lim_{n\to\infty} \left(\frac{L_n'}{n}\right),$$

which coincides with (8.21) (and $\ell_f'$). If, on the other hand, $\lambda_{\tau'} \neq 0$, then the second term in (8.22) is non-degenerate; the limit in (8.22) holds only in distribution. If we define $L_\infty'$ to be the right side of (8.22), then $L_T'/T \Rightarrow L_\infty'$; and since $\mathbf{E}[f(Z_\infty)] = \ell_f$, comparison with (8.21) verifies that $\mathbf{E}[L_\infty'] = \ell_f'$. □

Let us compare $\{L_T'/T, T > 0\}$ and $\{(L_n/\tau_n)', n > 0\}$ as estimators of $\ell_f'(\theta)$, in light of this result. At first glance, $L_T'/T$ has an advantage: it is unbiased over finite horizons (in the sense that $\mathbf{E}[L_T'/T] = \mathbf{E}[L_T/T]'$) because its denominator does not depend on $\theta$. On the other hand, $(L_n/\tau_n)'$ is easier to evaluate, at least in simulation, because it is easier to collect statistics at an event epoch than at a fixed time. More significantly, comparison of Theorems 8.3 and 8.4 shows that $\{(L_n/\tau_n)'\}$ converges under weaker conditions. We may summarize the difference like this: while $\{(L_n/\tau_n)'\}$ requires only that the *increments* $[\tau_n' - \tau_{n-1}']$ remain stable as $n \to \infty$, $\{L_T'/T\}$ requires stability of the epoch derivatives $\tau_n'$ themselves, so that $\lambda_{\tau'} = 0$ and (8.19) holds.

Below, we discuss conditions under which $\tau_n'/n \to 0$ as $n \to \infty$. But first we state a corollary to Theorem 8.4 that makes part (ii) of that theorem more useful. Recall (from Chapter 1, Section 1.3) that the property of *uniform integrability* justifies the interchange of limit and expectation. (This holds even if the convergence is in distribution rather than in probability.) From this we get

**Corollary 8.1.** Under the conditions of Theorem 8.4, for almost every $\theta \in \Theta$, if $\{L_T'/T, T > 0\}$ is uniformly integrable, then

$$\lim_{T\to\infty} \mathbf{E}[L_T'/T] = \ell_f'(\theta);$$

i.e., $\{L_T'/T\}$ is *asymptotically unbiased.*

This result validates an estimation procedure in which one averages independent replications of $L'_T/T$ for a large but finite horizon $T$. As *both* the number of replications and the time horizon increase, the estimator converges to $\ell'_f$. The condition of uniform integrability can be difficult to verify but is reasonable in practice.

We now turn to conditions under which $\tau'_n/n \to 0$ (i.e., $\delta_n(a_{n+1})/n \to 0$) with $\{\delta_n\}$ as defined in Section 8.3.1. From the recursion that defines it, it is clear that the growth of $\{\delta_n\}$ is tied to the way clock "perturbations" $\{\psi_\alpha(X_\theta(\alpha,k))\}$ are passed from one event to another. If perturbations are periodically flushed from the system, $\delta_n/n$ should converge to zero; but if perturbations continually build up, the limit is typically non-zero.

To make this distinction precise, we need the notion of *relevance* introduced in Chapter 4. Let $\mathcal{R}(\alpha)$ denote the set of $\alpha$-relevant events. Recall (from Definition 4.1) that $\alpha \in \mathcal{R}(\alpha)$; and that if $\beta_1 \in \mathcal{R}(\alpha)$ and if there are states $s$ and $s'$ with $p(s';s,\beta_1) > 0$ and $\beta_2 \in \mathcal{N}(s';s,\beta_1)$ then $\beta_2 \in \mathcal{R}(\alpha)$ also. A consequence of this definition is that if some $\beta$ is *not* in $\mathcal{R}(\alpha)$, then every triggering indicator (Definition 2.1) $\eta(\beta,j;\alpha,k)$ is zero (Lemma 4.1). From Lemma 2.1 and the definition of $\delta$ and $\psi$, we have the representation

$$\delta_{n-1}(a_n) = \frac{d\tau_n}{d\theta} = \sum_{i=1}^{n} \psi_{a_i}(X_\theta(a_i,k_i))\eta(a_n,k_n;a_i,k_i),$$

where $k_i$ is the number of instances of the event $a_i$ among $a_1,\ldots,a_i$. Thus, the events in $\mathcal{R}(\alpha)$ are those whose timing is affected by changes in the $\alpha$-clocks $\{X_\theta(\alpha,k), k = 1,2,\ldots\}$. If $\beta \notin \mathcal{R}(\alpha)$ no perturbation $\psi_\alpha(X_\theta(\alpha,k))$ is ever passed to any $\delta_n(\beta)$.

For any $A \subseteq \mathbf{A}$, let $\mathcal{R}(A) = \cup_{\alpha\in A}\mathcal{R}(\alpha)$. Let $A_\theta$ be the set of events whose clock distributions depend on $\theta$; i.e., the set of events $\alpha$ for which $\psi_\alpha$ is not identically zero. Write $\mathcal{R}(\theta)$ for $\mathcal{R}(A_\theta)$. In some cases, $A_\theta = \mathbf{A}$, hence $\mathcal{R}(\theta) = \mathbf{A}$, also. But it is often natural, when $\theta$ has a physical meaning, for $A_\theta$ to be a strict subset of $\mathbf{A}$; and, in this case, $\mathcal{R}(\theta)$ *may* be strictly contained in $\mathbf{A}$. If $\mathcal{R}(\theta) = \mathbf{A}$, then every perturbation is (potentially) passed to every other event; if $\mathcal{R}(\theta)$ is properly contained in $\mathbf{A}$, then perturbations are occasionally lost. More precisely, we have

**Theorem 8.5.** Suppose that $\{Z_t, t \geq 0\}$ is irreducible and is regenerative with respect to the sequence $\{\sigma_k, k \geq 0\}$ of exits from a one-clock state $s^*$. Suppose that $\lim_{n\to\infty}\delta_n(a_{n+1})/n$ exists. Then the limit is zero if and only if $\mathcal{R}(\theta) \neq \mathbf{A}$.

**Proof.** The statement of the theorem makes no reference to derivatives. The functions $\{\psi_\alpha\}$ in the definition of $\{\delta_n\}$ need not represent derivatives in any sense, so long as $\psi_\alpha \equiv 0$ if and only if $\alpha \notin A_\theta$.

We claim that if $\mathcal{R}(\theta) \neq \mathbf{A}$ then $\alpha^*$, the unique element of $\mathcal{E}(s^*)$, is not in $\mathcal{R}(\theta)$. In other words, we claim that if $\alpha^* \in \mathcal{R}(\theta)$, then $\mathcal{R}(\theta) = \mathbf{A}$. If $\alpha^* \in \mathcal{R}(\theta)$, then $\mathcal{R}(\alpha^*) \subseteq \mathcal{R}(\theta)$ (relevance is transitive), so to verify our claim we need only show that $\mathcal{R}(\alpha^*) = \mathbf{A}$. Pick any event $\beta$ and any state $s$ with $\beta \in \mathcal{E}(s)$. Irreducibility implies the existence of a sequence of events $\alpha_1, \ldots, \alpha_m$ and states $s_1, \ldots, s_m$ such that $\alpha_i \in \mathcal{E}(s_i)$, $i = 1, \ldots, m$, and $p(s_1; s^*, \alpha^*)p(s_2; s_1, \alpha_1) \cdots p(s; s_m, \alpha_m) > 0$. Since $\mathcal{E}(s^*) = \{\alpha^*\}$, $\alpha_1$ is in $\mathcal{N}(s_1; s^*, \alpha^*)$, so $\alpha_1 \in \mathcal{R}(\alpha^*)$. Applying the definition of relevance recursively, we conclude that $\beta \in \mathcal{R}(\alpha^*)$; thus, every event is in $\mathcal{R}(\alpha^*)$.

As before, let $\{K_k\}$ satisfy $\tau_{K_k} = \sigma_k$; then $a_{K_k} = \alpha^*$ because each regeneration point is an exit from $s^*$. If $\mathcal{R}(\theta) \neq \mathbf{A}$, then $\alpha^* \notin \mathcal{R}(\theta)$ so every $\delta_n(\alpha^*)$ is zero. In particular, the infinite sequence $\{\delta_{K_k-1}(a_{K_k}), k = 1, 2, \ldots\}$ is identically zero. It follows that $\liminf_{n\to\infty} |\delta_n(a_{n+1})| = 0$; so, if $\lim_{n\to\infty} \delta_n(a_{n+1})/n$ exists it can only be zero.

Suppose, on the other hand, that the limit exists and is non-zero. Then the limit through the subsequence $\{\delta_{K_k-1}(a_{K_k})\}$ also exists and is non-zero. Consequently, it is not possible to have $\{\delta_{K_k-1}(\alpha^*)\}$ identically zero; i.e., $\alpha^* \in \mathcal{R}(\theta)$. □

Combining this result with Theorem 8.3 (which guarantees the existence of $\lim_{n\to\infty} \delta_{n-1}(a_n)/n = \lim_{n\to\infty} \tau'_n/n)$ we get conditions for the strong consistency (a.e. on $\Theta$) of $\{L'_T/T, T > 0\}$.

Without a one-clock state, $\{\delta_n(a_{n+1})\}$ may fail to converge. As a trivial but illustrative example, consider a GSMP with just one state, $s$, and $\mathcal{E}(s) = \{\alpha, \beta\}$. Suppose that $A_\theta = \{\alpha\}$, $\psi_\alpha > 0$, and $\psi_\beta \equiv 0$. Then $\lim_{n\to\infty} \delta_n(\beta)/n = 0$ and $\lim_{n\to\infty} \delta_n(\alpha)/n > 0$, so $\lim_{n\to\infty} \delta_n(a_{n+1})$ does not exist. More generally, consider an open queueing network with two or more external arrival events which are in the event list of *every* state. If any one of these events is in $A_\theta$, $\{\delta_n(a_{n+1})\}$ typically fails to converge, because perturbations on one arrival stream are never passed to another arrival stream.

### 8.3.4 Some Clocks Exponential

The assumption of a one-clock state can be relaxed (while preserving an easily identified regenerative structure) if some of the clock distributions $\{F_\alpha, \alpha \in \mathbf{A}\}$ are exponential. Of course, in the extreme case where all clocks are exponential, $\{Z_t, t \geq 0\}$ is Markov and regenerates at every exit from (or return to) a fixed state. Here, we consider a more interesting intermediate case.

For every $s \in \mathbf{S}$, let $\mathcal{E}_{\exp}(s)$ be the (possibly empty) subset of $\mathcal{E}(s)$ consisting of those events with exponential clock distributions. Consider a state $s^*$ for which there is an event $\alpha^*$ such that

$$\mathcal{E}(s^*) = \mathcal{E}_{\exp}(s) \cup \{\alpha^*\}; \tag{8.23}$$

in other words, there is at most one non-exponential clock active in state $s$. This generalizes the notion of a one-clock state. Suppose that such a state exists and that, in addition, for all $s' \in \mathbf{S}$,

$$p(s'; s^*, \alpha^*) > 0 \Rightarrow p(s'; s^*, \beta) = 0, \quad \text{for all } \beta \in \mathcal{E}(s^*), \beta \neq \alpha^*. \tag{8.24}$$

This condition makes it possible to determine which transitions out of $s^*$ are due to $\alpha^*$ by observing only the states $\{Z_t, t \geq 0\}$. Just after such a transition, any "old" events have exponential clocks, so $\{Z_t, t \geq 0\}$ regenerates. If (8.23) and (8.24) hold, and if, a.s., $\alpha^*$ triggers a transition out of $s^*$ infinitely many times, then $\{Z_t, t \geq 0\}$ is a regenerative process. The epochs $\{\sigma_k, k \geq 0\}$ of transitions out of $s^*$ due to $\alpha^*$ form a sequence of regeneration points. Propositions 8.1-8.4 hold in this case with obvious modifications where needed.

Results for $\{(Y_n, c_n), n \geq 0\}$ and the augmented process $\{\mathbf{X}_n, n \geq 0\}$ do not, however, carry over quite as readily. Suppose, for example, that $\alpha_1, \alpha_2 \in \mathcal{E}_{\exp}(s^*)$, that $Y_n = s^*$, and that the $n+1$st event is $\alpha^*$. It is *not* the case (as it was with a one-clock state) that $c_{n+1}$ depends on $(Y_n, c_n)$ only through the fact that $Y_n = s^*$. For example, $c_{n+1}(\alpha_1) > c_{n+1}(\alpha_2)$ if and only if $c_n(\alpha_1) > c_n(\alpha_2)$, since $c_{n+1}(\alpha_i) = c_n(\alpha_i) - t^*(Y_n, c_n)$, $i = 1, 2$. Thus, following a transition out of $s^*$, $c_{n+1}$ still depends on $c_n$; the set $\Sigma^*$ is no longer a regeneration set for $\{(Y_n, c_n)\}$.

We can get around this difficulty if we modify the evolution of a GSMP by *resampling* residual clock times at each transition out of $s^*$ triggered by $\alpha^*$. To stress the difference, let us denote the sequence so obtained by $\{(Y_n, \tilde{c}_n), n \geq 0\}$, where $\tilde{c}_n(\alpha)$ is the (modified) residual clock time following the $n$th transition. We outline a construction based on this point of view and note that it preserves regeneration under (8.23), but omit details.

The evolution of $\{(Y_n, \tilde{c}_n)\}$ is the same as that of $\{(Y_n, c_n)\}$, except when $\alpha^*$ triggers a transition out of $\alpha^*$. Following such a transition, we set each $\tilde{c}_{n+1}(\alpha)$, $\alpha \in \mathcal{E}_{\exp}(s^*)$, equal to a new sample from the distribution $F_\alpha(x) = 1 - \exp(-x\mu_\alpha)$. This is permissible because the *ordinary* residual clock time $c_{n+1}(\alpha)$ would have this distribution following the transition. (This is the "memoryless" property of the exponential distribution.) With this modification, under condition (8.23), $\{(Y_n, \tilde{c}_n)\}$ regenerates at exits from $s^*$ due to $\alpha^*$. The set $\{(s, c) : s = s^*, a^*(s, c) = \alpha^*\}$ is a regeneration set if it is recurrent.

While this modification of the clocks leads to regeneration of $\{(Y_n, \tilde{c}_n)\}$, it is not sufficient when we incorporate the sequence $\{\Delta_n\}$; indeed, it is not even clear what $\Delta_n$ means when we resample clocks. To ensure regeneration, we require that

$$(\mathcal{E}(s^*) - \{\alpha^*\}) \cap \mathcal{R}(\theta) = \emptyset; \tag{8.25}$$

in particular, for every $\alpha \in \mathcal{E}_{\exp}$, $\alpha \neq \alpha^*$, we require $\psi_\alpha \equiv 0$. We also modify the evolution of $\{\Delta_n\}$ (replacing it with $\{\tilde{\Delta}_n\}$), but only at exits from $s^*$ triggered by $\alpha^*$. At such a transition, rather than following the usual recursion, we set

$$\tilde{\Delta}_{n+1}(\alpha) = \begin{cases} \psi_\alpha(\tilde{c}_{n+1}(\alpha)), & \alpha \in \mathcal{N}(Y_{n+1}; s^*, \alpha^*); \\ 0, & \text{otherwise.} \end{cases}$$

(This amounts to restarting the sequence, since this is essentially the rule used for initializing $\Delta_0$.) Except in this case, $\{\tilde{\Delta}_n\}$ evolves like $\{\Delta_n\}$. It is not to hard to see, with this modification, that $\tilde{\mathbf{X}}_n = (Y_n, \tilde{c}_n, \tilde{\Delta}_n)$ is a Markov chain, that it regenerates at transitions out of $s^*$ triggered by $\alpha^*$, and that, when (8.25) holds, $\tilde{\Delta}_n(a_{n+1}) = \tau'_{n+1} - \tau'_n$ Applying earlier arguments, *mutatis mutandis*, we have

**Theorem 8.6.** The conclusions of Theorems 8.5 and 8.6 hold if the assumption of a one-clock state is relaxed to (8.23)-(8.25).

It is important to note that the modifications described above are only needed to make the previous proofs go through. They in no way entail a change in the algorithm used to compute stochastic derivatives or to simulate the GSMP. The change in construction from $\{(Y_n, c_n)\}$ to $\{(Y_n, \tilde{c}_n)\}$ is purely conceptual; Theorem 8.6 applies to the "ordinary" derivative estimates defined earlier.

## 8.4 Estimation Through Cycles

In Sections 8.3.2 and 8.3.3, we used regenerative structure as a tool to establish the strong consistency of steady-state derivative estimators, but we did not make explicit use of the regenerative cycles in the estimates we considered. The sequences $\{[L_n/\tau_n]', n > 0\}$ and $\{L'_T/T, T > 0\}$ are defined without reference to regeneration. But at the end of Section 8.2.1, we pointed out that the ratio formula (8.4) for the steady-state mean of a regenerative process suggests an approach to derivative estimation that makes explicit use of cycles. Letting $\tilde{T}_k$ be $\sigma_{k+1} - \sigma_k$, the length of the $k$th cycle, and letting $\tilde{L}_k$ be the integral of $f(Z_t)$ over the $k$th cycle, we posed the question of when

$$\frac{\sum_k \tilde{L}'_k}{\sum_k \tilde{T}_k} - \frac{\sum_k \tilde{L}_k}{\sum_k \tilde{T}_k} \frac{\sum_k \tilde{T}'_k}{\sum_k \tilde{T}_k} \overset{?}{\rightarrow} \frac{\mathbf{E}[\tilde{L}_k]'}{\mathbf{E}[\tilde{T}_k]} - \frac{\mathbf{E}[\tilde{L}_k]}{\mathbf{E}[\tilde{T}_k]} \frac{\mathbf{E}[\tilde{T}_k]'}{\mathbf{E}[\tilde{T}_k]} = \ell'_f(\theta). \tag{8.26}$$

When this holds, the regenerative method of simulation output analysis provides a way of a estimating confidence interval for the steady-state derivative. We show how the ordinary regenerative method extends to (8.26) after considering the validity of (8.26).

### 8.4.1 Convergence

As noted earlier, a sufficient condition for (8.26) is that $\tilde{L}_k$ and $\tilde{T}_k$ be unbiased, in the sense that $\mathbf{E}[\tilde{L}'_k] = \mathbf{E}[\tilde{L}_k]'$ and $\mathbf{E}[\tilde{T}'_k] = \mathbf{E}[\tilde{T}_k]'$; this is (8.8). The problem is that this rarely holds, even when $\mathbf{E}[L'_n] = \mathbf{E}[L_n]'$ and $\mathbf{E}[L'_T] = \mathbf{E}[L_T]'$, for all $n$ and $T$. In general, $\tilde{L}_k$ is discontinuous in $\theta$ because its limits of integration $\sigma_{k-1}$ and $\sigma_k$ are. These differ from $\tau_n$ and $T$ because they depend on the sequence of states $\{Y_n\}$, and this sequence typically changes with $\theta$. Even a very small change in $\theta$ can change some $Y_i$, and in so doing, introduce a jump in $\sigma_k$.

Consider, for example, a GSMP with a one-clock state $s^*$, $\mathcal{E}(s^*) = \{\alpha^*\}$. To simplify the discussion, suppose $\sigma_0 \equiv 0$; i.e., the process starts just after an exit from $s^*$. In this case, $\sigma_1$ is the epoch of the first exit from $s^*$. Suppose that, initially, the $n-1$st state $Y_{n-1}$ is $s$, and $\mathcal{E}(s) = \{\alpha^*, \beta\}$. From this state, suppose the GSMP proceeds as follows:

$$s \xrightarrow{\beta} s^* \xrightarrow{\alpha^*} s',$$

for some $s'$. Then $\sigma_1 = \tau_{n+1}$; the first regeneration occurs upon exit from $s^*$. But suppose, now, that through a small change in $\theta$ the events $\beta$ and $\alpha^*$ change order. Suppose the GSMP satisfies (C2). The perturbed GSMP follows

$$s \xrightarrow{\alpha^*} s'' \xrightarrow{\beta} s',$$

for some $s''$. Condition (C2) ensures that, after the order change, the process returns to the same state, $s'$. Even so, the regeneration point has been eliminated, because now $\alpha^*$ occurs in $s$ rather than $s^*$. The epochs of occurrence of $\alpha^*$ and $\beta$ are continuous across the order change, but the order change introduces a jump in $\sigma_1$, hence also in $\tilde{L}_1$. As we have noted in several places, the presence of non-negligible discontinuities precludes unbiasedness.

Returning to (8.26), we see that there is still hope: unbiasedness is only a sufficient condition. Comparing the left side of (8.26) with the derivative of $\ell_f(\theta) = \mathbf{E}[\tilde{L}_k]/\mathbf{E}[\tilde{T}_k]$ (namely, the right side of (8.6)), we find that the precise condition for (8.26) to hold is

$$\frac{\mathbf{E}[\tilde{L}'_k] - \mathbf{E}[\tilde{L}_k]'}{\mathbf{E}[\tilde{L}_k]} = \frac{\mathbf{E}[\tilde{T}'_k] - \mathbf{E}[\tilde{T}_k]'}{\mathbf{E}[\tilde{T}_k]}, \tag{8.27}$$

if $\mathbf{E}[\tilde{L}_k] \neq 0$. In words, (8.27) requires that the *relative* biases in $\tilde{L}'_k$ and $\tilde{T}'_k$ "cancel out". On the surface, (8.27) seems unlikely indeed. Why should such a convenient cancellation occur? In fact, attempts to verify (8.27) directly have been stymied, except in "toy" examples for which each of the terms can be evaluated analytically. Nevertheless, it turns out that (8.27) is an inevitable consequence of the combination of (C2) with a one-clock state; and verification of (8.26) is much simpler than (8.27) suggests:

**Theorem 8.7.** Under the conditions of Theorem 8.3, (8.26) holds.

**Proof.** To simplify the argument, suppose that $\sigma_0 \equiv 0$; this can always be arranged by discarding the initial segment $\{Z_t, 0 \leq t \leq \sigma_0\}$, if necessary. Notice that

$$\sum_{k=1}^{i} \tilde{L}_k = \int_0^{\tau_{K_i}} f(Z_t)dt \quad \text{and} \quad \sum_{k=1}^{i} \tilde{T}_k = \tau_{K_i};$$

$$\begin{aligned}\sum_{k=1}^{i} \tilde{L}'_k &= \frac{d}{d\theta}\sum_{k=1}^{i} \tilde{L}_k \\ &= \frac{d}{d\theta}\int_0^{\tau_{K_i}} f(Z_t)dt \\ &= L'_{K_i};\end{aligned}$$

and

$$\sum_{k=1}^{i} \tilde{T}'_k = \tau'_{K_i}.$$

Thus,

$$\frac{\sum_{k=1}^{i} \tilde{L}'_k}{\sum_{k=1}^{i} \tilde{T}_k} - \frac{\sum_{k=1}^{i} \tilde{L}_k}{\sum_{k=1}^{i} \tilde{T}_k}\frac{\sum_{k=1}^{i} \tilde{T}'_k}{\sum_{k=1}^{i} \tilde{T}_k} = \left(\frac{L_n}{\tau_n}\right)'_{n=K_i}; \tag{8.28}$$

in other words, the estimator based on cycles (the left side) is just the (random) subsequence of $\{[L_n/\tau_n]', n > 0\}$ obtained as the index $n$ runs through $\{K_i, i > 1\}$. Since any subsequence of a convergent sequence is convergent with the same limit, the left side of (8.28) converges to $\ell'_f$ whenever $\{[L_n/\tau_n]', n > 0\}$ does. □

To summarize, (8.26) and (8.27) always hold for regenerative GSMPs with a one-clock state, satisfying (C2) and the mild regularity conditions of Theorem 8.3. An intuitive argument helps make the cancellation in (8.27) more natural. When a change in the order of events eliminates or creates a regeneration point, we can expect, with great generality, that either two cycles merge or one cycle splits in two. Multiple merges and splits occur with negligible, higher-order probability. When this is the case, the bias $\mathbf{E}[\tilde{L}'_k] - \mathbf{E}[\tilde{L}_k]'$ is the product of two terms: the "rate" at which cycles merge or split (due to a parameter change), and the expected change in $\tilde{L}_k$ due to a merge or split. (This is exactly the logic behind the smoothing methods of Chapter 7, where we evaluated jump rates and jump sizes.) The bias in $\tilde{T}'_k$ has a similar interpretation. But when two cycles merge or one splits, the expected increment in $\tilde{L}_k$ is just $\pm\mathbf{E}[\tilde{L}_k]$, the contribution of one cycle. The expected increment in $\tilde{T}_k$ is similarly $\pm\mathbf{E}[\tilde{T}_k]$. Thus, when we divide on each side of (8.27), we are left with just the merge-or-split "rate" on each side. For this (heuristic) reason, the cancellation in (8.27) is not so surprising after all.

### 8.4.2 Confidence Intervals

We can now use (8.26) to develop confidence interval estimates for steady-state derivatives. Let us first review, briefly, the standard regenerative method for steady-state estimation via

$$\frac{\sum_k \tilde{L}_k}{\sum_k \tilde{T}_k} \to \frac{\mathbf{E}[\tilde{L}_k]}{\mathbf{E}[\tilde{T}_k]} = \ell_f. \tag{8.29}$$

Our only assumptions are that the underlying process $\{Z_t, t \geq 0\}$ is regenerative and that $\mathbf{E}[\tilde{L}_k^2 + \tilde{T}_k^2] < \infty$. Given a level of precision $0 < \delta < 1$, the goal is to evaluate, from $\{(\tilde{L}_k, \tilde{T}_k), k = 1, \ldots, n\}$, quantities $h_n^l$ and $h_n^r$ such that

$$\lim_{n\to\infty} P(h_n^l \leq \ell_f \leq h_n^r) = 1 - \delta;$$

then $[h_n^l, h_n^r]$ is an asymptotically exact 100(1-$\delta$)% confidence interval for $\ell_f$.

The ordinary central limit theorem states that if $\{X_i, i > 0\}$ are i.i.d. with mean $m_X$ and variance $0 < \sigma_X^2 < \infty$, and if $\bar{X}_n = \sum_{i=1}^n X_i/n$, then $\sqrt{n}(\bar{X}_n - m_X)$ is asymptotically normal (mean zero, variance $\sigma_X^2$) as $n \to \infty$. In other words, for any $-\infty < z < \infty$,

$$\lim_{n\to\infty} P\left(\sum_{i=1}^n (X_i - m_X)/\sigma_X\sqrt{n} \leq z\right) = \Phi(z), \tag{8.30}$$

where

$$\Phi(z) = \frac{1}{\sqrt{2\pi}} \int_{-\infty}^{z} \exp(-x^2/2)dx$$

is the standard (zero mean, unit variance) normal distribution. The conclusion (8.30) can also be denoted by $\sqrt{n}(\bar{X}_n - m_X) \Rightarrow N(0, \sigma_X^2)$, where $N(0, \sigma_X^2)$ is the mean zero normal distribution with variance $\sigma_X^2$ and $\Rightarrow$ is convergence in distribution. From this limit it follows that, given precision $\delta$ and letting $z_\delta = \Phi^{-1}(1 - \delta/2)$, the interval

$$\left[\bar{X}_n - \frac{z_\delta \sigma_X}{\sqrt{n}}, \bar{X}_n + \frac{z_\delta \sigma_X}{\sqrt{n}}\right] \tag{8.31}$$

includes $m_X$ with probability approaching $1-\delta$ as $n \to \infty$. If we denote the sample variance for $\{X_i, i = 1, \ldots, n\}$, $n > 1$, by

$$s_X(n) = \frac{1}{n-1} \sum_{i=1}^{n} (X_i - \bar{X}_n)^2,$$

then $s_X(n) \to \sigma_X^2$ as $n \to \infty$, a.s.; and (8.30) and (8.31) remain valid if the true standard deviation $\sigma_X$ is replaced with the estimate $\sqrt{s_X(n)}$.

In (8.29), we have a *ratio* of sample means so the preceding analysis is not immediately applicable; we must transform the data. Let $X_i = \tilde{L}_i - \ell_f \tilde{T}_i$, $i = 1, 2, \ldots$. Then $\{X_i\}$ is an i.i.d. sequence with mean zero and variance

$$\sigma^2 = \mathbf{Var}[\tilde{L}_i] - 2\ell_f \mathbf{Cov}[\tilde{L}_i, \tilde{T}_i] + \ell_f^2 \mathbf{Var}[\tilde{T}_i];$$

thus, $\sqrt{n}\bar{X}_n \Rightarrow N(0, \sigma^2)$. Let $\bar{L}_n$ and $\bar{T}_n$ be sample means of $\{\tilde{L}_i, i = 1, \ldots, n\}$ and $\{\tilde{T}_i, i = 1, \ldots, n\}$. Substituting into (8.31) and rearranging terms, we find that

$$\frac{\bar{L}_n}{\bar{T}_n} - \frac{z_\delta \sigma}{\mathbf{E}[\tilde{T}_i]\sqrt{n}} \le \ell_f \le \frac{\bar{L}_n}{\bar{T}_n} + \frac{z_\delta \sigma}{\mathbf{E}[\tilde{T}_i]\sqrt{n}},$$

with probability approaching $1-\delta$ as $n \to \infty$. Thus, this provides an asymptotically exact 100(1-$\delta$)% confidence interval for $\ell_f$. In practice, $\sigma / \mathbf{E}[\tilde{T}_i]$ is estimated from $\{(\tilde{L}_k, \tilde{T}_k), k = 1, \ldots, n\}$ in forming this interval.

Confidence intervals for $\ell_f'$ based on (8.26) are not fundamentally different, but require a more elaborate transformation of the data. Reiman and Weiss [66], in considering derivative estimates based on *likelihood ratios*, prove the necessary limit theorem. We review their result, which is directly applicable to our estimates as well.

We assume that $\{(\tilde{L}_k, \tilde{T}_k, \tilde{L}_k', \tilde{T}_k'), k > 0\}$ is an i.i.d. sequence, and that each component has finite second moment. Using a bar to denote sample means (as above), let

$$\hat{r}_1(n) = \frac{\bar{L}_n}{\bar{T}_n}, \quad \hat{r}_2(n) = \frac{\bar{L}_n'}{\bar{T}_n}, \quad \hat{r}_3(n) = \frac{\bar{T}_n'}{\bar{T}_n};$$

and

$$r_1 = \frac{\mathbf{E}[\tilde{L}_k]}{\mathbf{E}[\tilde{T}_k]}, \quad r_2 = \frac{\mathbf{E}[\tilde{L}_k']}{\mathbf{E}[\tilde{T}_k]}, \quad r_3 = \frac{\mathbf{E}[\tilde{T}_k']}{\mathbf{E}[\tilde{T}_k]}.$$

Clearly, $r_1 = \ell_f$ and $r_2 - r_1 r_3 = \ell_f'$. For $k = 1, 2, \ldots$, define $X_1(k) = \tilde{L}_k - r_1 \tilde{T}_k$, $X_2(k) = \tilde{L}_k' - r_2 \tilde{T}_k$, and $X_3(k) = \tilde{T}_k' - r_3 \tilde{T}_k$. Then $\{X(k) = (X_1(k), X_2(k), X_3(k)), k > 0\}$ is an i.i.d. sequence, and, letting $\bar{X}_n$ denote the sample mean,

$$\sqrt{n}\bar{X}_n \Rightarrow N(0, A),$$

where the covariance matrix $A$ is given by $A_{ij} = \mathbf{Cov}(X_i(k), X_j(k))$, $i,j = 1,2,3$. If we define

$$\tilde{r}(n) = \sqrt{n}(\hat{r}_1(n) - r_1, \hat{r}_2(n) - r_2, \hat{r}_3(n) - r_3),$$

then $\tilde{r}(n) \Rightarrow N(0,B)$, with $B_{ij} = A_{ij}/(\mathbf{E}[\tilde{T}_k]^2)$, $i,j = 1,2,3$.

Let $\hat{r} = (\hat{r}_1, \hat{r}_2, \hat{r}_3)$ and $r = (r_1, r_2, r_3)$. Our derivative estimate based on $n$ cycles is $\hat{r}_2(n) - \hat{r}_1(n)\hat{r}_3(n)$. If we let $g : \mathbf{R}^3 \to \mathbf{R}$ be the function $g(x,y,z) = y - xz$, then our estimate is $g(\hat{r}_n)$, which converges almost surely to $g(r) = \ell'_f$, because $g$ is continuous and $\hat{r}_n \to r$. Using a Taylor series argument (e.g., Cramér [14], p.354) it follows that

$$\sqrt{n}[g(\hat{r}) - \ell'_f] = \sqrt{n}[g(\hat{r}) - g(r)] \Rightarrow N(0, \sigma_g^2),$$

where

$$\sigma_g^2 = \sum_{1 \le i,j \le 3} g'_i(r) B_{ij} g'_j(r),$$

and $g'_i(r)$, $i = 1,2,3$, is the $i$th partial derivative of $g$, evaluated at $r$. Thus, the error $\sqrt{n}[g(\hat{r}) - \ell'_f]$ is asymptotically normal with mean zero and variance

$$\sigma_g^2 = r_3^2 B_{11} + B_{22} + r_1^2 B_{33} - 2r_1 r_3 B_{12} + 2r_1 B_{13} - 2r_1 B_{23}.$$

This asymptotic variance can be estimated from $\{(\tilde{L}_k, \tilde{T}_k, \tilde{L}'_k, \tilde{T}'_k), k = 1, \ldots, n\}$ using the sample covariance matrix of $\hat{r}$ in place of $B$. Substituting the variance estimate for $\sigma^2$ and $g(\hat{r}_n)$ for $\bar{X}$ in (8.31) produces an approximate 100(1-$\delta$)% confidence interval that is exact as $n \to \infty$.

* * *

## Notes and Comments

Much of the early work on derivative estimation via perturbation analysis considered steady-state performance. This may be, in part, a consequence of the general tendency in queueing theory and in simulation to focus on steady state. Another incentive for looking at steady state is that it makes it easier to compare experimental results with analytic results for tractable systems; most analytic results apply to steady state. But we believe that a satisfactory treatment of derivative estimation over an infinite horizon must be preceded by a thorough investigation of the finite-horizon theory.

Cao [6] established the convergence of throughput sensitvity estimates as the time horizon grows in the special case of closed Jackson networks. Zazanis [85], and Zazanis and Suri [88] established strong consistency for waiting times in the GI/G/1 queue. Other approaches to this problem can be found in Zazanis [86], Wardi [81], Hu [56], and Chen and Yao [11]. Hu [57] shows how stochastic convexity can be used to establish strong consistency in specially structured networks. In studying queues with discrete service times, Wardi [82] points out that the question of strong consistency may be ill-posed: even in non-pathological cases, the steady state waiting time may not be differentiable. Cao, Gong, and Wardi [8] give an example.

Zazanis and Suri [88] were the first to consider perturbation analysis estimates over regenerative cycles. Specific regenerative estimators are studied in Glasserman [25,29]. General conditions for estimates over cycles to be strongly consistent are given in Heidelberger, Cao, Zazanis, and Suri [48], where the cancellation in (8.27) is also discussed. This cancellation remains somewhat baffling and is probably not the best way of looking at the question of consistency; Theorem 8.7 provides a better way. The intuitive explanation that follows the theorem is validated through explicit calculation in Heidelberger et al. [48] for the special case of an M/G/1 queue.

This chapter follows the approach to steady state derivative estimation developed in Glasserman, Hu, and Strickland [34], using the general conditions (Theorem 8.1) of Hu and Strickland [58]. In [34], a less "obvious" regenerative structure is also considered.

Most of the general results on regenerative processes can be found in the books by Asmussen [1] and Ross [68], but are also widely scattered throughout the applied probability literature. Some specific results on regeneration of GSMPs can be found in Glynn [40], Haas and Shedler [47], and Shedler [75], where Proposition 8.4 appears. Asmussen [1] includes a detailed treatment of Harris recurrent Markov chains. The regeneration we get from (8.9) is relatively simple compared to what the general theory can handle. Following an exit from a one-clock state, $(Y_{n+1}, c_{n+1})$ is independent of $(Y_n, c_n)$; but it would be enough for this to hold only with a certain "probability". See pages 150-152 of [1]

The regenerative method of simulation is developed in Crane and Iglehart [15,16,17]. These papers include a detailed treatment of confidence interval estimation. As noted earlier, the modified procedure for derivative estimates—involving the difference of ratios—is from Reiman and Weiss [66]. Since likelihood ratio derivative estimates are typically unbiased over regenerative cycles, in that setting the extension from finite horizon to steady state is relatively straightforward (but see Glynn [39] for a more subtle use of regeneration).

# Bibliography

[1] Asmussen, S., *Applied Probability and Queues,* John Wiley and Sons, New York, 1987.

[2] Barlow, R.E., and Proschan, F., *Statistical Theory of Reliability and Life Testing*, Holt, Rinehart and Winston, New York, 1975.

[3] Bratley, P., Fox, B.L., and Schrage, L.E., *A Guide to Simulation*, Springer, New York, 1983.

[4] Brémaud, P., *Point Processes and Queues,* Springer, New York, 1981.

[5] Cao, X.R., "Convergence of Parameter Sensitivity Estimates in a Stochastic Experiment," *IEEE Transactions on Automatic Control,* **AC-30,** pp.845-853, 1985.

[6] Cao, X.R., "Realization Probability in Closed Jackson Queueing Networks and its Application", *Advances in Applied Probability*, **19**, pp.708-738, 1987.

[7] Cao, X.R., "A Sample Performance Function of Closed Jackson Queueing Networks", *Operations Research*, **36**, pp.128-136, 1988.

[8] Cao, X.R., Gong, W.B., and Wardi, Y., "Non-Differentiable Performance Functions of Queueing Systems with Discrete Random Service Times," Digital Equipment Corporation, Marlboro, Massachusetts, 1990.

[9] Cao, X.R., and Ho, Y.C., "Estimating Sojourn Time Sensitivity in Queueing Networks Using Perturbation Analysis", *Journal of Optimization Theory and Applications*, **53**, 353-375, 1987.

[10] Cassandras, C.G., and Strickland, S.G., "On-Line Sensitivity Analysis of Markov Chains", *IEEE Transactions on Automatic Control*, **AC-34**, pp.76-86, 1989.

[11] Chen, H., and Yao, D.D., "Derivatives of the Expected Delay in the GI/G/1 Queue", 1990, *Journal of Applied Probability,* to appear.

[12] Chung, K.L., *A Course in Probability Theory,* Academic Press, Orlando, Florida, 1974.

[13] Çinlar, E., *Introduction to Stochastic Processes*, Prentice-Hall, Englewood Cliffs, New Jersey, 1975.

[14] Cramér, H., *Mathematical Methods of Statistics*, Princeton University Press, Princeton, New Jersey, 1958.

[15] Crane, M.A., and Iglehart, D.L., "Simulating Stable Stochastic Systems I: General Multiserver Queues", *Journal of the Association for Computing Machinery,* **21**, pp.103-113, 1974.

[16] Crane, M.A., and Iglehart, D.L., "Simulating Stable Stochastic Systems II: Markov chains", *Journal of the Association for Computing Machinery,* **21**, pp.114-123, 1974.

[17] Crane, M.A., and Iglehart, D.L., "Simulating Stable Stochastic Systems III: Regenerative Processes and Discrete Event Simulation", *Operations Research*, **23**, pp.33-45, 1975.

[18] Dieudonné, J.A., *Foundations of Modern Analysis*, Academic Press, New York, 1960.

[19] Fox, B.L., and Glasserman, P., "Estimating Derivatives via Poisson's Equation," Technical Report, Mathematics Department, University of Colorado, Denver, 1990.

[20] Fox, B.L., and Glynn, P.W., "Discrete-Time Conversion for Simulating Semi-Markov Processes", *Operations Research Letters*, **5**, pp.191-196, 1986.

[21] Fox, B.L., and Glynn, P.W., "Discrete-Time Conversion for Simulating Finite-Horizon Markov Processes", *SIAM Journal on Appl. Math.*, 1990.

[22] Fu, M.C., "Convergence of a Stochastic Approximation Algorithm for the GI/G/1 Queue Using Infinitesimal Perturbation Analysis", *Journal of Optimization Theory and Applications*, **65**, pp.149-160, 1990.

[23] Glasserman, P., "Sensitivity of Sample Values Not Generated By Inversion", *Journal of Optimization Theory and Applications*, **52**, pp.487-493, 1987.

[24] Glasserman, P., *Equivalence Methods in the Perturbation Analysis of Queueing Networks,* Ph.D. Thesis, Division of Applied Sciences, Harvard University, 1988.

[25] Glasserman, P., "Infinitesimal Perturbation Analysis of a Birth and Death Process," *Operations Research Letters*, **7**, pp.43-49, 1988.

[26] Glasserman, P., "Performance Continuity and Differentiability in Monte Carlo Optimization", *Proceedings of the Winter Simulation Conference*, M. Abrams, P. Haigh, and J. Comfort, eds., pp.518-524, 1988.

[27] Glasserman, P., "Structural Conditions for Perturbation Analysis Derivative Estimation: Finite-Time Performance Indices", AT&T Bell Laboratories, Holmdel, New Jersey, 1989.

[28] Glasserman, P., "Structural Conditions for Perturbation Analysis of Queueing Systems", 1989, *Journal of the Association for Computing Machinery*, to appear.

[29] Glasserman, P., "The Limiting Value of Derivative Estimates Based on Perturbation Analysis", *Communications in Statistics: Stochastic Models*, **6**, pp.229-257, 1990.

[30] Glasserman, P., "Derivative Estimates from Simulation of Continuous-Time Markov Chains", 1989, revision to appear in *Operations Research.*

[31] Glasserman, P., "Discrete-Time 'Inversion' and Derivative Estimation for Markov Chains", *Operations Research Letters*, **9**, no.5, 1990.

[32] Glasserman, P., and Gong, W.B., "Derivative Estimates From Discontinuous Realization: Smoothing Techniques", *Proceedings of the Winter Simulation Conference*, E.A. MacNair, K.J. Musselman, P. Heidelberger, eds., pp.381-389, 1989.

[33] Glasserman, P., and Gong, W.B., "Smoothed Perturbation Analysis for a General Class of Discrete Event Systems", *IEEE Transactions on Automatic Control*, **AC-35**, no.11, 1990.

[34] Glasserman, P., Hu, J.Q., and Strickland, S.G., "Strong Consistency of Sample Path Derivative Estimates", Technical Report, Division of Applied Sciences, Harvard University, 1990.

[35] Glasserman, P., and Yao, D.D., "Monotonicity in Generalized Semi-Markov Processes," AT&T Bell Laboratories, Holmdel, New Jersey, 1989.

[36] Glasserman, P., and Yao, D.D., "Generalized Semi-Markov Processes: Antimatroid Structure and Second-Order Properties," AT&T Bell Laboratories, Holmdel, New Jersey, 1990.

[37] Glynn, P.W., "Construction of Process Differentiable Representations for Parametric Families of Distributions", Technical Report, University of Wisconsin Mathematics Research Center, Madison, Wisconsin, 1986.

[38] Glynn, P.W., "Stochastic Approximation for Monte Carlo Optimization", *Proceedings of the Winter Simulation Conference*, J. Wilson, J. Henriksen, S. Roberts, eds., pp.356-364, 1986.

[39] Glynn, P.W., "Likelihood Ratio Gradient Estimation: An Overview", *Proceedings of the Winter Simulation Conference*, A. Thesen, H. Grant, W. David Kelton, eds., pp.366-374, 1987.

[40] Glynn, P.W., "A GSMP Formalism for Discrete Event Systems" *Proceedings of the IEEE*, **77**, pp.14-23, 1990.

[41] Glynn, P.W., and Iglehart, D.L., "Simulation Methods for Queues: An Overview", *Queueing Systems*, **3**, pp.221-256, 1988.

[42] Gong, W.B., "Smoothed Perturbation Analysis of Discrete Event Dynamic Systems", Ph.D. thesis, Division of Applied Sciences, Harvard University, 1987.

[43] Gong, W.B., "Smoothed Perturbation Analysis Algorithm for a G/G/1 Routing Problem," *Proceedings of the Winter Simulation Conference*, M. Abrams, P. Haigh, and J. Comfort, eds., pp.525-531, 1988.

[44] Gong, W.B., Cassandras, C.G., Kallmes, M.H., and Wardi, Y., "A New Class of Gradient Estimators for Queueing Systems with Real-Time Constraints," *29th IEEE Conference on Decision and Control*, to appear, December 1990.

[45] Gong, W.B., and Ho, Y.C., "Smoothed (Conditional) Perturbation Analysis of Discrete Event Dynamical System", *IEEE Transactions on Automatic Control*, **32**, pp.858-866, 1987.

[46] Gong, W.B., and Glasserman, P., "Perturbation Analysis of the M/G/1/K Queue", *Proceedings of the 27th IEEE Conference on Decision and Control*, pp.1114-1118, 1988.

[47] Haas, P.J., and Shedler, G.S., "Regenerative Generalized Semi-Markov Processes", *Communications in Statistics: Stochastic Models*, **3**, pp.409-438, 1987.

[48] Heidelberger, P., Cao, X.R., Zazanis, S. and Suri, R., "Convergence Properties of Infinitesimal Perturbation Analysis Estimates", *Management Science,* **34**, pp.1281-1302, 1988.

[49] Ho, Y.C., "Performance Evaluation and Perturbation Analysis of Discrete Event Dynamic Systems", *IEEE Transactions on Automatic Control*, **AC-32**, pp.563-572, 1987.

[50] Ho, Y.C., Eyler, M.A., and Chien, T.T., "A Gradient Technique for General Buffer Storage Design in a Serial Production Line", *International Journal of Production Research*, **17**, pp.557-580, 1979.

[51] Ho, Y.C., and Cao, X.R., "Optimization and Perturbation Analysis of Queueing Networks," *Journal of Optimization Theory and Applications,* **40**, pp.559-582, 1983.

[52] Ho, Y.C., and Hu, J.Q., "An Infinitesimal Perturbation Analysis Algorithm for a Multi-Class G/G/1 Queue," *Operations Research Letters*, **9**, pp.35-44, 1990.

[53] Ho, Y.C., and Li, S. "Extensions of perturbation analysis of discrete-event dynamic systems", *IEEE Transactions on Automatic Control*, **AC-33**, pp.427-438, 1988.

[54] Ho, Y.C., and Yang, P.Q., "Equivalent Network, Load Dependent Servers and Perturbation Analysis—An Experimental Study", *Proc. Int. Conf. Teletraffic Analysis and Computer Performance Evaluation*, North-Holland, Amsterdam, 1986.

[55] Holtzman, J.M., "On Using Perturbation Analysis to do Sensitivity Analysis: Derivatives vs. Differences", *Proceedings of the 28th IEEE Conference on Decision and Control*, pp.2018-2023, 1989.

[56] Hu, J.Q., "Stong Consistency of Infinitesimal Perturbation Analysis for the G/G/1 Queue", 1990, *Management Science*, to appear.

[57] Hu, J.Q., "Convexity of Sample Path Performances and Strong Consistency of Infinitesimal Perturbation Analysis Estimates", 1990, *IEEE Transactions on Automatic Control*, to appear.

[58] Hu, J.Q., and Strickland, S.G., "General Conditions for Strong Consistency of Sample Path Derivative Estimates", 1990, *Applied Mathematics Letters*, to appear.

[59] Kelly, F.J., *Reversibility and Stochastic Networks*, John Wiley and Sons, New York, 1979.

[60] Kiefer, J., and Wolfowitz, J., "Stochastic Estimation of the Maximum of a Regression Function", *Annals of Mathematical Statistics*, **23**, pp.462-466, 1952.

[61] Kushner, H.J., and Clark, D.S., *Stochastic Approximation Methods for Constrained and Unconstrained Systems*, Springer, New York, 1978.

[62] L'Ecuyer, P., Giroux, N., and Glynn, P.W., "Stochastic Optimization by Simulation: Some Experiments With a Simple Steady-State Queue", Technical Report, Département d'Informatique, Université Laval, Ste-Foy, Québec, 1989.

[63] Li, S., and Ho, Y.C., "Sample Path and Performance Homogeneity of Discrete Event Dynamic Systems", *Automatica*, **25**, pp.907-916, 1989.

[64] Matthes, K., "Zur Theorie der Bedienungsprozesse", *Third Prague Conference on Information Theory, Statistical Decision Functions and Random Processes*, Prague, 1962.

[65] Prabhu, N.U., *Stochastic Processes*, Macmillan, New York, 1965.

[66] Reiman, M.I. and Weiss, A., 'Sensitivity analysis for simulations via likelihood ratios', *Operations Research*, **37**, pp.830-844, 1989.

[67] Robbins, H., and Monro, S., "A Stochastic Approximation Method", *Annals of Mathematical Statistics*, **22**, pp.400-407, 1951.

[68] Ross, S.M., *Stochastic Processes*, John Wiley and Sons, New York, 1983.

[69] Rubinstein, R., "Sensitivity Analysis and Performance Extrapolation for Computer Simulation Models," *Operations Research*, **37**, pp.72-81, 1989.

[70] Rudin, W., *Principles of Mathematical Analysis*, McGraw-Hill, New York, 1964.

[71] Rudin, W., *Real and Complex Analysis*, McGraw-Hill, New York, 1966.

[72] Schassberger, R., "Insensitivity of Steady-State Distributions of Generalized Semi-Markov Processes, I." *Annals of Probability*, **5**, pp.87-99, 1977.

[73] Schassberger, R., "Insensitivity of Steady-State Distributions of Generalized Semi-Markov Processes, II." *Annals of Probability*, **6**, pp.85-93, 1978.

[74] Schassberger, R., "Insensitivity of Steady-State Distributions of Generalized Semi-Markov Processes with Speeds" *Advances in Applied Probability*, **10**, pp.836-851, 1978.

[75] Shedler, G.S., *Regeneration and Networks of Queues*, Springer, New York, 1987.

[76] Suri, R., "Implementation of Sensitivity Calculations on a Monte-Carlo Experiment", *Journal of Optimization Theory and Applications*, **40**, pp.625-630, 1983.

[77] Suri, R., "Infinitesimal Perturbation Analysis for General Discrete Event Systems", *Journal of the Association for Computing Machinery,* **34**, pp.686-717, 1987.

[78] Suri, R., and Leung, Y.T., "Single Run Optimization of Discrete Event Simulations: An Empirical Study Using the M/M/1 Queue", *IIE Transactions*, to appear.

[79] Suri, R., and Zazanis, M., "Perturbation Analysis Gives Strongly Consistent Sensitivity Estimates for the M/G/1 Queue', *Management Science*, **34**, pp.39-64, 1988.

[80] Vakili, P., and Ho, Y.C., "Alternative Representation and Perturbation Analysis in a Routing Problem" *Proceedings of the 28th IEEE Conference on Decision and Control*, pp.1082-1083, 1989.

[81] Wardi, Y., "Interchangeability of Expectation and Differentiation of Waiting Times in GI/G/1 Queues", Technical Report, School of Electrical Engineering, Georgia Institute of Technology, 1988.

[82] Wardi, Y., "Consistency of Perturbation Analysis for Non-Markovian Queueing Networks", Memorandum TMYW01-89, School of Electrical Engineering, Georgia Institute of Technology, 1989.

[83] Wardi, Y., Gong, W.B., Cassandras, C.G., and Kallmes, M.H., "A New Class of Perturbation Analysis Algorithms for Piecewise Continuous Sample Performance Functions," Technical Report, School of Electrical Engineering, Georgia Institute of Technology, 1990.

[84] Whitt, W., "Continuity of Generalized Semi-Markov Processes", *Mathematics of Operations Research*, **5**, pp.494-501, 1980.

[85] Zazanis, M.A., *Statistical Properties of Perturbation Analysis Estimates*, Ph.D. Thesis, Division of Applied Sciences, Harvard University, 1986.

[86] Zazanis, M.A., "Weak Convergence of Sample Path Derivatives for the Waiting Time in a Single-Server Queue", *Proceedings of the 25th Allerton Conference*, pp.297-304, 1987.

[87] Zazanis, M.A., "Compensators and Derivative Estimation for Queueing Systems", *Proceedings of the 26th Allerton Conference*, pp.549-555, 1988.

[88] Zazanis, M.A., and Suri, R., "Perturbation Analysis of the GI/G/1 Queue", Technical Report, IE/MS Department, Northwestern University, Evanston, Illinois, 1986.

[89] Zazanis, M.A., and Suri, R., "Comparison of Perturbation Analysis with Conventional Sensitivity Estimates for Stochastic Systems", Technical Report, IE/MS Department, Northwestern University, Evanston, Illinois, 1985.

# Index